Wave Propagation in Petroleum Engineering

WAVE PROPAGATION IN PETROLEUM ENGINEERING

Modern applications to drillstring vibrations, measurement-while-drilling, swab-surge, and geophysics

Wilson C. Chin

Gulf Publishing Company
Houston, London, Paris, Zurich, Tokyo

To

Victoria, Jessica, and Jonathan

Gulf Publishing Company
Book Division
P.O. Box 2608
Houston, Texas 77252-2608

10 9 8 7 6 5 4 3 2 1

Library of Congress Cataloging-in-Publication Data

Chin, Wilson C.
Wave propagation in petroleum engineering / Wilson C. Chin.
p. cm.
Includes bibliographical references and index.
ISBN 0-88415-169-7
1. Petroleum engineering. 2. Waves. I. Title.
TN871.C4958 1994
622′.338′01531133—dc20 93-48179
CIP

Contents

Preface .... **ix**

Part I
Fundamental Principles

1. Overview and Basic Ideas **3**

Goals of the book, 3
The classical wave equation, 5
Functional representation, 9
Separation of variables and eigenfunction expansions, 10
Standing versus propagating waves, 18
Laplace transforms, 22
Fourier transforms, 29
External forces versus boundary conditions, 33
Point force and dipole wave excitation, 46
First order partial differential equations, 50
References, 53

2. Kinematic Wave Theory **55**

Whitham's theory in nondissipative media, 56
Simple attenuation modeling, 62
Kinematic wave theory in homogeneous and dissipative media, 65
High-order kinematic wave theory, 69
Effect of low-order nonuniformities, 76
Three-dimensional kinematic wave theory, 82
References, 87

3. Elementary Examples from Classical Mechanics **88**

Example 3-1: Lateral vibration of simple beams, 88
Example 3-2: Acoustic waves in waveguides, 91
Example 3-3: Gravity-capillary waves in deep water, 102

Example 3-4: Fluid-solid interaction, waves on elastic membranes, 106
Example 3-5: Problems in hydrodynamic stability, 111
References, 115

Part II
Drillstring Vibrations

4. Typical Downhole Vibration Environment 119

What is wave motion?, 119
Drillstring vibration modes, 121
Long-standing vibrations issues, 125
Practical applications, 127
Elastic line model of the drillstring, 129
Objectives and discussion plan, 132
References, 134

5. Axial Vibrations 136

Pioneering axial vibration studies, 137
Governing differential equation, 139
Conventional separation of AC/DC solutions, 143
Boundary conditions: old and new ideas, 146
Global energy balance, 155
Simple solution for rate-of-penetration, 159
Finite difference modeling, 163
Complete formulation and numerical solution, 170
Modeling pipe-to-collar area changes, 174
Example Fortran implementation, 176
Drillstring and formation imaging, 183
References, 200

6. Lateral Bending Vibrations.. 204

Why explain this drilling paradox?, 204
Lateral vibrations in deep water operations, 205
Bending vibrations in directional control, 205
A downhole paradox: the case of the vanishing waves, 206

Why drillstrings fail at the neutral point, 211
Surface detection of downhole bending disturbances, 224
Linear boundary value problem formulation, 226
Finite difference modeling, 229
Example Fortran implementation, 233
Nonlinear interaction between axial and lateral bending vibrations, 235
References, 236

7. Torsional and Whirling Vibrations 239

Torsional vibrations, 240
Example Fortran implementation, 247
Whirling motions, 248
References, 251

8. Coupled Axial, Torsional, and Lateral Bending Vibrations 254

Importance to PDC bit dynamics, 255
Coupled axial, torsional, and bending vibrations, 255
Example 8-1: Simple desktop experiment, 256
Notes on the coupled model, 256
Coupled axial, torsional, and bending equations, 257
General Fortran implementation, 263
Examples 8-1 and 8-2: Example calculations for bit bounce, stick-slip, rate-of-penetration, and drillstring precession, 267
Precessional instabilities, 271
Drillstring vibrations in horizontal wells, 274
References, 275

Part III
Modern Petroleum Engineering Applications

9. Mud Dynamics in the Borehole Annulus 281

Governing Lagrangian equations, 282
Governing Eulerian equations, 290
Transient finite difference modeling, 296
Swab-surge modeling, 298

Steady state annular flow modeling, 301
References, 316

10. Measurement-While-Drilling 318

Basic MWD system components, 319
Candidate transmission technologies, 319
Mud pulse telemetry: the acoustic source, 321
Mud pulse telemetry: the transmission channel, 329
References, 335

11. Geophysical Applications 337

Classical wave modeling: eikonal methods and ray tracing, 338
Fermat's principle of least time (via calculus of variations), 341
Fermat's principle revisited via kinematic wave theory, 344
Modeling wave dissipation, 345
Ray tracing over large space-time scales, 349
Subtle high-order effects, 352
Travel time modeling, 356
References, 362

12. Wave and Current Interaction in the Ocean 364

Recapitulation, 364
Damped waves in deep water, 365
Waves in finite depth water, 366
Sources of hydrodynamic loading, 367
Instabilities due to heterogeneity, 367
References, 371

Cumulative References 372
Index 390

Preface

Wave motions abound in petroleum engineering. They manifest themselves in drillstring vibrations that haunt and intimidate drillers; in measurement-while-drilling, or MWD, where mud pulse telemetry delivers real-time downhole information to rig hands and geologists . . .

In dynamic swab and surge, where propagating waves can induce wellbore instabilities; in oil and gas flowlines, where accurate wave speed and attenuation modeling is difficult; in surface seismics and crosswell tomography, useful in delineating underground structural features . . .

In borehole geophysics, where high frequency acoustic signals reveal important clues about the neighboring stratigraphy; and in the design of offshore structures, where ocean wave loadings on platforms and cables can create crippling damage.

In an increasingly competitive business environment, how do petroleum engineers—*tomorrow's managers*—stay abreast of the technologies that they will need to oversee in the future? How do we become competitive through interdisciplinary study and technology transfer?

Physical systems support complicated wave motions, but even the simplest of these require years of graduate study in applied mathematics. This has transformed petroleum engineering education into "how-to" recipes, thus denying students the opportunity to grow with and contribute to new technologies.

This book discusses the broad subject of *wave propagation in petroleum engineering*. Physical features common to specific but different disciplines will be studied, and the mathematical and numerical methods required to model their similar areas are developed once and for all.

Although wave propagation modeling is inherently mathematical, our approach is applications oriented. The pertinent analytical and computational techniques are introduced only in the context of the physical problem at hand, and then, only as much as is required to solve the problem.

Mathematical modeling and numerical analysis are developed only where they are stimulated by necessity and engineering need. Importantly, both are presented in a practical, down-to-earth, readable manner that is equally rigorous and scientifically precise. Extensive references are provided for readers interested in additional information.

Wave models are usually *equation-based;* e.g., the classical equation $\partial^2 u/\partial t^2 - c^2 \partial^2 u/\partial x^2 = 0$ is standard in math courses. But many wave motions do not satisfy this equation, and students often develop the mistaken impression that studying it alone leaves them with all of the required tools.

In reality, many wave motions satisfy formulations that are quite different. Cases in point: water waves following Laplace's elliptic equation, lateral drillstring vibrations satisfying a fourth-order beam equation, and geophysical waves governed by a system of equations with combined normal and shear stress effects.

Even when basic phenomena do satisfy $\partial^2 u/\partial t^2 - c^2 \, \partial^2 u/\partial x^2 = 0$, a study of the boundary conditions and forcing functions required in modeling a range of applications consumes months. Until the early 1970s, the study of wave propagation meant a conglomeration of fragmented and specialized disciplines.

Inasmuch as Newton's $\mathbf{F} = m\mathbf{a}$ has revolutionized classical mechanics, the emergence of *kinematic wave theory* (KWT) two decades ago has unified much of traditional wave mechanics. For once, students can grasp *the big picture* easily, allowing equation-based methods to cover those necessary details that certainly will vary with particular applications.

KWT is useful in many areas of petroleum engineering. For example, the eikonal method used in seismics is restricted to nonattenuative media; while this is unimportant in surface seismics, where reflections act over thin earth layers, improved methods can be devised using KWT for crosswell tomography where rays act over thousands of feet in attenuative media.

Using extended eikonal methods, again via KWT, it is possible to explain exactly *why* drillstrings fail at the neutral point, and *why* the severe lateral vibrations that induce this failure cannot be seen at the surface. This solves a long-standing mystery well known to drillers.

The use of wave equation based finite difference methods in seismic migration and travel-time estimation has raised confusion regarding numerical solutions that lead to unphysical results. KWT again resolves the differences, when dissipation is properly accounted for in the high-order terms.

In this book, classical and kinematic wave theories are developed in a hands-on manner with minimal formalism, so that they are understandable to undergraduate students. Applications results are discussed in a manner likely to be of value to practitioners, researchers, and software developers.

For instance, in the example on drillstring failure, we give equations derived from KWT that show how axial stress variations alter wavenumber density. This induces refractive effects that cause lateral waves to focus and trap at the neutral point, fluctuate violently, and produce high-cycle fatigue.

And we go further. Algorithms are given which simulate these instabilities. Because of the research nature of our subjects, we sometimes speculate; here, we suggest, with full mathematical justification, the detection of downhole lateral vibrations by monitoring surface axial vibrations.

We will also show how dynamic wave loadings acting on offshore platform legs and ocean-immersed cables and pipelines can be calculated using KWT in a relatively simple manner, without the need to perform the detailed perturbation analyses required in water wave modeling.

Our most important contribution is an integrated perspective of wave propagation. For example, the concept of monopole and dipole modeling in describing different excitations is developed from physical considerations using simple-to-understand analogies related to transversely vibrating strings.

These ideas are useful in modeling MWD pulsers and also in simulating the kinematics of rotating tricone bits as they affect rate-of-penetration and drillstring dynamics. The monopole and dipole ideas employed here, developed in earthquake seismology to simulate source point effects, apply almost immediately upon simple vocabulary changes.

For example, the symmetric stress field due to an underground nuclear blast is much like the symmetric pressure field about a negative pressure MWD pulser. Antisymmetric stresses, by contrast, model positive pressure valves and mud sirens. Like waves propagate and evolve similarly; their solution methods are almost identical.

Jump conditions, dreaded and suspicious subjects even in mathematics, are discussed rigorously, but using only ideas requiring no more than calculus. The use of Dirac delta functions and their derivatives is introduced, but only as a formalism which facilitates recipe-type solutions based on operational Laplace and Fourier transforms.

Our methodologies are not new. For example, many were developed in earthquake seismology and physical acoustics; KWT, originally devised to study water waves, has been used in optics, boundary layer instability, and hydraulic jump formation.

Our algorithms for drillstring vibrations, MWD, and swab-surge were developed in aerodynamics and geophysics. This book is introductory in the sense that it develops the physical principles and mathematical tools needed to model petroleum applications in a readable manner.

But it is also an advanced applications book. The subject matter, literature reviews, and research results described are state-of-the-art, and much appears in print for the first time. We hope to define promising areas for future industry research. There is no other book covering the applications discussed here, and at the same time delving into mathematics and algorithms.

It goes without saying that writing this book, and putting together its pieces over the years, has been fun, challenging, and intellectually rewarding. Gratitude is extended to all my colleagues and friends who have shaped and molded my thinking about wave mechanics. In particular, I thank:

Professor Marten T. Landahl, who taught me wave mechanics, and supervised my Ph.D. thesis at the Massachusetts Institute of Technology, almost two decades ago;

Mr. Larry J. Leising, the Schlumberger veteran who, for better or for worse, introduced me to the petroleum industry, and MIT colleague who fascinated me with mundane oilfield problems such as mud pulse telemetry, drillstring vibrations, and swab and surge;

Professor Gerald B. Whitham at the California Institute of Technology, inventor of sonic boom analysis and kinematic wave theory, who allowed me, over twenty years ago, to audit his over-subscribed classes despite my lack of prerequisites;

Dean Stanley Ebner at the University of Houston, who provided computer time, facilities, and students when I needed them the most;

The author's colleagues and friends, and Mr. Harry D. Smith, Jr., Manager of Research, Halliburton Energy Services, who provided the supportive environment and facilities this author drew upon during the final writing and revision of this manuscript; and last, but not least,

Mr. William J. Lowe, Editor-in-Chief, Book Division, Gulf Publishing Company, and Ms. Joyce Alff and Ms. Melissa Beck, Editors, who have always lent open hearts and minds to new technologies waiting to happen.

Wilson C. Chin
Houston, Texas

Part I
Fundamental Principles

1
Overview and Basic Ideas

What is *wave propagation*? Commonplace examples include obvious yet mathematically complicated events such as the breaking of water waves on beaches, the vibrations of tuning forks, and children singing in hallways.

Wave propagation, a.k.a. *vibration theory*, *hyperbolic equations*, and *asymptotic analysis*, has been studied mathematically for three hundred years. Its literature is vast. Its developers include the greatest minds in physics. Libraries of books have been written on esoteric aspects of this subject.

Wave propagation encompasses many disciplines: optics, acoustics, structural vibrations, aeroelasticity, electromagnetism, hydrodynamic stability, laminar flow transition, underwater sound, philharmonic hall design, plasma stability, earthquake seismology, geophysics, and spiral galactic instabilities.

How, then, does the student of petroleum engineering (whose work ultimately embraces drilling vibrations, seismics, and MWD, among other evolving technologies) build an adequate technical foundation without an overwhelming amount of graduate work?

Goals of the book. Despite the seriousness of the material and the mathematics presented in this volume, the present book is intended to be an introductory textbook, but in a sense not usually taken in formal courses.

We will provide the student with a *flavor* for the formal constructive techniques usually taught in advanced courses, so that he acquires enough familiarity with the jargon to read professional papers and books.

This is accomplished by offering a synopsis of the conventional mathematics taught in partial differential equations courses, covering standard solution methods, and their strengths and weaknesses. We introduce pertinent analysis areas to give students resources to pursue additional study.

Our objective is not the teaching of classical mathematics *per se*. Thorough understanding can only be achieved through intensive and diligent exercise. We only introduce these methods, but in such a manner as to demonstrate their powerful capabilities or their severe limitations.

This objective allows us to communicate *the big picture*, so that the reader can at once grasp an appreciation of what has been made possible by generations of eminent scholars.

But since it turns out that many of the wave propagation issues encountered in modern petroleum engineering do not draw directly upon these techniques, this overview suffices insofar as providing the basic foundation needed for more specialized analysis.

The physical problems particular to our rapidly changing industry are reviewed for each of the titled applications, and mathematical and numerical models are developed to handle the specialized circumstances these specific applications demand.

By taking this approach, the requirements for new analysis methods and the mathematical issues associated with their implementation appear naturally, and without intimidating theorems, proofs, and corollaries.

Modern subjects such as finite difference modeling, multiple-scaling, stationary phase, monopole and dipole source properties, Fermat's principle of least time, kinematic wave theory, and group velocity are introduced and developed rigorously ... *but in a simple and intuitive way.*

This approach to introducing specialized subjects in the simplest manner has posed the greatest challenge. But this philosophy reflects the trials and tribulations of my own learning process, as I took advantage of the diverse opportunities the Oil Patch offered, after a false start in aerospace engineering.

Before we embark on wave propagation, we state clearly what this book does *not* cover. We will not discuss electromagnetic waves; we omit references to resistivity and electromagnetic logging instruments. We will not discuss spectral analysis, since good books on the subject are available. And we will not develop seismic theory, since many techniques are overly specialized.

However, limited discussion will focus on *new* ideas aimed at improving the formulation process itself. Thus, we will revisit Fermat's principle of least time, examine the role played by attenuation in eikonal analysis, and introduce kinematic wave theory, which allows direct use of measured results while circumventing the equation-based modeling process.

We will not discuss ultrasonics and borehole geophysics, although the ideas developed here for crosswell seismics no doubt apply, nor will we consider hydrodynamic stability in general, although some limited applications are given. With these disclaimers in place, we begin our study of waves.

THE CLASSICAL WAVE EQUATION

Many students learn *wave propagation* by way of the *undamped classical wave equation* (e.g., see Hildebrand, 1948, or Tychonov and Samarski, 1964)

$$\partial^2u/\partial t^2 - c^2\, \partial^2u/\partial x^2 = 0 \tag{1.1}$$

for u(x,t), where x is the propagation direction and t denotes time. We may think of Equation 1-1 as describing the *transverse vibrations* of a string.

If so, u(x,t) is the *displacement* from equilibrium, and c is the *disturbance speed* ($c^2 = T/\rho_l$, where T is *tension* and ρ_l is *lineal mass density*). Examples include waves on violin and guitar strings. In this book, new terminology is introduced intuitively through discussion and highlighted in context by *italics.*

The *domain of x* can be *finite*, *semi-infinite*, or *infinite*, depending on the application. But time is always zero or positive, and future events must never affect present and past motions. This *principle of causality* governs *equations of evolution*, particularly *hyperbolic equations* such as Equation 1-1.

Not all equations are *causal*. The *elliptic equations* used in *reservoir flow simulation* deal with *domains of influence and dependence* crafted so that every point affects and is influenced by every other point (Chin, 1993).

In Equation 1-1, $\partial/\partial t$ and $\partial/\partial x$ represent *partial derivatives*, or derivatives of functions of several variables relative to the particular independent variable shown, with all others held fixed. We may also express partial derivatives using *subscript notation*,

$$u_{tt} - c^2\, u_{xx} = 0 \tag{1.2}$$

Its general solution was obtained two hundred years ago as

$$u(x,t) = f(x+ct) + g(x-ct) \tag{1.3}$$

To understand why this holds, we apply the *chain rule* of calculus, stating that the x-derivative of h{p(x)} is {h'(p)}{dp(x)/dx}, where primes (') denote differentiation with respect to p. Since $u_x = f' + g'$, $u_{xx} = f'' + g''$, $u_t = c f' - cg'$, and $u_{tt} = c^2 f'' + c^2 g''$, substitution in Equation 1-2 proves Equation 1-3.

Note: Two densities, the *lineal* mass density ρ_l and the mass density *per unit volume*, are used in this book. The first is appropriate to vibrating string problems satisfying $\rho_l\, \partial^2u/\partial t^2 - T\, \partial^2u/\partial x^2 = 0$ where the tension T has units of force. The second applies to drillstring vibrations and borehole acoustics problems satisfying $\rho\, \partial^2u/\partial t^2 - E\, \partial^2u/\partial x^2 = 0$ where E, either Young's modulus or the bulk modulus, has units of force per unit area.

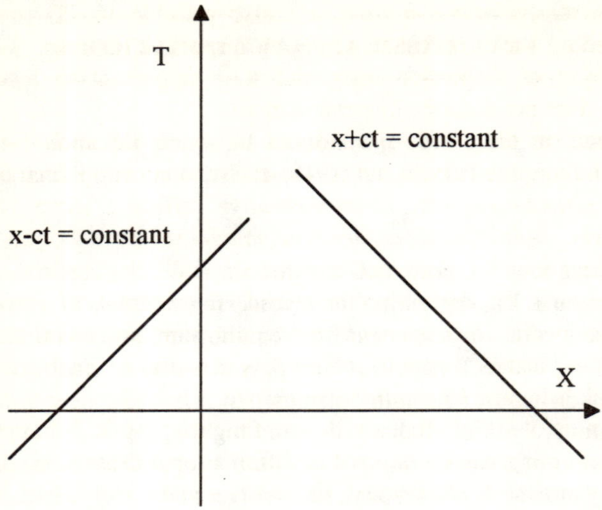

Figure 1-1. The characteristic plane.

Equation 1-3 contains much information. If we consider the u(x,t) = f(x+ct) contribution, we observe that u(x,t) must be constant if x+ct is constant. But "x+ct = constant" is just the straight line in Figure 1-1. Since time must increase, the *argument* x+ct must represent *left-going* waves.

Similarly, u(x,t) = g(x-ct) is constant along *trajectories* with x-ct = constant. Thus, g(x-ct) represents *right-going* waves. Depending upon the application, wave solutions may be both *up-going* and *down-going*.

Equation 1-3 states that solutions of Equation 1-1 can be constructed from general families of left and right-going waves. Lines for which x+ct and x-ct are constant are known as *characteristics* or *rays*; their coordinates represent the *natural* or *canonical* variables describing the wave propagation.

In linear theory, characteristics within each wave family are parallel. The *method of characteristics*, used analytically and numerically, solves problems by tracing values of the dependent variable along rays. We will use this method in Chapter 2 to obtain closed form solutions for use in *kinematic wave theory*.

From Figure 1-1, the slope c in the *characteristic x-t plane*, is the *wave* or *sound speed*; e.g., the line x-ct = constant has the slope c which is obviously the *propagation velocity* with which *disturbances* travel. In general, a disturbance propagates with speed c to both the left and the right.

Equation 1-1 states that c is the *only* available wave speed. In many problems, other speeds are possible. In 3D *waveguides*, additional speeds are possible. Waves with different lengths may travel at different speeds: an initially confined *wave group* may *disperse* and lose its identity.

Such *dispersive wave motions* are also possible in 1D problems; for example, bending waves on beams represent dispersive motions. We will deal with the subject of dispersion later, but it is important to recognize that Equation 1-1 does not describe dispersion at all.

And since, as discussed, disturbances satisfying Equation 1-1 propagate with shape and amplitude both intact, the classical equation does not describe *dissipative*, *attenuative*, or *nonconservative* effects, terms often used interchangeably. It applies to *conservative*, *undamped* wave motions only.

Also, Equation 1-1 does not contain *variable coefficients*: the motion occurs in *uniform* or *homogeneous* media, in contrast to *nonuniform* or *heterogeneous* media. *Heterogeneous media* may contain *spatial* and *temporal inhomogeneities*, that is, variable coefficients in x and t. Such general waves are easily modeled using *kinematic wave theory*.

In Equation 1-3, "+" appears in one function, while "-" appears in the other. Different conventions are used in different applications; $u(x,t) = f(t+x/c) + g(t-x/c)$ may appear in one context, but $u(x,t) = g(-t - x/c) - f(-t + x/c)$ may be more convenient in others. Sometimes f and g are reversed for convenience.

To solve the general *initial value problem* on an *infinite unbounded domain*, subject to *initial conditions* prescribing $u(x,0)$ and the *velocity* $u_t(x,0)$ at $t = 0$, we have *D'Alembert's solution*,

$$u(x,t) = \{u(x-ct,0) + u(x+ct,0)\}/2 + \{1/(2c)\} \int_{x-ct}^{x+ct} u_t(x',0)\, dx' \tag{1.4}$$

Consider an initial prescribed displacement with vanishing velocity. The integral term disappears, leaving $u(x,t) = \{u(x-ct,0) + u(x+ct,0)\}/2$, showing through the "+" and "-" sign differences that half of the initial wave will propagate to the left, while the remaining half propagates to the right.

Reflection properties. Equation 1-3 is significant to wave propagation in many facets of petroleum engineering. Unfortunately, math courses do not address engineering application, and the consequences of Equation 1-3 often remain buried in the obscurity of subscripts and theorems.

We have already alluded to Equation 1-1 as a model for *transverse* string vibrations. It is under this interpretation that much of partial differential equations is taught at the undergraduate and graduate level.

As a prelude to drillstring vibrations and seismics, we also envision $u(x,t)$ as the *longitudinal elastic displacement* of a one-dimensional bar, in which case $c^2 = E/\rho$ where E is Young's modulus and ρ is the *material density* or *mass density per unit volume*. The *normal stress* σ satisfies *Hooke's law* $\sigma = E\, \partial u/\partial x$, where $\partial u/\partial x$ represents the *strain*.

Let us start with $u(x,t) = f(x+ct) + g(x-ct)$, and calculate, using the chain rule, the space derivative $u_x = f'(x+ct) + g'(x-ct)$, and the timewise derivative $u_t = cf'(x+ct) - c\,g'(x-ct)$. Now assume that $x = 0$ represents the *termination end* of a finite or semi-infinite one-dimensional system. How are f and g related, and what are the physical implications?

Example 1-1: Rigid end termination. Suppose that our termination end can be modeled by a *rigid wall*, so that the *displacement* satisfies $u(0,t) = 0$. Then, it is clear that the choice

$$u(x,t) = f(x+ct) - f(-x+ct) \tag{1.5}$$

solves the problem, since $u(0,t) = f(ct) - f(ct)$ yields $u(0,t) = 0$, for any choice of the function f. By differentiating, the end *velocity* $u_t(0,t)$ is also zero.

At $x = 0$, the derivative $u_x(x,t) = f'(x+ct) + f'(-x+ct)$ equals $2f'(ct)$, so that *stress doubles at a rigid end* relative to the incident $f'(ct)$ value. Also, from Equation 1-5, an incident *displacement* pulse $+f(x+ct)$ reflects from $x = 0$ as $-f(-x+ct)$: it reflects with the same shape and size, but "flips over" with a sign change. Since both terms in $u_x(x,t) = f'(x+ct) + f'(-x+ct)$ have like signs, incident *stress* pulse reflects with the same sign.

Example 1-2: Stress-free end. On the other hand, consider a *stress-free* boundary condition satisfying $u_x(0,t) = 0$. Obviously, the choice

$$u(x,t) = f(x+ct) + f(-x+ct) \tag{1.6}$$

solves the problem, since $u_x(x,t) = f'(x+ct) - f'(-x+ct)$ vanishes at $x = 0$. Note that $u(0,t) = 2\,f(ct)$, so that the displacement, and hence the velocity $u_t(0,t) = 2\,cf'(ct)$ doubles (relative to the incident $cf'(ct)$ value) at a stress-free boundary.

Unlike Example 1-1, a *free end* causes an incident *displacement* wave $f(x+ct)$ to reflect with the same sign as $f(-x+ct)$, again with its shape and size the same. But since the terms in $u_x(x,t) = f'(x+ct) - f'(-x+ct)$ have unlike signs, an incident *stress* pulse reflects with a sign change.

For both rigid and free ends, incident pulse *shapes* remained unchanged. But *shape invariance* is not the norm, even for our *nondispersive* Equation 1-1; *distortions* can arise, as we will show, from non-standard boundary conditions.

How might our results be relevant to drillstring vibrations? Suppose our bit were modeled as a zero displacement condition at $x = 0$ (this crude model is *not* used later). Then, it is clear that an approaching wave must double in stress at the bit as it reflects, increasing its tendency to break rock.

Note on acoustics. Equation 1-1 applies to *acoustics* problems, where *axial* or *longitudinal vibrations* of *fluid columns* are of interest (see Chapter 9). The only change is the new definition $c^2 = B/\rho$, where B is the bulk modulus of the fluid. For air, the stress-free boundary condition models long wave sound emission by an *open pipe* into a *large empty space* such as a room.

New acoustics students often find difficulty envisioning waves reflecting at "nothing". After all, since ping-pong balls pass through and out, why shouldn't sound waves? *They don't:* in fluid mechanics, acoustic and hydraulic flow problems differ distinctly and satisfy different equations.

By *acoustics*, we mean wave motions arising from the *compressibility* of the medium, whereas by *hydraulics*, we refer to the *incompressible constant density* flow associated with *material transport* "from here to there". These subjects and their subtleties will be discussed later.

FUNCTIONAL REPRESENTATION

In mathematics, functions can be represented by different series, each useful for particular applications. Why might one wish to represent any function in anything more than classical *Taylor series*?

The answer is simple: the required functional representation is determined by the *boundary value problem* one wishes to solve and the *geometry* of the *problem domain*. Since the possibilities for problem formulations are numerous, there is no shortage of representations, e.g., Legendre, Bessel, and Hankel series. Refer to Hildebrand (1948) for a complete discussion on series.

Fortunately, for the one-dimensional analyses considered in this book, no more than classical Taylor and *Fourier series* are required. We now discuss these more conventional functional representation techniques.

Taylor series. Suppose we wish to represent $y = f(x)$ using a power series in x. We recall from calculus that we may expand f(x) about any value of x, say x_0, where the value of the function and its derivatives are known.

This assumes that the latter exist, which may not be the case; $y = x^{-1/2}$, for example, is infinite and has *singular* (*infinite*) derivatives at $x = 0$, and therefore no Taylor series representation there. But away from $x = 0$, this representation applies. If f(x) and its derivatives $f^n(x)$ exist at $x = x_0$, then its Taylor series expansion is

$$\begin{aligned} f(x) &= f(x_0) + f'(x_0)(x-x_0) + 1/2\, f''(x_0)(x-x_0)^2 + \ldots \\ &= \Sigma\, f^n(x_0)(x-x_0)^n /n! \end{aligned} \quad (1.7)$$

where superscripts indicate the order of the derivative.

Fourier series. Now let us represent $y = f(x)$, defined on $0 < x < L$, using Fourier series. Because its values on $-L < x < 0$ are not significant, we can *assume* f(x) to be *even* with respect to $x = 0$. Since $f(x) = f(-x)$, we can write

$$f_e(x) = A_0 + \Sigma\, A_n \cos n\pi x/L \quad (1.8)$$

provided

$$A_0 = \{\int f_e(x)\, dx\}/L \tag{1.9}$$

$$A_n = \{2 \int f_e(x) \cos n\pi x/L\, dx\}/L \tag{1.10}$$

Alternatively, if f(x) is *assumed* to be *odd* with f(x) = - f(-x), we can write

$$f_o(x) = \Sigma\ B_n \sin n\pi x/L \tag{1.11}$$

$$B_n = \{2 \int f_o(x) \sin n\pi x/L\, dx\}/L \tag{1.12}$$

In the above, the subscripts *e* and *o* indicate *even* and *odd*. Also, in Equations 1-9 to 1-12, the limits of integration extend from x = 0 to L, and the summation extends from n = 1 to ∞ (limits are omitted for brevity).

Consider any function g(x) defined on the *wider* interval -L < x < L. From g(x) = {g(x) + g(-x)}/2 + {g(x) - g(-x)}/2, it is clear that the first term {g(x) + g(-x)}/2 remains unchanged, and that the second term {g(x) - g(-x)}/2 reverses in sign, if in each case x is replaced by -x. Since {g(x) + g(-x)}/2 is *even*, while {g(x) - g(-x)}/2 is *odd*, it follows from the above arguments that

$$g(x) = C_0 + \Sigma\, C_n \cos n\pi x/L + \Sigma\ D_n \sin n\pi x/L \tag{1.13}$$

$$C_0 = \{\int \{g(x) + g(-x)\}/2\, dx\}/L \tag{1.14}$$

$$C_n = \{2 \int \{g(x) + g(-x)\}/2 \cos n\pi x/L\, dx\}/L \tag{1.15}$$

$$D_n = \{2 \int \{g(x) - g(-x)\}/2 \sin n\pi x/L\, dx\}/L \tag{1.16}$$

Examples of commonly used Fourier series may be found in mathematical tables and references.

SEPARATION OF VARIABLES AND EIGENFUNCTION EXPANSIONS

Here we demonstrate how Fourier series are used to solve *boundary value problems.* As suggested, equations such as Equation 1-1 possess numerous solutions, which must be fixed by *auxiliary constraints* called *boundary* and *initial conditions.* We will emphasize *constructive techniques,* but not dwell upon *existence* and *uniqueness* issues in this book.

Separation of variables is the best known of all partial differential equation solution methods. Unfortunately, for most practical purposes, it is also the least useful. Nonetheless, in many graduate level courses, it is the standard by which student comprehension and ingenuity is based.

We introduce the technique using several examples marked by increasing levels of difficulty. We also take the opportunity to introduce the mathematical and engineering terminology used in solving vibration problems. All of our examples are defined on the *finite domain* $0 < x < L$.

Example 1-3: String with pinned ends and general initial conditions. Consider Equation 1-1, which does not include *damping* and *external excitation*. In particular, we solve $\partial^2u/\partial t^2 - c^2\, \partial^2u/\partial x^2 = 0$ for the dependent variable $u(x,t)$ subject to the boundary conditions $u(0,t) = 0$ and $u(L,t) = 0$, plus *general initial conditions* $u(x,0) = \phi(x)$ and $u_t(x,0) = \psi(x)$. We assume *separable product solutions* of the form

$$u(x,t) = X(x)T(t) \tag{1.17}$$

where $X(x)$ and $T(t)$ are unknown. Substitution into Equation 1-1 leads to $X''(x)/X(x) = (1/c^2)\, T''(t)/T(t)$. Since the left side, a function of x alone, equals a function of t alone, each side must in turn equal the constant $-\lambda$. The result

$$X''(x)/X(x) = (1/c^2)\, T''(t)/T(t) = -\lambda \tag{1.18}$$

implies two *ordinary* differential equations,

$$X''(x) + \lambda\, X(x) = 0 \tag{1.19}$$

$$T''(t) + c^2\lambda\, T(t) = 0 \tag{1.20}$$

Here, λ is a positive constant (zero and negative choices lead to trivial solutions). The *assumption* (or, *Ansatz*) $u(x,t) = X(x)T(t)$ and the boundary conditions $u(0,t) = X(0)T(t) = 0$ and $u(L,t) = X(L)T(t) = 0$ show that $X(0) = X(L) = 0$. Thus, the solution $X(x) = D_1\cos(x\sqrt{\lambda}) + D_2\sin(x\sqrt{\lambda})$ requires $X(0) = D_1 = 0$ and $X(L) = D_2\sin(L\sqrt{\lambda}) = 0$.

$D_2 = 0$ would yield a zero solution, which violates initial conditions, so we consider $\sin(L\sqrt{\lambda}) = 0$; this requires $L\sqrt{\lambda} = n\pi$, where $n = 1, 2, 3, \ldots$ may be arbitrary. To note these possibilities, we introduce the subscripted quantity

$$\lambda_n = (\pi n/L)^2 \tag{1.21}$$

to which we associate the functions

$$X_n(x) = C_n\sin(\pi nx/L) \tag{1.22}$$

$$T_n(t) = A_n\cos(\pi nct/L) + B_n\sin(\pi nct/L) \tag{1.23}$$

noting that the time function is obtained by solving the equation for $T(t)$.

Since *a* solution takes the form $u_n(x,t) = X_n(x)T_n(t) = \{A_n\cos(\pi nct/L) + B_n\sin(\pi nct/L)\}\sin(\pi nx/L)$, where we have assumed $C_n = 1$ without loss of generality, the more complete superposition solution satisfies

$$u(x,t) = \Sigma u_n(x,t)$$
$$= \Sigma \{A_n \cos(\pi nct/L) + B_n \sin(\pi nct/L)\} \sin(\pi nx/L) \quad (1.24)$$

where the summation is taken from n = 1 to ∞.

We have already used *boundary conditions* to select a *sine* expansion for u(x,t), but the series coefficients are still undetermined. These can be fixed by applying *initial conditions.* The substitution $u(x,0) = \phi(x) = \Sigma u_n(x,0) = \Sigma A_n \sin(\pi nx/L)$, together with Equations 1-11 and 1-12, lead to

$$A_n = \{2 \int \phi(x) \sin n\pi x/L \, dx\}/L \quad (1.25)$$

Similarly, $u_t(x,0) = \psi(x) = \Sigma \partial u_n(x,0)/\partial t = \Sigma(\pi nc/L) B_n \sin(\pi nx/L)$ yields

$$B_n = \{2 \int \psi(x) \sin n\pi x/L \, dx\}/(\pi nc) \quad (1.26)$$

Since A_n and B_n are known, the solution for u(x,t) is completely determined. This *formal construction* does not necessarily produce a solution, e.g., if the derived series diverges or if it is not differentiable.

Example 1-4: String with distributed forces. Now let us consider the *inhomogeneous wave equation*, namely,

$$\partial^2 u/\partial t^2 - c^2 \partial^2 u/\partial x^2 = f(x,t) \quad (1.27)$$

where the right side may represent the acceleration of a *concentrated* force, a *distributed* force, or both. Again, we consider the boundary conditions $u(0,t) = 0$ and $u(L,t) = 0$, and the initial conditions $u(x,0) = \phi(x)$ and $u_t(x,0) = \psi(x)$.

Separation of variables again shows that sin πnx/L is the appropriate function satisfying boundary conditions. But since f(x,t) is nonzero and arbitrary, we take a modified approach and write

$$u(x,t) = \Sigma u_n(t) \sin \pi nx/L \quad (1.28)$$

To be consistent, we also expand the functions φ(x), ψ(x) and f(x,t) in similar series so that, for example, we obtain the Fourier series pair

$$f(x,t) = \Sigma f_n(t) \sin \pi nx/L \quad (1.29a)$$

$$f_n(t) = \{2\int f(x,t) \sin n\pi x/L \, dx\}/L \quad (1.29b)$$

where t is a parameter. Likewise, we obtain,

$$\phi(x) = \Sigma \phi_n \sin \pi nx/L \quad (1.30a)$$

$$\phi_n = \{2\int \phi(x) \sin n\pi x/L \, dx\}/L \quad (1.30b)$$

$$\psi(x) = \Sigma \psi_n \sin \pi nx/L \quad (1.31a)$$

$$\psi_n = \{2\int \psi(x) \sin n\pi x/L \, dx\}/L \qquad (1.31b)$$

Substitution into Equation 1-27 leads to Σ (sin $\pi nx/L)\{-c^2(\pi n/L)^2 u_n(t) - u_n''(t) + f_n(t)\} = 0$. This is satisfied if the coefficients within the { } brackets vanish, that is, if the functions $u_n(t)$ obey the ordinary differential equation

$$u_n''(t) + c^2(\pi n/L)^2 u_n(t) = f_n(t) \qquad (1.32)$$

Now, using Equation 1-28, our initial conditions can be written as $u(x,0) = \phi(x) = \Sigma\, u_n(0) \sin \pi nx/L = \Sigma\, \phi_n \sin \pi nx/L$ and $u_t(x,0) = \psi(x) = \Sigma\, du_n(0)/dt \sin \pi nx/L = \Sigma\, \psi_n \sin \pi nx/L$ where we have designated

$$u_n(0) = \phi_n \qquad (1.33)$$

and

$$du_n(0)/dt = \psi_n \qquad (1.34)$$

As in the previous example, ϕ_n and ψ_n are known from Fourier expansions (see Equations 1-24 to 1-26). This provides the initial conditions needed to integrate Equation 1-32.

If spatial variations in $f(x,t)$ are rapid, numerous series terms are required to provide adequate spatial resolution. This is the case with concentrated *delta function* loads, for which alternative solution methods are presented later.

Example 1-5: Alternative boundary conditions. We consider anticipated changes for other boundary condition classes. If we had posed the derivative model $u_x(0,t) = u_x(L,t) = 0$, the solution would take the form $u(x,t) \propto \cos n\pi x/L$, with $n = 0, 1, 2, 3, \ldots$, so that $u_x(x,t) \propto \sin n\pi x/L$ vanishes at $x = 0, L$.

And if we had posed the boundary conditions $u(0,t) = u_x(L,t) = 0$, we would have taken $u(x,t) \propto \sin n\pi x/L$, which satisfies $u(0,t) = 0$. Then, $u_x(x,t) \propto \cos n\pi x/L$ leads to $u_x(L,t) \propto \cos n\pi = 0$; hence, $n = 1/2, 3/2, 5/2$, and so on. Similar remarks apply to the problem requiring $u_x(0,t) = u(L,t) = 0$.

From these examples, we observe recurring types of functions and operations. The functions sin $n\pi x/L$ and cos $n\pi x/L$ which arose as a result of our boundary condition classes and Fourier series manipulations are known as *eigenfunctions, normal modes,* or *mode shapes*.

The λ_n shown in Equation 1-21 are known as the *eigenvalues* or *characteristic values* of the system. If we write the cos $(\pi nct/L)$ function in Equation 1-23 in terms of a more familiar cos ωt, we have $\omega_n = \pi nc/L$, where the ω_n represent the *natural frequencies* of the system.

Example 1-6: Mixed boundary conditions. Again, eigenfunctions allow us to fit boundary conditions to series expansions. We have dealt with sine and cosine series so far, but others are also used in mathematics, e.g., series containing Bessel, Hankel, Hermite, Laguerrre and Legendre functions.

A given function may be expanded in any of these series; the nature of the boundary value problem, however, determines the exact type used. Once the appropriate *sequence of eigenfunctions* is identified for a particular problem, we can express any *piecewise continuous function* in the corresponding series, since the coefficients depend only on *integrated (not differentiated) properties* (see Equations 1-9 to 1-12).

This is important because many variables in petroleum applications contain embedded *discontinuities*, for which spatial derivatives do not exist. The relaxed piecewise continuity requirement allows us to obtain analytical solutions for practical problems. Series containing discontinuities cannot be differentiated, of course, but they can be integrated.

Numerous wave applications are associated with discontinuities. For example, *discontinuities in pressure* characterize the acoustic source for positive pressure MWD pulsers, while *velocity discontinuities* model negative pressure pulsers. Discontinuities in axial displacement, i.e., *displacement sources*, model the kinematics of rotating drillbits in drillstring vibrations.

In order for series representations to be possible, the eigenfunction sequence must *complete*. This requirement, often construed by engineers as mathematical formality, is actually quite important to problems such as those cited, where internal discontinuities originate at the source point.

It also turns out that a function with a finite number of discontinuities (e.g., multiple MWD signal generators in a multiplexed system) *can* be represented by eigenfunction series in a *least-squares* sense. But in order that this representation be *easy* to accomplish, the sequence must be *mutually orthogonal*. For our purposes, it suffices to say that orthogonality is defined by special integral requirements that must be satisfied by eigenfunction pairs.

It is necessary to introduce these abstract ideas because not all classes of boundary conditions yield to simple analyses such as those presented earlier. For example, in axial drillstring vibrations, both surface and downhole boundary conditions may take the *mixed* form $\alpha u_x + \beta u_t + \lambda u = 0$.

For such problems, the boundary conditions turn out to depend on the eigenvalue k, and the eigenfunctions will *not* be mutually orthogonal. Although for the boundary condition just given, the eigenfunctions are complete, it is difficult to compute the coefficients of the eigenfunction series.

Thus, obtaining a solution to any initial value problem for such boundary conditions using series can prove to be elusive. Usually, the solution is obtained circuitously, e.g., using *Laplace transforms*, or numerical simulations, as we will when we model *rate-of-penetration* in drillstring vibrations.

For a comprehensive discussion, refer to Morse and Feshbach (1953); there, alternative but complicated solution methods are given, which apply when orthogonality breaks down. Leighton (1967) also gives mathematical discussions on orthogonality and completeness.

Example 1-7: Problems without initial conditions. Very often, vibrations may have been maintained over large enough time scales, that the effects of attenuation have rendered the influence of initial conditions insignificant.

Thus we consider *problems without initial conditions*, or *dynamically steady* wave motions where oscillations at constant frequency are maintained. These formulations are useful because the required labor is reduced, and the results, unlike series expansions, are amenable to physical interpretation.

Because steady state formulations lead to usable results more rapidly than general solutions, we can increase the complexity of the underlying physical model. Equation 1-1, for example, describes undamped motions only, and arises as an approximation to the *dissipative formulation*

$$\partial^2 u/\partial t^2 + \zeta\, \partial u/\partial t - c^2\, \partial^2 u/\partial x^2 = 0 \tag{1.35}$$

We will solve both Equations 1-1 and 1-35, using dynamically steady assumptions, and introduce simple concepts from complex variables (Equation 1-35 is discussed at length in Chapter 5). Again, the undamped equation is

$$\partial^2 u/\partial t^2 - c^2\, \partial^2 u/\partial x^2 = 0 \tag{1.36}$$

We consider $0 \le x \le \infty$, and prescribe the end displacement $u(0,t) = A \cos \omega t$, where ω and the amplitude A are real. The resulting waves clearly propagate to the right. We solve this problem using several elementary methods, demonstrating the advantages and disadvantages of each.

Example 1-7a: Naive approach. Students often wonder why *complex exponentials* are used at all. Since practical solutions must ultimately appear in real terms, why should imaginary numbers be used in the solution process?

We attempt to solve the problem using real functions only, pretending that complex functions do not exist. Knowing that trigonometric functions yield simplifications, we assume the separable solution for a *real* function u(x,t) as

$$u(x,t) \cong A \cos \omega t\, U(x) \tag{1.37}$$

where the *real* function U(x) is to be determined. The $\cos \omega t$ in Equation 1-37 is motivated by its appearance in the boundary condition $u(0,t) = A \cos \omega t$.

If we substitute Equation 1-37 into Equation 1-36, we obtain the one-dimensional *Helmholtz equation*

$$U''(x) + \omega^2/c^2\, U(x) = 0 \tag{1.38}$$

whose solution is $U(x) = C_1 \sin \omega x/c + C_2 \cos \omega x/c$. We may write $u(x,t) \cong C_3 \cos \omega t \sin \omega x/c + C_4 \cos \omega t \cos \omega x/c$, on replacing AC_1 and AC_2 by C_3 and C_4.

The identities cos ωt sin ωx/c = 1/2 {sin ω (x/c - t) + sin ω (x/c + t)} and cos ωt cos ωx/c = 1/2 {cos ω(x/c - t) + cos ω(x/c + t)} allow us to recast our solution as the sum of four terms, in particular, $u(x,t) \cong 1/2\ C_3 \sin \omega\ (x/c + t) + 1/2\ C_4 \cos \omega(x/c + t) + 1/2\ C_3 \sin \omega\ (x/c - t) + 1/2\ C_4 \cos \omega(x/c - t)$.

Now, the sin ω (x/c + t) and cos ω(x/c + t) terms represent left-going waves, and the physical solution is right-going. But if we set the left-going coefficients C_3 and C_4 to zero, the right-going terms also vanish. Something is logically inconsistent, and we will see why using a reformulated approach.

Example 1-7b: Correct approach. Let us replace the "obvious" choice assumed in Equation 1-37 by the separable product

$$u(x,t) \cong e^{i\omega t}\, U(x) \tag{1.39}$$

where both u(x,t) and U(x) may be *complex* functions. If we substitute into Equation 1-36, we again obtain Equation 1-38. Now, we take its solution in the form $U(x) = D_1 e^{i\omega x/c} + D_2 e^{-i\omega x/c}$ so that $u(x,t) \cong D_1 e^{i\omega(x/c+t)} + D_2 e^{i\omega(-x/c+t)}$.

Since the D_1 term is a left-going wave, we discard it, setting $D_1 = 0$; this leaves $u(x,t) \cong D_2 e^{i\omega(-x/c+t)}$, which satisfies $u(0,t) \cong D_2 e^{i\omega t}$. Because we have prescribed u(0,t) = A cos ωt, we select $D_2 = A$, so that $u(x,t) \cong Ae^{i\omega(t-x/c)}$.

Next we use the substitution $u = u^r + i\, u^i$, and the complex identity $e^{i\theta} = \cos \theta + i \sin \theta$, where $i = \sqrt{(-1)}$. Since A was assumed to be real, the real part of our complex solution yields $u^r(x,t) \cong A \cos \omega(t - x/c)$, while the imaginary part leads to $u^i(x,t) \cong A \sin \omega(t - x/c)$.

The real (right-going) part $u^r(x,t) \cong A \cos \omega(t - x/c)$ clearly satisfies the boundary condition $u^r(0,t) \cong A \cos \omega t$, and solves the boundary value problem. But our procedure gives an additional "free" solution, $u^i(x,t) \cong A \sin \omega(t - x/c)$, which satisfies $u^i(0,t) = A \sin \omega t$. Thus, *complex variables* produce simplifications when properly employed.

Example 1-7c: Faster approach. We could have written the solution immediately from first principles. Since the waves *must* be right-going, u(x,t) *must* be a function of "x-ct", that is, the dimensionless quantity ω(t - x/c). The equation $u(0,t) \cong A \cos \omega t$ motivates us to replace ωt, which does not contain x, by ω(t - x/c), which does. This yields $u(x,t) \cong A \cos \omega(t - x/c)$.

Example 1-8: Dissipative wave solution. Solutions to Equation 1-35 are more complicated, and analysis shows that the damping $\zeta\ \partial u/\partial t$ affects both amplitude and *phase* (that is, the ω(t - x/c) and ω(t + x/c) in the above solutions). Equations 1-35 and 1-39 yield the complex differential equation

$$U''(x) + \{(\omega^2 - i\omega\zeta\,)/c^2\}\, U(x) = 0 \tag{1.40}$$

for a complex U(x). Exponential rather than trigonometric functions are again used to satisfy boundary conditions, and we attempt solutions of the form U(x) = $e^{\alpha x}$. Substitution in Equation 1-40 leads to

$$\alpha^2 = -\omega^2/c^2 + i\omega\zeta/c^2 \tag{1.41}$$

When $\zeta = 0$, the solutions to Equation 1-41 are $\alpha = i\,\omega/c$ and $-\,i\,\omega/c$. The assumption $u(x,t) \cong e^{i\omega t}\,U(x)$ produces the solutions $e^{i(\omega t+\omega x/c)}$ and $e^{i(\omega t-\omega x/c)}$ as before, the first of which is eliminated since it is left-going.

In general, we can express any *complex number* X + i Y in the *polar form* $X + i\,Y = R(\cos\Theta + i\sin\Theta)$, where $R = \sqrt{(X^2 + Y^2)}$ and $\Theta = \tan^{-1}\{Y/X\}$. From *De Moivre's Theorem* in complex variables, it follows that $\{X + i\,Y\}^{1/2} = \{R(\cos\Theta + i\sin\Theta)\}^{1/2} = R^{1/2}\{\cos(\Theta+2k\pi)/2 + i\sin(\Theta+2k\pi)/2\}$, k = 0, 1.

For dissipative problems, the solution to Equation 1-41 is required. Setting $X = -\,\omega^2/c^2$ and $Y = \omega\zeta/c^2$, the two square roots of α obtained are

$$\alpha_1 = (\omega^4/c^4 + \omega^2\zeta^2/c^4)^{1/4} \tag{1.42a}$$

$$\times\ [\cos\{\tfrac{1}{2}\arctan(-\zeta/\omega)\} + i\sin\{\tfrac{1}{2}\arctan(-\zeta/\omega)\}]$$

$$\alpha_2 = (\omega^4/c^4 + \omega^2\zeta^2/c^4)^{1/4} \tag{1.42b}$$

$$\times\ [\cos\{\tfrac{1}{2}\arctan(-\zeta/\omega) + \pi\} + i\sin\{\tfrac{1}{2}\arctan(-\zeta/\omega) + \pi\}]$$

Following Example 1-7, only the root corresponding to the right-going wave is physically meaningful. In the zero damping limit when $\zeta = 0$, our dissipative solution should reduce to the earlier one.

We choose as the *principal branch* of the $\tan^{-1}$ function the one yielding $\tan^{-1}(-\,\zeta/\omega) = \pi$ when $\zeta = 0$, so that $\alpha_1 = (\omega/c)\,[\cos\{\tfrac{1}{2}\pi\}\} + i\sin\{\tfrac{1}{2}\pi\}] = i\omega/c$ and $\alpha_2 = (\omega/c)\,[\cos\{\tfrac{1}{2}\pi + \pi\} + i\sin\{\tfrac{1}{2}\pi + \pi\}] = -\,i\omega/c$ as before.

Substitution in $u(x,t) \cong e^{i\omega t}\,U(x) = e^{i\omega t}\,e^{\alpha x}$ clearly shows that the α_1 solution "$e^{i\omega(t+x/c)}$" is left-going, while the α_2 solution "$e^{i\omega(t-x/c)}$" is the desired right-going wave. Only the α_2 root is relevant, with

$$\alpha_2 = (\omega^4/c^4 + \omega^2\zeta^2/c^4)^{1/4}\cos\{\tfrac{1}{2}\arctan(-\zeta/\omega) + \pi\}$$
$$+\, i\,(\omega^4/c^4 + \omega^2\zeta^2/c^4)^{1/4}\sin\{\tfrac{1}{2}\arctan(-\zeta/\omega) + \pi\} \tag{1.42c}$$

If Equation 1-42c is introduced into $u(x,t) \cong e^{i\omega t}\,U(x) = e^{i\omega t}\ e^{\alpha x}$, with a factor A to handle the boundary condition, we obtain the complex function

$$u(x,t) = A\exp\{\,(\omega^4/c^4 + \omega^2\zeta^2/c^4)^{1/4}\cos\{\tfrac{1}{2}\arctan(-\zeta/\omega) + \pi\}\}x$$
$$\mathrm{X}\ \exp i\{\omega t + (\omega^4/c^4 + \omega^2\zeta^2/c^4)^{1/4}\sin\{\tfrac{1}{2}\arctan(-\zeta/\omega) + \pi\}x\,\} \tag{1.43}$$

whose real part is

$$u^r(x,t) = A\exp\{(\omega^4/c^4 + \omega^2\zeta^2/c^4)^{1/4}\cos\{\tfrac{1}{2}\arctan(-\zeta/\omega) + \pi\}\}x$$
$$\times\ \cos\{\omega t + (\omega^4/c^4 + \omega^2\zeta^2/c^4)^{1/4}\sin\{\tfrac{1}{2}\arctan(-\zeta/\omega) + \pi\}x\} \quad (1.44)$$

Equation 1-44 clearly satisfies $u(0,t) = A \cos \omega t$. While this is reminiscent of the $u^r(x,t) \cong A \cos \omega(t - x/c)$ in Example 1-7, there are obvious differences: an exponential damping with x, plus a complicated cosine argument showing that dissipation affects the wave phase "$\omega(t - x/c)$" obtained in the undamped case.

Thus, in attenuative media, waves undergo shape distortions as well as changes to propagation speed. From Equation 1-44, the exact speed is

$$dx/dt = -\omega/\{(\omega^4/c^4 + \omega^2\zeta^2/c^4)^{1/4}\sin\{\tfrac{1}{2}\arctan(-\zeta/\omega) + \pi\}\} \quad (1.45)$$

We emphasize that all of our results apply to Equation 1-35 only, and that this damping model is applicable to a limited number of engineering problems.

STANDING VERSUS PROPAGATING WAVES

In Examples 1-3 and 1-4 for the *bounded* domain $0 \leq x \leq L$, we obtained closed form solutions for problems with and without external excitation. In Examples 1-7 and 1-8, we examined excitation sources located at $x = 0$, and constructed right-going waves that propagated to infinity. More general kinds of dissipation models will be studied using *kinematic wave theory* methods to be discussed in Chapter 2.

Standing waves. On bounded domains, superpositions of left and right-going waves may combine to form *standing waves* that do not move either left or right, although their amplitudes will locally vary with time.

The simplest example is produced by plucking a guitar string: the string executes back-and-forth motions which appear geometrically similar, whose amplitudes increase and decrease with time. Since the waves travel nowhere, and do not transport energy, they are called *standing waves.*

Standing waves contain systems of *nodes* or *minima*, where amplitudes vanish, and *anti-nodes* or *maxima,* where amplitudes are largest, which are of engineering interest. In machine vibrations, anti-nodes indicate where shock absorbers *should* be placed. Nodes represent locations where transducers *should not* be placed: they yield no information for *transfer function* analysis.

Propagating waves. On the other hand, the problems discussed in Examples 1-7 and 1-8 illustrate *propagating waves.* Here, the waves are not stationary, since a reflective right boundary does not exist. *Wave energy* is propagated over a distance, and the waves are said to be *propagating waves*.

How are propagating and standing waves related in bounded systems? All waves start in a highly transient state and propagate outward. Finite boundaries interrupt this propagation, reflecting part of the energy back toward the source; repeated indefinitely, standing waves appear after sufficient time.

Combined standing and propagating waves. A wave system can be standing *and* propagating. Consider a semi-infinite string ($0 < x < \infty$) attached to a constraint at $x = 0$. If a point excitation at $x = x^* > 0$ oscillates sinusoidally at constant frequency, then $0 < x < x^*$ contains a standing wave, while $x > x^*$ hosts the propagating wave.

Characterizing propagating waves. We have considered two simple wave models so far, the undamped and damped wave equations in *canonical* form. Here we will *characterize* propagating waves in greater detail.

We will consider the undamped transverse vibrations of a string satisfying $\rho_l\, \partial^2u/\partial t^2 - T\, \partial^2u/\partial x^2 = 0$, and dissipative waves satisfying $\rho_l\, \partial^2u/\partial t^2 + \gamma\, \partial u/\partial t - T\, \partial^2u/\partial x^2 = 0$ (note that $c^2 = T/\rho_l$).

Here, ρ_l, γ, and T denote lineal mass density, damping factor, and tension (a formal derivation is given in the next section). In going from the undamped to the damped model, we uncovered new effects such as shape distortion and changes to wave speed, but many more effects await discovery.

In order to understand wave propagation more fully, let us consider the undamped model $\rho_l\, \partial^2u/\partial t^2 - T\, \partial^2u/\partial x^2 = 0$, and assume a *uniform, plane, monochromatic* wave solution $u(x,t) = A \sin (kx - \omega t)$, where A is the *wave amplitude*, k is the *wavenumber*, and ω is the *frequency*.

The wavenumber is the number of *wave crests* in a length 2π, whereas the frequency is the number of oscillations in a time 2π. Direct substitution of $u(x,t) = A \sin k \{x - (\omega/k)\, t\}$ in the wave equation yields $\rho_l\omega^2 = Tk^2$. Thus, the solution for the *speed* $\omega/k = \pm \sqrt{(T/\rho_l)}$ shows that both right and left-going waves are correctly found as part of the substitution; also, a sinusoidal wave component stays sinusoidal as it propagates.

On the other hand, suppose we consider the more complicated situation in which the string rests on an elastic base, so that our vibrations now satisfy the equation $\rho_l\, \partial^2u/\partial t^2 + \kappa\, u - T\, \partial^2u/\partial x^2 = 0$. Here, the Greek letter κ (kappa) describes the elastic constant of the foundation.

The substitution $u(x,t) = \sin (kx - \omega t)$ now leads to $\omega^2 = (T/\rho_l)k^2 + \kappa/\rho_l$, so that the speed $\omega/k = \pm (\sqrt{\{(T/\rho_l)k^2 + \kappa/\rho_l\}})/k$ depends on k. Thus, we find that waves having different lengths propagate with different speeds. Such waves are said to be *dispersive*, as opposed to being *nondispersive*.

The equations $\rho_l\omega^2 = Tk^2$ and $\omega^2 = (T/\rho_l)k^2 + \kappa/\rho_l$ are *dispersion relations* because they describe wave dispersion. Because a *general disturbance* consists of wave components with different lengths, it will lose its identity after a finite time in a dispersive medium. Dispersion in *MWD communications channels* is undesirable, since information packets must be periodically reassembled; in mud pulse telemetry, dispersion is much less a problem than is attenuation.

In time-dependent problems, wave properties *may* be proportional to $e^{i\omega t}$ or sin $(\omega t+\phi)$, where ϕ is a *phase angle*, with the *angular frequency* ω being constant. Since ω cycles appear in 2π units of time, the time per cycle or *period* is $2\pi/\omega$, and the *number of cycles per unit time* or *frequency* is $f = \omega/2\pi$.

The units of frequency f are *hertz* (Hz), where *1 Hz is 1 cycle per second,* while the units of angular frequency are *radians per second.* If a periodic dependence on space (x) exists, e.g., sin $(kx-\omega t)$, or sin $k(x-ct)$ where k is the *wavenumber* and $c = \omega/k$ is the *phase velocity,* the repetition length or *wavelength* is $\lambda = 2\pi/k = 2\pi c/\omega = c/(\omega/2\pi)$ or $\lambda = c/f$ (e.g., if c = 4,800 ft/sec and f = 12 Hz, then λ = 400 ft).

Once the dispersion relation for a system is known, either analytically or experimentally, the relationship between k and ω is uniquely defined; in other words, when k is specified, ω can be computed, and vice-versa. As we will show in Chapter 2, *the dispersion relation contains the entire physical description of the wave propagation.*

Two measures of wave speed are considered by physicists and engineers, the *phase velocity* C_p cited above, and the *group velocity* C_g,

$$C_p = \omega/k \tag{1.46}$$

$$C_g = \partial\omega/\partial k \tag{1.47}$$

For the classical equation $\rho_l\, \partial^2u/\partial t^2 - T\, \partial^2u/\partial x^2 = 0$, both phase and group velocities *happen* to be identical. For $\rho_l\, \partial^2u/\partial t^2 + \kappa\, u - T\, \partial^2u/\partial x^2 = 0$, though, phase and group velocities differ.

The distinction between phase and group velocities was discovered by physicists at the turn of the century. It turns out that *wave energy* (among other quantities of physical interest) propagates with group velocity. Phase velocity, in the scheme of nature, is unimportant kinematically and dynamically.

In drillstring vibrations, severe downhole lateral oscillations cause catastrophic failures and twistoffs. Yet, they *cannot* be detected from the surface, even in vertical wells where borehole wall damping is insignificant. This paradox is easily explained using group velocity arguments in Chapter 6.

In the foregoing examples, the dispersion relations obtained were *real,* as opposed to *complex.* Let us now consider the damped wave equation $\rho_l\, \partial^2u/\partial t^2 + \gamma\, \partial u/\partial t - T\, \partial^2u/\partial x^2 = 0$, and attempt $u(x,t) = \sin(kx - \omega t)$ once more.

It is clear from the even and odd derivative combinations in the differential equation that our trigonometric functions will never cancel. This may have been anticipated from Equation 1-44, which indicates that dissipation affects phase nontrivially.

As before, we instead attempt a complex substitution. The choice $u(x,t) = A \exp i(kx-\omega t)$ leads to $\rho_l \omega^2 = Tk^2 - i \gamma\omega$, which is a *complex dispersion relation*. So, ω must be likewise complex; we consider the *complex frequency*

$$\omega(k) = \omega_r(k) + i\, \omega_i(k) \tag{1.48}$$

Direct substitution now yields $\rho_l (\omega_r^2 - \omega_i^2 + i\, 2\omega_r\omega_i) = Tk^2 - i\gamma(\omega_r + i\,\omega_i) = Tk^2 + \gamma(\omega_i - i\,\omega_r)$. Next, taking real parts gives $\rho_l (\omega_r^2 - \omega_i^2) = Tk^2 + \gamma\,\omega_i$, whereas taking imaginary parts gives $\rho_l\, 2\omega_r\omega_i = -\gamma\,\omega_r$ or $\omega_i = -\gamma/(2\rho_l)$.

As Equation 1-44 suggests, the *real dispersion relation* $\rho_l (\omega_r^2 - \omega_i^2) = Tk^2 + \gamma\,\omega_i$ is affected by dissipation. The *imaginary frequency* $\omega_i = -\gamma/(2\rho_l)$, for *our* damped equation, is a constant independent of k, but in general this is not so physically. If the attenuation is weak, with $|\omega_i/\omega_r| << 1$, and γ small, the real dispersion relation can be approximated by the undamped result $\rho_l\,\omega_r^2 = Tk^2$.

There are more wave properties to elaborate upon. First, observe that in all of the equations considered so far, none contain *variable coefficients*, also known as *nonuniformities*, *inhomogeneities*, or *heterogeneities*.

Certainly, we can envision a problem where the lineal mass density ρ_l varies with x, and the tension T varies with time. If these variations are weak, we can explicitly write $\rho_l(\varepsilon x)\, \partial^2 u/\partial t^2 - T(\varepsilon t)\, \partial^2 u/\partial x^2 = 0$, and expect that the *uniform plane wave* result $\rho_l\,\omega_r^2 = Tk^2$ obtained earlier applies locally.

That is, we physically expect that $\rho_l(\varepsilon x)\,\omega_r^2 = T(\varepsilon t)\, k^2$ holds to leading order, which can be demonstrated by formal *WKB asymptotic methods*. The new dispersion relation now describes a *slowly varying wavetrain*. For waves in dissipative media, the *inhomogeneous* dispersion relation is

$$\omega(k) = \omega_r(k,x,t) + i\, \omega_i(k,x,t) \tag{1.49}$$

where $\omega_r(k,x,t)$ and $\omega_i(k,x,t)$ may be obtained analytically, as we have done here, or experimentally from laboratory results.

Classical wave theory is *equation-based* in that partial differential equation models are studied for physical systems at hand; for example, the model $\rho_l\, \partial^2 u/\partial t^2 + \gamma\, \partial u/\partial t - T\, \partial^2 u/\partial x^2 = 0$ is assumed and analyzed.

If the consequences of a model do not agree with experiment, another is proposed, and its consequences studied. This approach can be tedious, since the simplest equations often elude convenient solution.

Kinematic wave theory, which deals with *propagating waves*, provides an important and powerful alternative to physical modeling. The dispersion functions $\omega_r(k,x,t)$ and $\omega_i(k,x,t)$ are often available experimentally; for example, they may be derived from wave speed and attenuation data.

KWT provides the formalism through which equation based modeling can be circumvented, allowing immediate use of laboratory results in practical problems. This is clear intuitively: $\omega_r(k,x,t)$ and $\omega_i(k,x,t)$ already embody the physics, and in this sense, equation modeling is superfluous.

The word *kinematic* in kinematic wave theory must not be taken lightly: it indicates that the propagation properties for assumed *linear systems* do *not* depend upon the wave amplitude in the leading description.

When *nonlinear* or *finite amplitude effects* (as opposed to *infinitesimal*) are important, analyses similar to those given above can be performed, but the dispersion relations obtained now contain a dependence on wave amplitude.

For example, we can write $\omega = \omega_r(k,x,t,A^2)$ for nondissipative systems. The formalism extending KWT to dissipative nonlinear systems, presented in Whitham (1974) for low-order systems, was further extended by Chin (1976, 1981) to high-order systems. Some general results are presented in Chapter 2.

We return to our survey of classical techniques and continue our discussion by introducing Laplace transforms. This technique is useful in drillstring vibrations modeling, and in mud pulse telemetry modeling, because it greatly simplifies source description through *delta function* formulations.

LAPLACE TRANSFORMS

Laplace transforms provide powerful tools for the solution of boundary value problems. The foundation underlying this classical technique, developed by the English engineer Heaviside a century ago, is firmly rooted in the theory of complex variables, which we will not attempt to reproduce here.

Churchill (1941, 1958) provides readable expositions on the method, treating problems in a down-to-earth manner. Here, we will give an overview only, preferring to demonstrate through examples how solutions are obtained.

Let us derive the equation satisfied by the transverse displacements u(x,t) of a string acting under its own weight,

$$\partial^2 u/\partial t^2 - c^2\, \partial^2 u/\partial x^2 = -g \tag{1.50}$$

where g is the acceleration due to gravity. We apply Newton's law **F** = ma in the vertical direction to the mass element ρ_l dx shown in Figure 1-2, where ρ_l is the lineal mass density and dx is a small length.

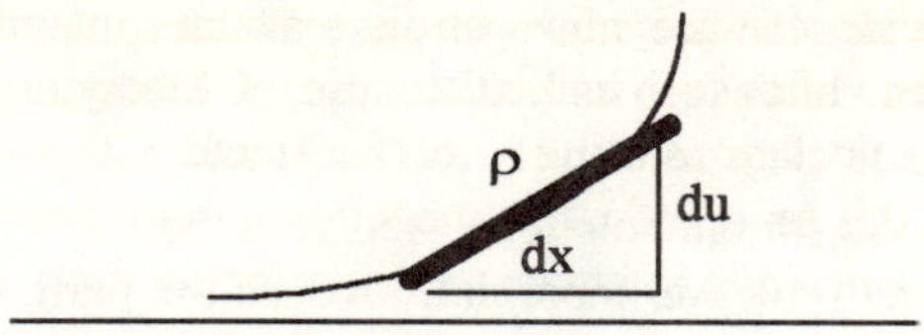

Figure 1-2. Transverse vibrations of a string.

Along the string segment, the slope $\Delta u/\Delta x$ must equal the vertical-to-horizontal force ratio V/H. For small displacements, $H \cong T$, where T is the constant tension acting along the string. Thus, the vertical force V acting at one side of the control element equals $T\Delta u/\Delta x$ or $T\, \partial u/\partial x$.

At the opposite side, the force is $T\, \partial u/\partial x + (T\, \partial u/\partial x)_x\, dx$, so that the incremental force equals $(T\partial u/\partial x)_x\, dx$. Since the weight is $-\rho_l g\, dx$, we obtain $\rho_l\, \partial^2u/\partial t^2 - T\, \partial^2 u/\partial x^2 = -\rho_l g$ which, upon division, yields $\partial^2u/\partial t^2 - c^2\, \partial^2u/\partial x^2 = -g$. In the *free fall* limit with $T = 0$, the familiar $u \cong -1/2\, gt^2$ result reappears.

Example 1-9: String falling under its own weight. We solve Equation 1-50 subject to the initial elevation $u(x,0) = 0$ and the static condition $\partial u(x,0)/\partial t = 0$. The string is rigidly attached to a wall $x = 0$, that is, $u(0,t) = 0$; at infinity, its geometric slope is zero, with $\partial u/\partial x(\infty,t) = 0$.

The reason for our recipe will be apparent shortly. We multiply each side of Equation 1-50 by e^{-st} (with $s > 0$), and integrate with respect to time from $t = 0$ to ∞. Integration by parts shows that $\int e^{-st}\, \partial^2u/\partial t^2\, dt = s^2 \int e^{-st}\, u(x,t)\, dt - su(x,0) - \partial u(x,0)/\partial t$, where we have omitted the limits $(0,\infty)$ for brevity.

The definite integral on the right side is known as the *Laplace transform* of the function u(x,t), and is denoted by

$$L\{u(x,t)\} = U(x,s) = \int_0^\infty e^{-st}\, u(x,t)\, dt \tag{1.51}$$

The time integration has eliminated the appearance of t, replacing it by the *positive* parameter s (this choice ensures that U(x,s) exists).

The key advantage with Laplace transforms is a practical one: once U(x,s) is available, where x is regarded as a constant parameter, the solution u(x,t) can be obtained by "table look-up". Sophisticated *Bromwich contour integration* using *complex variables* provides another powerful option (e.g., see Carrier, Krook, and Pearson, 1966).

The former is preferable, since there is no shortage of transform tables. It goes without saying that the opposite process, finding transforms for given polynomial, trigonometric, and transcendental functions, follows similar lines.

Advances in software now permit real-time integration using *symbolic manipulation*; products such as Mathematica™, Macsyma™, and Maple™ easily handle tedious calculations using recursive algebras.

We proceed with our solution. Applying the $\int e^{-st} \ldots dt$ operation to each term of Equation 1-50, we importantly reduce the partial differential equation to the ordinary one

$$c^2U''(x,s) - s^2U(x,s) = g/s \tag{1.52}$$

with primes denoting x derivatives, where we have simplified intermediate results using initial conditions. Similarly, the boundary conditions transform into $U(0,s) = 0$ and $U'(\infty,s) = 0$.

This ordinary differential equation can be solved using standard techniques, and the result is the superposition of *homogeneous* and *particular solutions* (Leighton, 1967). The solution takes the form $U(x,s) = C_1e^{-sx/c} + C_2e^{+sx/c} - g/s^3$. Application of transformed boundary conditions leads to

$$U(x,s) = -g\,\{1 - e^{-(x/c)s}\}/s^3 \tag{1.53}$$

and table look-up yields the *falling string* solution

$$u(x,t) = -g(2cxt - x^2)/2a^2,\ x \leq ct;\ \text{and}\ - gt^2/2,\ x \geq ct \tag{1.54}$$

The first part of Equation 1-54 shows that the string closest to the wall always remains attached, satisfying $u(0,t) = 0$, and that its shape at any instant is parabolic. The second part for the far end is, consistently, *free fall*.

Example 1-10: Semi-infinite string with a general end support. Here, we will ignore gravity, and consider the boundary value problem defined by

$$\partial^2u/\partial t^2 - c^2\, \partial^2u/\partial x^2 = 0;\ x > 0,\ t > 0 \tag{1.55}$$

the initial conditions $u(x,0) = 0$ and $\partial u(x,0)/\partial t = 0$, and the end model

$$M\, \partial^2u(0,t)/\partial t^2 + \zeta\, \partial u(0,t)/\partial t + ku(0,t) = T\, \partial u(0,t)/\partial x \tag{1.56}$$

Thus we study the coupled motion of the restraint and the string, when the *mass-spring-damper* model is excited by the vertical end force $T\, \partial u(0,t)/\partial x$. A schematic of the problem, showing the mass M, the spring constant k, and the dashpot dissipation factor ζ, is given in Figure 1-3a.

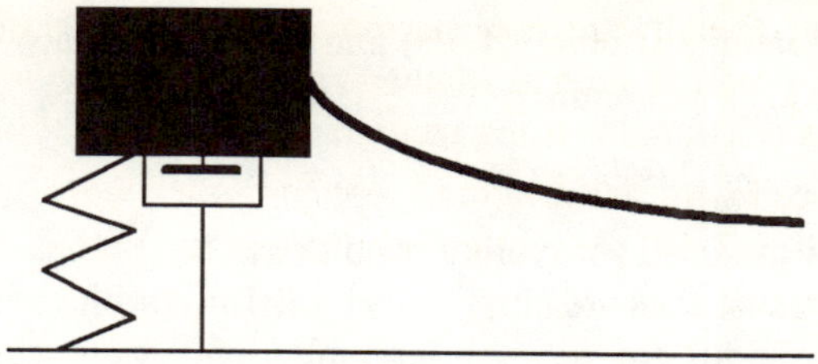

Figure 1-3a. Transverse string coupled to mass-spring-damper.

Earlier we indicated that solutions to the undamped wave equation, which satisfy u(x,t) = f(ct-x) + g(ct+x), propagate with their form unchanged: once a propagating wave enters a one-dimensional *waveguide*, it propagates at a speed c with its shape and size intact.

This is not to say that an initial waveform will never distort. It *can* even when dissipation and dispersion are neglected in the partial differential equation, even over short space and time scales: *shape distortion can be introduced at boundaries for certain classes of auxiliary conditions.*

Here, we will demonstrate some important capabilities of Laplace transforms, and show how transforms, together with the general solution u(x,t) = f(ct-x) + g(ct+x), can yield powerful mathematical results.

A second objective is physically oriented: what happens to a wave of a given shape on reflection from the end boundary condition given above? Earlier we showed that, for simple end conditions, a wave may reflect with either a 0° *or* a 180° phase change.

But not so in general: a sinusoidal incident wave can, on reflection, acquire non-sinusoidal "DC" changes. As we will show in Chapter 5, this property is useful in modeling drillstring rate-of-penetration and bit bounce.

In this exercise, we will assume that the functional form of the incident wave g(ct+x) heading toward x = 0 is given. We will solve for the reflected wave f(ct-x), and then the complete solution u(x,t), so that the transverse oscillations are completely determined everywhere for all time.

The general solution is difficult if we attack the partial differential equation directly. Instead, we solve the ordinary differential equation at x = 0 exactly, and use u(x,t) = f(ct-x) + g(ct+x), with the *Convolution Theorem*, to *analytically continue* the solution into x > 0.

We substitute u(x,t) = f(ct-x) + g(ct+x) into $M\, \partial^2 u(x,t)/\partial t^2 + \zeta\, \partial u(x,t)/\partial t + ku(x,t) = T\, \partial u(x,t)/\partial x$, and then set x = 0. This leads to

$$Mc^2 f''(\xi) + (\zeta c+T) f'(\xi) + kf(\xi) = -Mc^2 g''(\xi) - (\zeta c-T) g'(\xi) - kg(\xi) \qquad (1.57)$$

where ξ is a dummy variable. If F(s) and G(s) are Laplace transforms of f and g with respect to ξ, we obtain

$$F(s) = -G(s) + 2TsG(s)/\{Mc^2s^2 + (\zeta c+T)s + k\} \tag{1.58}$$

where we have assumed zero initial conditions.

Laplace transforms are almost always defined with respect to the time, but they may be used for any *time-like* variables, e.g., x on semi-infinite domains, or ξ above.

It is clear that the first two terms easily invert to $f(\xi) = -g(\xi) + \ldots$, but the inverse of the remaining term is less obvious. However, it is clear that it can be expressed as the product of simpler functions of s.

The *Convolution Theorem* provides direct inversions when a transform can be factored into simpler transforms with known inverses. If $f_1(s)$ and $f_2(s)$ represent Laplace transforms for $F_1(t)$ and $F_2(t)$, then the inverse of $f_1(s)f_2(s)$, denoted by $L^{-1}\{f_1(s)f_2(s)\}$, is simply

$$L^{-1}\{f_1(s)f_2(s)\} = \int_0^t F_1(t-\tau)\, F_2(\tau)\, d\tau \tag{1.59}$$

Note that a function cannot be *arbitrarily* decomposed into the product of *any* two functions $f_1(s)$ and $f_2(s)$. From Equation 1-51, each of these functions *must* vanish as $s \to \infty$. If not, the functions $F_1(t)$ and $F_2(t)$ will *not* exist, and the required inversions will *not* appear in any published tables.

We will take $f_1(s) = G(s)$ and $f_2(s) = 2Ts/\{Mc^2s^2 + (\zeta c+T)s + k\}$, noting that $f_2(\infty) = 0$ as required. Then, the Convolution Theorem yields

$$f(\xi) = -g(\xi) + (2T/Mc^2)(a-b)^{-1} \int_0^\xi g(\xi^*)\, \{ae^{a(\xi-\xi^*)} - be^{b(\xi-\xi^*)}\}\, d\xi^* \tag{1.60}$$

where the constants "a" and "b" satisfy

$$a = \{-(\zeta c+T) - \sqrt{[(\zeta c+T)^2 - 4kMc^2]}\}/2Mc^2 \tag{1.61}$$

$$b = \{-(\zeta c+T) + \sqrt{[(\zeta c+T)^2 - 4kMc^2]}\}/2Mc^2 \tag{1.62}$$

Since $g(\xi)$ is prescribed, $f(\xi)$ is now known, and is easily obtained by integration. By substituting the required dummy variables in ct+x and ct-x, we construct the exact solution as

$$u(x,t) = g(ct+x) - g(ct-x) + \{2T/[Mc^2(a-b)]\} \int_0^{ct-x} g(\sigma) \{ae^{a(ct-x-\sigma)} - be^{b(ct-x-\sigma)}\} d\sigma \quad (1.63)$$

Note that M, ζ, k, T, c and ρ_l do not appear individually in the final solution, but implicitly through the parameters a and b. We also infer from the exponential terms in Equation 1-63 that "ac" and "bc" have dimensions of inverse time. Thus, $(ac)^{-1}$ and $(bc)^{-1}$ define time scales in addition to the period; these are analogous to those in electrical RC circuits.

In Equation 1-63, the g(ct+x) term represents the known, prescribed incident wave emerging from infinity, whereas the term -g(ct-x) represents the reflection at a rigid pinned connection (this is so because the sum g(ct+x) - g(ct-x) vanishes at x = 0). The last term is the *shape distortion* to the incoming wave due to the flexible support; this consists of a phase shift and a shape change.

Example 1-10a: Rectangular pulse. Figure 1-3b shows how a rectangular pulse is distorted at a *pure* elastic boundary (with $M = \zeta = 0$); its closed form solution is given by Graff (1975). We next determine the exact distortion effects for two important types of incident waves.

Example 1-10b: Impulse response. The time response of a system to an *impulse input* is important in *control theory* because it contains all of the system's dynamic characteristics, e.g., reflections, attenuation, and distortions.

We consider an ideal impulse, that is, an infinitely thin pulse of infinite height having bounded energy. This incident pulse, heading toward the reflector at x = 0, is "set off" at t = 0; it is described by the *Dirac delta function* $\delta(t)$ discussed in the next section. Following our recipes there, and the use of Equation 1-63 leads to the exact solution

$$u(x,t) = A\,\delta\{(t+x/c)/t_{ref}\} - A\,\delta\{(t-x/c)/t_{ref}\} + \{2ATt_{ref}/[Mc(a-b)]\}\{ae^{a(ct-x)} - be^{b(ct-x)}\} \quad (1.64)$$

where t_{ref} is a constant with dimensions of time. The $A\,\delta\{(t+x/c)/t_{ref}\}$ term is the incoming wave, while $-A\,\delta\{(t-x/c)/t_{ref}\}$ represents the reflected outgoing wave, had we assumed rigid displacement boundary conditions.

Two time scales $(ac)^{-1}$ and $(bc)^{-1}$ are introduced by the *mixed boundary condition* in Equation 1-56. Our solution shows that they are responsible for a *propagating exponential distortion* $\{ae^{a(ct-x)} - be^{b(ct-x)}\}$ having *nonzero* width.

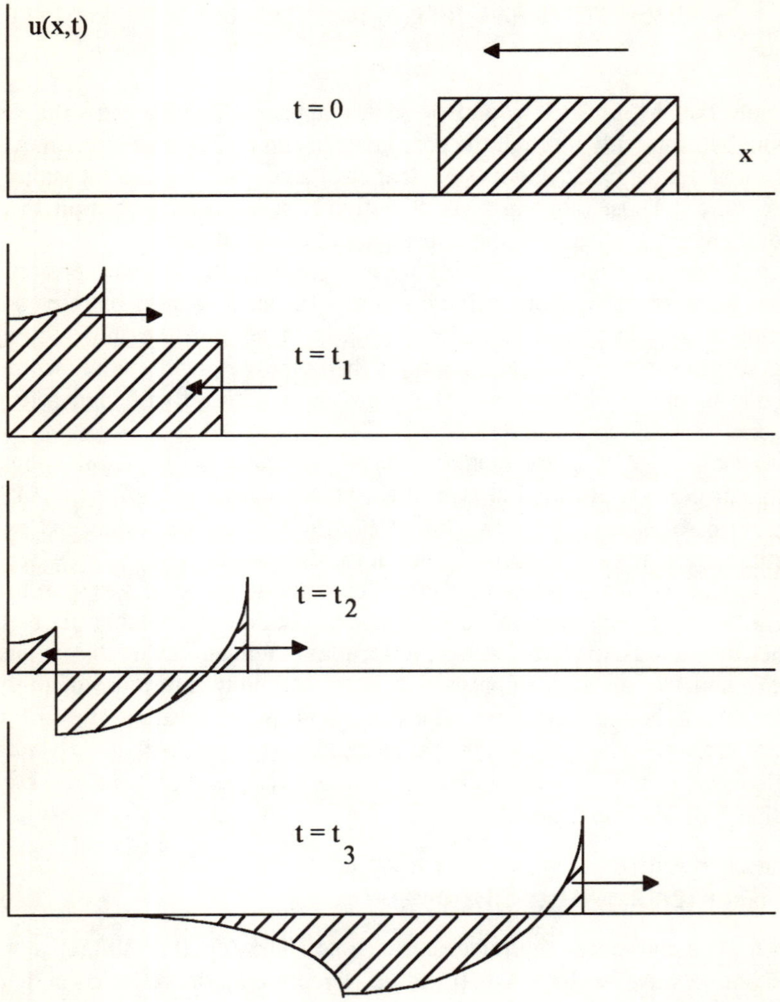

Figure 1-3b. Shape distortion due to elastic boundary.

Example 1-10c: Incident sinusoidal wavetrain. We consider the incident displacement wave A sin $\omega(ct+x)/c$, where ω is the *frequency*. The exact solution, continued next page, is

$$u(x,t) = A \sin \{\omega(ct+x)/c\} - A \sin \{\omega(ct-x)/c\} \tag{1.65}$$

$$+ 2ATb\{b \sin(\omega(ct-x)/c)$$
$$+ (\omega/c) \cos(\omega(ct-x)/c)\}/\{Mc^2(a-b)(b^2 + \omega^2/c^2)\}$$
$$- 2ATa\{a \sin(\omega(ct-x)/c)$$
$$+ (\omega/c) \cos(\omega(ct-x)/c)\}/\{Mc^2(a-b)(a^2 + \omega^2/c^2)\}$$
$$+\{2AT\omega/[Mc^3(a-b)]\}\{ae^{a(ct-x)}/(a^2 + \omega^2/c^2)$$
$$- be^{b(ct-x)}/(b^2 + \omega^2/c^2)\}$$

The first line represents the incoming wave and its pinned end reflection, where A is the wave amplitude. The second and third lines represent the *phase change* due to the support, while the fourth shows that a distortion with *exponential smearing* emerges that shifts the mean DC level.

FOURIER TRANSFORMS

We have seen how Laplace transforms can be used to solve wave problems with nontrivial boundary conditions. Another popular transformation is the *Fourier transform*, useful in solving problems on infinite domains. Consider a function u(x) defined on $-\infty < x < \infty$. Its Fourier transform $U(\omega)$ is defined by

$$U(\omega) = \int_{-\infty}^{+\infty} e^{i\omega x}\, u(x)\, dx \tag{1.66}$$

where ω is a parameter. If the transform $U(\omega)$ were known, then the original function u(x) can be recovered by the complementary integral

$$u(x) = \{1/(2\pi)\} \int_{-\infty}^{+\infty} e^{-i\omega x}\, U(\omega)\, d\omega \tag{1.67}$$

Example 1-11: Propagation of an initially static disturbance. Let us consider the classic initial value problem

$$\partial^2 u(x,t)/\partial t^2 - c^2\, \partial^2 u/\partial x^2 = 0 \tag{1.68}$$

$$u(x,0) = f(x) \tag{1.69}$$

$$\partial u(x,0)/\partial t = 0 \tag{1.70}$$

$$u,\ \partial u(x,t)/\partial x\ ,\ \partial u(x,t)/\partial t \rightarrow 0, \text{ as } \sqrt{(x^2 + t^2)} \rightarrow \infty \tag{1.71}$$

Next, introduce the Fourier transform

$$U(\omega,t) = \int_{-\infty}^{+\infty} e^{i\omega x}\, u(x,t)\, dx \tag{1.72}$$

where t is regarded as a parameter. If we multiply Equation 1-68 by $e^{i\omega x}$, integrate the result with respect to x over $(-\infty,+\infty)$ by parts, and apply the *regularity condition* in Equation 1-71, we obtain

$$U_{tt}(\omega,t) + c^2\omega^2\, U(\omega,t) = 0 \tag{1.73}$$

which is an ordinary differential equation in t. Equation 1-73 has the solution

$$U(\omega,t) = A(\omega) \sin \omega ct + B(\omega) \cos \omega ct \tag{1.74}$$

Equations 1-69 and 1-70 can be similarly transformed, with the result that

$$U(\omega,0) = F(\omega) \tag{1.75}$$

$$U_t(\omega,0) = 0 \tag{1.76}$$

where $F(\omega)$ is the Fourier transform of $f(x)$. It follows that the solution to Equation 1-74 satisfying Equations 1-75 and 1-76 is

$$U(\omega,t) = F(\omega) \cos \omega ct \tag{1.77}$$

Following Equation 1-67, we write the solution $u(x,t)$ as

$$u(x,t) = \{1/(2\pi)\} \int_{-\infty}^{+\infty} e^{-i\omega x}\, F(\omega) \cos \omega ct\, d\omega \tag{1.78}$$

which is the *exact* integral. A more enlightening interpretation is possible if we observe that

$$F(\omega) \cos \omega ct = 1/2\, F(\omega) \exp (i\omega ct) + 1/2\, F(\omega) \exp (-i\omega ct) \tag{1.79}$$

Thus, Equation 1-78 can be written as

$$u(x,t) = \{1/(2\pi)\} \int_{-\infty}^{+\infty} e^{-i\omega x}\, \{1/2\, F(\omega)e^{ic\omega t} + 1/2\, F(\omega)e^{-ic\omega t}\}d\omega \tag{1.80}$$

which involves products of transforms. We turn to the *Fourier Convolution Theorem*, which relates k(x) and u(x) to their transforms $K(\omega)$ and $F(\omega)$, i.e.,

$$\{1/(2\pi)\} \int_{-\infty}^{+\infty} e^{-i\omega t}\, K(\omega)\, U(\omega)\, d\omega = \int_{-\infty}^{+\infty} k(t-x)\, u(x)\, dx \tag{1.81}$$

The transform functions $e^{ic\omega t}$ and $e^{-ic\omega t}$ are associated with the *delta functions* (see next section) shown in Equations 1-82 and 1-83,

$$\int e^{i\omega x}\,\delta(x-ct)\,dx = e^{ic\omega t} \tag{1.82}$$

$$\int e^{i\omega x}\,\delta(x+ct)\,dx = e^{-ic\omega t} \tag{1.83}$$

Thus, the solution in Equation 1-80 can be written as

$$u(x,t) = 1/2 \int_{-\infty}^{+\infty} \delta(x-\xi-ct)\,f(\xi)\,d\xi + 1/2 \int_{-\infty}^{+\infty} \delta(x-\xi+ct)\,f(\xi)\,d\xi \tag{1.84}$$

which is easily evaluated to give

$$u(x,t) = 1/2\, f(x-ct) + 1/2\, f(x+ct) \tag{1.85}$$

This agrees with the general solution in Equation 1-4. Equation 1-85 shows that, at each point x, half of the initial $u(x,0) = f(x)$ disturbance propagates to the right, while the remaining half propagates to the left.

Example 1-12: Directional properties of special wave operators. In constructive solutions to $\partial^2 u(x,t)/\partial t^2 - c^2\, \partial^2 u/\partial x^2 = 0$, the independent variable combinations "x-ct" and "x+ct" arise all too frequently. This suggests that

$$\xi = x-ct \tag{1.86}$$

and

$$\eta = x+ct \tag{1.87}$$

may be more natural than x and t. This is, in fact, true: ξ and η are called *natural, characteristic,* or *canonical* variables.

Let us write $u(x,t) = U(\xi,\eta)$, and determine the equation satisfied by U in the coordinates (ξ,η). Application of the chain rule of calculus shows that

$$u_t(x,t) = U_\xi \xi_t + U_\eta \eta_t = -cU_\xi + cU_\eta \tag{1.88}$$

$$\begin{aligned} u_{tt}(x,t) &= -c(U_{\xi\xi}\xi_t + U_{\xi\eta}\eta_t) + c(U_{\eta\xi}\xi_t + U_{\eta\eta}\eta_t) \\ &= c^2 U_{\xi\xi} - 2c^2 U_{\xi\eta} + c^2 U_{\eta\eta} \end{aligned} \tag{1.89}$$

$$u_x(x,t) = U_\xi \xi_x + U_\eta \eta_x = U_\xi + U_\eta \tag{1.90}$$

$$\begin{aligned} u_{xx}(x,t) &= U_{\xi\xi}\xi_x + U_{\xi\eta}\eta_x + U_{\eta\xi}\xi_x + U_{\eta\eta}\eta_x \\ &= U_{\xi\xi} + 2U_{\xi\eta} + U_{\eta\eta} \end{aligned} \tag{1.91}$$

Substitution of Equations 1-88 to 1-91 in Equation 1-68 leads to

$$\partial^2 U(\xi,\eta)/\partial\xi\partial\eta = 0 \tag{1.92}$$

A first integral can be obtained as

$$\partial U(\xi,\eta)/\partial\xi = \alpha(\xi) \tag{1.93}$$

since $\partial^2 U(\xi,\eta)/\partial\xi\partial\eta = \partial\alpha(\xi)/\partial\eta = 0$ exactly. A second integral is

$$U(\xi,\eta) = \int \alpha(\xi)\, d\xi + \beta(\eta) \tag{1.94}$$

Thus, the general solution to Equation 1-92 takes the form

$$U(\xi,\eta) = A(\xi) + B(\eta) \tag{1.95}$$

where A and B are arbitrary *functions*, thereby reiterating the fact that "x-ct" and "x+ct" are natural independent variables.

This fact can be put to good use. Suppose that a complicated wave field u(x,t) is available experimentally, and that the component waves traveling in each direction are desired. For example, axial waves emanating from the drillbit may carry important lithological *information* up the drillstring.

If the up-going waves propagating along the characteristic ξ = x-ct = constant are required, a $\partial U(\xi,\eta)/\partial\xi$ operation differentiating u(x,t) with respect to ξ suppresses all waves that are functions of the down-going coordinate η = x+ct. This differentiation is accomplished using the fact that

$$\partial U(\xi,\eta)/\partial\xi = u_x x_\xi + u_t\, t_\xi \tag{1.96}$$

Equation 1-96 translates $\partial U(\xi,\eta)/\partial\xi$ requirements into those for $\partial u(x,t)/\partial t$ and $\partial u/\partial x$, that is,

$$\partial U(\xi,\eta)/\partial\xi = 1/2\, (u_x - 1/c\, u_t) \tag{1.97}$$

since Equations 1-86 and 1-87 imply that $x = (\xi+\eta)/2$ and $t = (\eta-\xi)/2c$.

The above u_x term states that spatial derivatives are required, while u_t indicates that time derivatives of u(x,t) are needed. The hardware requirements for *directional filtering* are, therefore, multiple transducers, time storage capabilities, analog-to-digital discretization, and microprocessors.

The exact schemes implementing Equation 1-97 can be drawn from *finite difference* representations of the derivatives at discrete points in space and time (see Chapter 5). What the derivatives in Equation 1-97 require in terms of percentages of wavelengths and periods depends on case-by-case study.

Directional filtering ideas are well known to modern aerodynamicists; in fact, they were used extensively at Boeing in the 1970s to model the effects of flap motions on total lift. In simulating transient compressible flow, *large* computational boxes were formerly used so that numerical errors generated during iterations do not reflect from close boundaries back onto wing surfaces.

Farfield conditions such as $u_x + 1/c\ u_t = 0$ or $u_x - 1/c\ u_t = 0$, known as *radiation conditions* in classical physics, were routinely used at box edges to filter disturbances that might propagate to undesired directions. They allow unidirectional wave simulation on *small* computationally inexpensive boxes, which otherwise create unacceptable reflections.

Directional filters are used in geophysics too. *Imaging algorithms* are tested on *synthetic data* generated by the 3D wave equation for finite reservoir volumes, where *outgoing wave conditions* are used at computational walls to eliminate spurious reflections (e.g., Bleistein (1984), Claerbout (1985a,b)).

EXTERNAL FORCES VERSUS BOUNDARY CONDITIONS

The distinction between *applied external forces* and *prescribed boundary conditions* is sometimes unclear since both are used to excite physical systems. Their differences will be clarified here, since both are needed to pursue MWD and drilling vibration applications.

Single point forces. To visualize the problem, we return to the *transverse displacements* u(x,t) of a stretched string, with lineal mass density ρ_l and tension T. Everyday guitar strings execute such oscillations, and such motions are ideal for illustrating key mathematical ideas. In the equation

$$\rho_l\ \partial^2u/\partial t^2 + \zeta\ \partial u/\partial t - T\ \partial^2u/\partial x^2 + \rho_l g = 0 \qquad (1.98)$$

we have introduced a term $\rho_l g$ proportional to *body force*, g being the acceleration due to gravity, and also, an *internal viscous dissipation* $\zeta\ \partial u/\partial t$. These do *not* affect any *jump properties* to be discussed, but they are retained in our discussion to show *why* they won't.

When an external point force of strength F(t) is applied to $x = x_s$, as shown in Figure 1-4, the *Dirac delta function* representation

$$\rho_l\ \partial^2u/\partial t^2 + \zeta\ \partial u/\partial t - T\ \partial^2u/\partial x^2 + \rho_l g = F(t)\ \delta(x-x_s) \qquad (1.99)$$

is used, where $\delta(x-x_s)$ is the *delta function*. We follow the usual convention, prefixing F with a "+" sign when the force points upward.

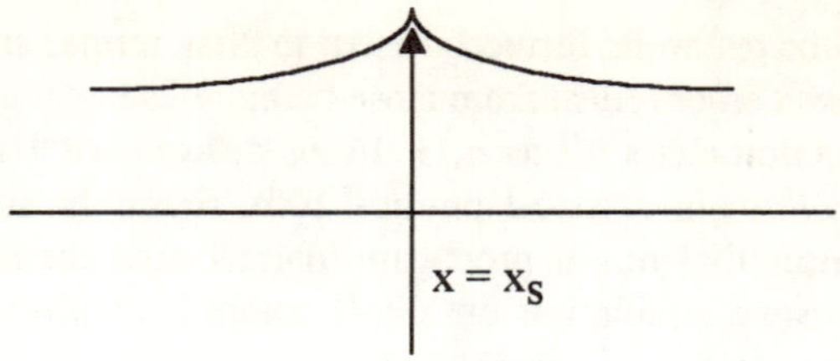

Figure 1-4. Single point force acting on a string.

The delta function is often used in (recipe-oriented) *operational* or *transform calculus*. For us, it suffices to list its two best known properties,

$$\int \delta(x-x_s)\, dx = 1 \tag{1.100}$$

and

$$\int P(x)\, \delta(x-x_s)\, dx = P(x_s) \tag{1.101}$$

where the $\int \ldots dx$ is carried out over any domain containing x_s, and P(x) may be any function. Books dealing with *distributions* are available, but we need not delve further, e.g., Lighthill (1959). If more than one force appears, multiple delta functions simply appear on the right side.

It is clear from Figure 1-4 that *F is responsible for an abrupt change in the slope* $\partial u/\partial x$. In particular, while the displacement u itself is *continuous* or *single-valued* through the point of contact (and hence, the speed and acceleration), the slope $\partial u/\partial x$ takes on *two* different values at the left and right: it is *double-valued.*

That is, a point force (or delta function) produces a *slope discontinuity* or a *jump* in the first *spatial* derivative of the dependent variable u. This jump is conventionally denoted by enclosing the discontinuous quantity in rectangular brackets; in this example, we have $[\partial u/\partial x]$.

The exact magnitude can be obtained from momentum considerations, but we instead obtain it by integrating Equation 1-99 from $x = x_s - \varepsilon$ to $x = x_s + \varepsilon$ where ε is a small positive number,

$$\rho_l \int \partial^2 u/\partial t^2\, dx + \zeta \int \partial u/\partial t\, dx$$
$$- T \int \partial^2 u/\partial x^2\, dx + \rho_l g \int dx = F(t) \int \delta(x-x_s)\, dx \tag{1.102}$$

Integration limits are omitted for brevity. It is clear that, if our string does not break, the acceleration $\partial^2 u/\partial t^2$ takes on the same value at either side of the force; thus, its integral vanishes identically as ε tends to zero (the same vanishing applies to the $\partial u/\partial t$ velocity and gravity terms, and the elastic force had we included it in our formulation).

Now, the Dirac delta function is always defined with the property that its areal sum is *exactly* unity (see Equation 1-100). Since the slope $\partial u/\partial x$ can and will change through $x = x_s$, as shown in Figure 1-4, we conclude that Equation 1-102 reduces to

$$[\partial u/\partial x]_{x=xs} = \partial u/\partial x \mid_{x=xs+\varepsilon} - \partial u/\partial x \mid_{x=xs-\varepsilon} = - F(t)/T \quad (1.103)$$

Thus, an applied force F(t) will produce a jump in slope in the amount -F(t)/T that is independent of ρ_l.

Suppose a hypothetical mechanical device creates internal slope discontinuities $[\partial u/\partial x]$ of magnitude -F(t)/T at $x = x_s$. Then this device is modeled by Equation 1-99. Since Figure 1-4 shows a single force, that is a single delta function, we refer to the excitation as a *monopole*.

Properties of point loads. The point load F(t) is associated with more properties, which are easily illustrated by the action of a bow on a violin string (by simply *writing* F(t), we imply that its magnitude *can* be controlled, and that adequate feedback and control makes this mechanically possible).

First, while F is responsible for a local discontinuity or jump in $\partial u/\partial x$, it does *not* directly dictate the exact value of $\partial u/\partial x$ or u, as a prescribed boundary condition would: *F is associated with differences in* $\partial u/\partial x$ *only.*

Second, waves heading towards $x = x_s$ from left or right will pass *through* x_s without change. Waves on a violin string pass *through* the bow contact point: they do not terminate there and reflect backwards. In other words, *point forces are transparent to propagating waves.*

Example 1-13: Boundary conditions versus forces. Consider a finite string of length L, rigidly attached to a wall at x = 0 so that u(0,t) = 0. We will discuss modeling options for a sinusoidal excitation at the end x = L.

The *boundary condition* $u(L,t) = A_2 \sin \omega t$ *might* be a reasonable model, if everything that can conceivably happen can be described by a prescribed displacement. But it would be restrictive if the actual excitation were due to an external point load entity close to x = L, and *another* constraint describing *another* feature characterizing x = L were needed: we would have "used up" the boundary condition at x = L, thus limiting ourselves in modeling options.

But by using a *forcing function* model instead, e.g., $\rho_l\, \partial^2u/\partial t^2 + \zeta\, \partial u/\partial t - T\, \partial^2 u/\partial x^2 + \rho_l g = A_3 \sin \omega t\, \delta(x-L+\delta)$, where $\delta > 0$, we open up more classes of modeling options because we have not yet "used up" the boundary condition at x = L. These others might be $\partial u(L,t)/\partial x = A_1 \sin \omega t$, $u(L,t) = A_2 \sin \omega t$, or a *radiation condition* $u(x,t) \approx u(x-ct)$ to model semi-infinite effects at a point.

Also, boundary conditions are not transparent: they isolate "left" and "right" events in this sense. Suppose that the force in Figure 1-4 were replaced by a rigid wall, and that the *boundary condition* excitation $u(x_s,t) = A_2 \sin \omega t$ was mechanically enforced at the wall x_s. Then the strings to the left and right of the wall will act independently; they will not "see" each other.

This wall separates events in $x < x_s$ from those in $x > x_s$. If any of these reflect back from finite boundaries, their motions terminate here and re-reflect at the wall, because the exact level of u has been prescribed.

Thus, boundary conditions for u or $\partial u/\partial x$ should be used only when the *exact magnitude* (of u or $\partial u/\partial x$) can be prescribed with certainty. Force models are used when, at most, *differences* in the first spatial derivative can be guaranteed, and when transparency to impinging waves is desired.

These ideas are crucial to many wave applications in petroleum engineering. They provide the flexibility needed in drilling vibrations to model drillbit kinematics *as well as* rock-bit interaction and penetration rate; and in MWD mud pulse telemetry, to model pulser source input *as well as* reflection from the drillbit and the mud pump.

Couples or dipoles. A single point force is known as a *monopole* source excitation. When two forces having equal magnitudes but opposite directions, separated only by a negligibly small distance, are applied, we term the combined effect a *dipole* or *couple*.

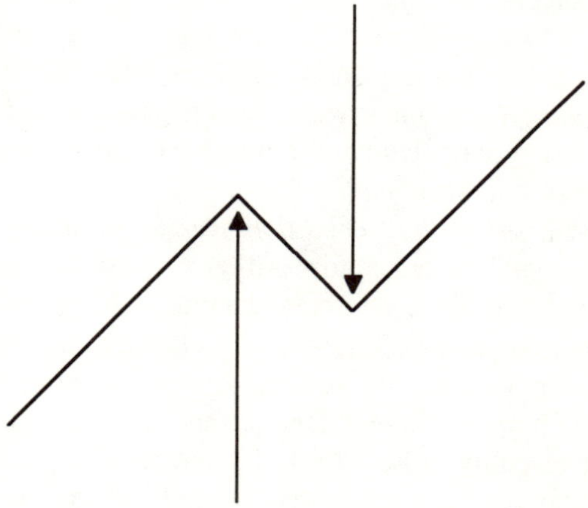

Figure 1-5. An external dipole loading.

A single force produces a slope discontinuity $[\partial u/\partial x]_{x=xs} = - F(t)/T$. From Figure 1-5, a dipole load leaves the slope $\partial u/\partial x$ continuous, but it introduces a jump in the displacement function u itself (and hence, in velocity and acceleration). That is, it induces a nonzero value of $[u(x_s)]$ through $x = x_s$, and the right side of

$$\rho_l\, \partial^2u/\partial t^2 + \zeta\, \partial u/\partial t - T\, \partial^2u/\partial x^2 + \rho_l g = F(t)\, \{\delta(x-x_s+\varepsilon) - \delta(x-x_s-\varepsilon)\} \qquad (1.104)$$

is responsible for it. More formally, the displacement u is discontinuous and double-valued, while the spatial derivative $\partial u/\partial x$ is continuous and single-valued.

To determine the magnitude of [u], it is not necessary to invoke esoteric theories on *distributions* or *generalized functions*. A less elegant "engineering derivation" actually suffices. We rewrite Equation 1-104 in the *conservation form* (i.e., "$\partial A/\partial t + \partial B/\partial x = 0$"),

$$\begin{aligned} &\partial(\rho_l\, \partial u/\partial t + \zeta\, u)/\partial t - \partial(T\, \partial u/\partial x - \rho_l g x)/\partial x \\ &\quad = F(t)\, \{\delta(x-x_s+\varepsilon) - \delta(x-x_s-\varepsilon)\} = F(t)\, \delta\,'(x-x_s)\, dx \\ &\quad = \partial\{F(t)\, \delta\, (x-x_s)\, dx\}/\partial x \end{aligned} \qquad (1.105)$$

where $\delta\,'(x-x_s) = \{\delta(x-x_s+\varepsilon) - \delta(x-x_s-\varepsilon)\}/dx$ and $dx = 2\varepsilon > 0$ is small. Kanwal (1983) discusses in detail the *unit dipole* $\delta'(x)$; Hudson (1980) discusses *three-dimensional dipoles* and their use in earthquake seismology. Next, we simplify the result so that

$$\partial(\rho_l\, \partial u/\partial t + \zeta\, u)/\partial t - \partial(T\, \partial u/\partial x - \rho_l g x + F(t)\, \delta\, (x-x_s)\, dx)/\partial x = 0 \qquad (1.106)$$

Now introduce a new dependent variable $\phi(x,t)$ defined by the transformations

$$\partial\phi/\partial x = \rho_l\, \partial u/\partial t + \zeta\, u \qquad (1.107)$$

and

$$\partial\phi/\partial t = T\, \partial u/\partial x - \rho_l g x + F(t)\, \delta\, (x-x_s)\, dx \qquad (1.108)$$

If we differentiate these with respect to x and t, respectively, we have

$$\partial^2\phi/\partial x^2 = \rho_l\, \partial^2u/\partial t\partial x + \zeta\, \partial u/\partial x \qquad (1.109)$$

$$\partial^2\phi/\partial t^2 = T\, \partial^2u/\partial x\partial t + F\,'(t)\, \delta\, (x-x_s)\, dx \qquad (1.110)$$

noting that the body force term disappears. Next, multiply the first equation by T and the second equation by ρ_l, to obtain

$$T\, \partial^2\phi/\partial x^2 = T\rho_l\, \partial^2u/\partial t\partial x + T\zeta\, \partial u/\partial x \qquad (1.111)$$

$$\rho_l\, \partial^2\phi/\partial t^2 = T\rho_l\, \partial^2u/\partial x\partial t + \rho_l\, F\,'(t)\, \delta\, (x-x_s)\, dx \qquad (1.112)$$

Subtraction leads to

$$\rho_l\, \partial^2\phi/\partial t^2 - T\, \partial^2\phi/\partial x^2 = \rho_l\, F\,'(t)\, \delta\, (x-x_s)\, dx - T\zeta\, \partial u/\partial x \qquad (1.113)$$

From the definition $\partial\phi/\partial t = T\,\partial u/\partial x - \rho_l g x + F(t)\,\delta(x-x_s)\,dx$ in Equation 1-108, we have

$$- T\zeta\,\partial u/\partial x = - \rho_l g x\zeta + \zeta F(t)\,\delta(x-x_s)\,dx - \zeta\,\partial\phi/\partial t \qquad (1.114)$$

which can be used to eliminate the $- T\zeta\,\partial u/\partial x$ term in Equation 1-113. Thus, the governing equation satisfied by $\phi(x,t)$ is

$$\rho_l\,\partial^2\phi/\partial t^2 - T\,\partial^2\phi/\partial x^2$$
$$= \rho_l F'(t)\,\delta(x-x_s)\,dx - \rho_l g x\zeta + \zeta F(t)\,\delta(x-x_s)\,dx - \zeta\,\partial\phi/\partial t \qquad (1.115)$$

or

$$\rho_l\,\partial^2\phi/\partial t^2 + \zeta\,\partial\phi/\partial t - T\,\partial^2\phi/\partial x^2$$
$$= - \rho_l g x\zeta + \rho_l F'(t)\,\delta(x-x_s)\,dx + \zeta F(t)\,\delta(x-x_s)\,dx$$
$$= - \rho_l g x\zeta + \{\rho_l F'(t) + \zeta F(t)\}\,dx\,\delta(x-x_s) \qquad (1.116)$$

Drawing upon the arguments leading to Equation 1-103, the force-like quantity $\{\rho_l F'(t) + \zeta F(t)\}\,dx$ must be responsible for a discontinuity $[\partial\phi/\partial x]$ in the first spatial derivative of ϕ. In particular, using Equation 1-103, we have

$$[\partial\phi/\partial x]_{x=xs} = \partial\phi/\partial x\,|_{x=xs+\varepsilon} - \partial\phi/\partial x\,|_{x=xs-\varepsilon}$$
$$= - \{\rho_l F'(t) + \zeta F(t)\}\,dx / T \qquad (1.117)$$

$$[\partial\phi/\partial x]_{x=xs} = - \{\rho_l F'(t) + \zeta F(t)\}\,dx / T \qquad (1.118)$$

which, on substitution of our original definition for $\partial\phi/\partial x$ in Equation 1-107, yields

$$[\rho_l\,\partial u/\partial t + \zeta u]_{x=xs} = - \{\rho_l F'(t) + \zeta F(t)\}\,dx / T \qquad (1.119)$$

Since $F(t)$, ρ_l and ζ are arbitrary, the derivative and undifferentiated terms lead to separate identities, $[\partial u/\partial t]_{x=xs} = - F'(t)\,dx/T$ and $[u]_{x=xs} = - F(t)\,dx/T$. But these are equivalent, the first being the derivative of the second; thus, we take

$$[u]_{x=xs} = - F(t)\,dx / T \qquad (1.120)$$

without loss of generality. We return to our original equation for $u(x,t)$, Equation 1-104, and equivalently rewrite it as

$$\rho_l\,\partial^2 u/\partial t^2 + \zeta\,\partial u/\partial t - T\partial^2 u/\partial x^2 + \rho_l g$$
$$= F(t)\,\{\delta(x-x_s+\varepsilon)-\delta(x-x_s-\varepsilon)\}$$
$$= F(t)dx\{\delta(x-x_s+\varepsilon)-\delta(x-x_s-\varepsilon)\}/(2\varepsilon)$$
$$= F(t)\,dx\,\delta'(x-x_s)$$
$$= -T\,[u]_{x=xs}\,\delta'(x-x_s) \qquad (1.121)$$

Hence, if a jump $[u]_{x=xs}$ in the dependent variable (as opposed to its spatial derivative) is desired, an external dipole excitation $-T\,[u]_{x=xs}\,\delta'(x-x_s)$

must be applied, where $\delta\,'(x\text{-}x_s)$ is the derivative of the delta function; here, the first spatial derivative will be continuous and single-valued. The resulting equation, as we will show, is easily solved using Laplace transforms.

We have discussed our delta function ideas having arguments in x, but it is clear that similar ideas apply to delta function *impulse* loads with arguments in t (we will give an example later). Then, the consequences would be expressed in terms of jumps in u or the velocity $\partial u/\partial t$.

Returning to our discussion of Figure 1-5, note how, when the discontinuity $[u(t)]_{x=xs}$ itself is prescribed, the formulation using $\phi(x,t)$ is actually simpler. We show this by combining Equations 1-116 and 1-120,

$$\rho_l\, \partial^2\phi/\partial t^2 + \zeta\, \partial\phi/\partial t - T\, \partial^2\phi/\partial x^2$$

$$= -\rho_l g x \zeta + \{\rho_l\, F\,'(t)\, dx + \zeta\, F(t)\, dx\}\, \delta\,(x\text{-}x_s)$$

$$= -\rho_l g x \zeta - T\, \{\rho_l\, [u]_{x=xs} + \zeta\, [u]_{x=xs}\}\, \delta\,(x\text{-}x_s) \qquad (1.122)$$

which contains a simple delta function and not its derivative $\delta\,'(x\text{-}x_s)$.

The ϕ transform renders the mathematical problem *force-like* or *monopole-like*, since Equation 1-122 is similar in appearance to Equation 1-99. Force-like models are easily solved numerically, as we will demonstrate in Chapter 5.

Multiple forces and higher order moments. Higher-order forces and couples may be constructed to model different kinds of external excitations. In Figure 1-6, we show two equal forces pointing upwards, and one force with the combined magnitude of the two opposing loads pointing downwards.

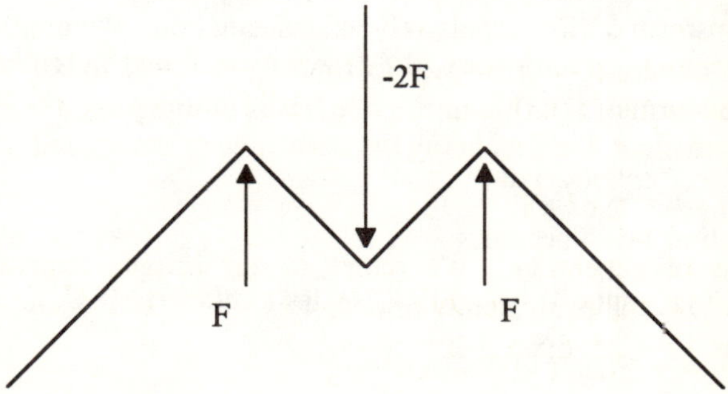

Figure 1-6. An unconventional force arrangement.

This force configuration introduces a slope discontinuity similar to that of a single force acting alone; however, the net force here vanishes, while a single force is always nonzero. The use of this or other load models will be dictated by physical requirements. Other load models are easily conceptualized.

Symmetries and anti-symmetries. Force loadings are associated with additional properties. To fix ideas, refer back to Equation 1-99, and consider an infinite string. A single-force or monopole will create a jump $[\partial u/\partial x]$, but the jump is such that the two $\partial u/\partial x$'s at its sides are equal and opposite; also, as proven earlier, they will propagate in opposite directions.

A double-force dipole load will create a jump [u], but the jump is such that the two u's at its sides are equal and opposite; also, they will propagate in opposite directions. Modeling decisions will also depend on required symmetries or anti-symmetries (in the dependent variable or its derivative) desired at the source point.

Impulse response. Engineers talk freely about *impulse responses.* We hear statements that the system response to an impulse excitation completely characterizes the dynamics of the system. Often, explanations are given in terms of Green's functions and similar abstractions, but it is possible to develop these ideas without higher math.

What is an impulse? To develop the ideas, we return to the single-force model for u(x,t), with a general time-dependent F(t),

$$\rho_l\, \partial^2 u/\partial t^2 + \zeta\, \partial u/\partial t - T\, \partial^2 u/\partial x^2 + \rho_l g = F(t)\, \delta(x-x_s) \qquad (1.123)$$

Consider first the case when F(t) is the timewise *unit delta function* $\delta(t)$, so that

$$\rho_l\, \partial^2 u/\partial t^2 + \zeta\, \partial u/\partial t - T\, \partial^2 u/\partial x^2 + \rho_l g = \delta(t)\, \delta(x-x_s) \qquad (1.124)$$

Since we are concerned with initial value problems, we take the time Laplace transform of this impulsively loaded equation. We multiply throughout by e^{-st} and integrate with respect to time from $t = 0$ to ∞. We define the Laplace transform of u(x,t), omitting the limits of integration for brevity, by

$$U(x,s) = \int e^{-st}\, u(x,t)\, dt \qquad (1.125)$$

Then, Equation 1-124 becomes

$$\rho_l\{s^2U(x,s) - su(x,0) - u_t(x,0)\} + \zeta\{sU(x,s) - u(x,0)\} - T\, U_{xx}(x,s) + \rho_l g/s$$
$$= \int e^{-st}\, \delta(t)\, \delta(x-x_s)\, dt = \delta(x-x_s) \qquad (1.126)$$

or

$$U_{xx}(x,s) - \{(\rho_l s^2 + \zeta s)/T\}U$$
$$= -\delta(x-x_s)/T - \{(\rho_l s+\zeta)/T\}u(x,0) - (\rho_l/T)u_t(x,0) + (\rho_l g/T)/s$$
$$= -\delta(x-x_s)/T - \ldots \qquad (1.127)$$

Equation 1-127 is an ordinary differential equation for U(x,s) in x. It can also be solved by transform methods, depending on the type of spatial domain.

If the problem is defined over $-\infty < x < \infty$, the Fourier transform can be used; if it is defined on $x > 0$ only, the Laplace transform in x can be invoked.

It is not necessary to actually solve Equation 1-127 in order to obtain general results. We simply assume that the solution to the *homogeneous equation* $U_{xx}(x,s) - \{(\rho_l s^2 + \zeta s)/T\}U = 0$ can be obtained using standard methods, as $U_1(x,s)$ and $U_2(x,s)$.

Then, the method of *variation of parameters* from the theory of second-order ordinary differential equations (e.g., see Leighton (1967)) guarantees that a *particular solution* for Equation 1-127 exists in the form

$$U_p(x,s;\delta(t)) = -\int_{x_o}^{x} \{ \ldots \} \, \delta(\lambda - x_s)/T \, d\lambda + \ldots \tag{1.128}$$

where the $\delta(t)$ in $U_p(x,s;\delta(t))$ emphasizes the fact that Equation 1-128 applies to the impulsive excitation $\delta(t)$ only, "..." refers to terms associated with initial conditions, and the curly bracket denotes

$$\{ \ldots \} = |U_1(\lambda,s)U_2(x,s) - U_1(x,s)U_2(\lambda,s)| / \\ |U_1(\lambda,s)U_{2,x}(\lambda,s) - U_{1,x}(\lambda,s)U_2(\lambda,s)| \tag{1.129}$$

Using the *sifting property* of the delta function (see Equation 1-101), Equation 1-128 simplifies to

$$U_p(x,s;\delta(t)) = - |U_1(x_s,s)U_2(x,s) - U_1(x,s)U_2(x_s,s)| / \\ \{T \, |U_1(x_s,s)U_{2,x}(x_s,s) - U_{1,x}(x_s,s)U_2(x_s,s)| \} + \ldots \tag{1.130}$$

whose time-domain inverse we denote $u_p(x,t)$.

Over large times, the effects of viscous dissipation will damp out all traces of the initial conditions $u(x,0)$ and $u_t(x,0)$, so that the " ..." will no longer enter the solution. Then, the particular solution $u_p(x,t)$ becomes the sole response of the impulse $\delta(t)$ when the effects of gravity are neglected.

Now it is clear that had we carried out our analysis for a more general time-dependent force F(t) on the right side of Equation 1-127, as opposed to $\delta(t)$, we would have obtained instead the particular solution

$$U_p(x,s;F(t)) = - F^*(s) \, |U_1(x_s,s)U_2(x,s) - U_1(x,s)U_2(x_s,s)| / \\ \{T \, |U_1(x_s,s)U_{2,x}(x_s,s) - U_{1,x}(x_s,s)U_2(x_s,s)| \} \tag{1.131}$$

where $F^*(s)$ is the Laplace transform of F(t). Interestingly, we can rewrite Equation 1-131 as the product of two Laplace transforms

$$U_p(x,s;F(t)) = - F^*(s) \, U_p(x,s;\delta(t)) \tag{1.132}$$

where we have used the impulse response in Equation 1-130.

We return to our *Convolution Theorem* (see Equation 1-59) which states that the time domain function corresponding to the product $f_1(s)\ f_2(s)$ of two Laplace transforms $f_1(s)$ and $f_2(s)$ with inverses $f_1(t)$ and $f_2(t)$ is the integral

$$\int_0^t f_1(\tau)\ f_2(t-\tau)\ d\tau \tag{1.133}$$

Accordingly, the inverse solution u(x,t) to $U_p(x,s;F(t))$ in Equation 1-132 must take the form

$$\int_0^t F(\tau)\ u_p\{t-\tau;\delta(t)\}\ d\tau \tag{1.134}$$

where u_p, again, is the undamped time domain response due to an impulse.

This completes the proof: *once $u_p(t)$ is known or available, the complete transient response to a more general excitation F(t) can be written using Equation 1-134.*

Of course, generating a true impulse experimentally is a difficult challenge. As an exercise, the reader should repeat these steps using an initial test waveform *other* than an impulse, and determine the superposition integral analogous to Equation 1-134.

The above conclusion is quite general, and applies to all linear systems. It is not restricted to the wave operator, although we chose that route to make the argument concrete. Observe that, from the variation of parameters argument onwards, *any* linear differential operator may have preceded the discussion.

On the subtle meaning of impulse. When we state that the dynamical response of a physical system to any excitation is determined completely by its impulse response, we must understand that this refers to specific differential equations and excitations. For example, if we are considering

$$\rho_l\ \partial^2 u/\partial t^2 + \zeta\ \partial u/\partial t - T\ \partial^2 u/\partial x^2 + \rho_l g = F(t)\ \delta(x-x_s) \tag{1.135}$$

as the model for an entity u(x,t), the actual excitation F *must* experimentally create a jump discontinuity in $\partial u/\partial x$ in order for us to use its u(x,t) response as the impulse response.

Those $u_p(x,t)$'s so obtained can be used to produce more general solutions *for u(x,t)* using the convolution integral in Equation 1-134. For the transverse string modeled by Equation 1-135, the impulsive strike of a guitar pick or a violin bow, as suggested in Figure 1-4, might do.

Example 1-14: Incorrect use of impulse response. Suppose Joe Engineer uses a small clamp-on excitation source that actually implements the *dipole* load in Figure 1-5. Because the "sharp source" acts over a short distance, he (mistakenly) believes that he has excited the system with a point impulse.

His measurements for $u_p(x,t)$, and their application to Equation 1-134 for u(x,t), would be useless: the convolution result is based upon the assumption of a true $\delta(x-x_s)$ source that creates jumps in $\partial u/\partial x$, but this is clearly not the case from Figure 1-5 (the dipole here, rather than being $\delta(x-x_s)$, is actually $\delta'(x-x_s)$, which does *not* lead to Equation 1-134).

If Joe Engineer *insists* on using the loading of Figure 1-5, then in what sense *are* his measurements meaningful? Since the $[\partial u/\partial t]$ or [u] suggested by Figure 1-5 *really* represent jumps in the spatial derivative $\partial\phi/\partial x$ (see Equation 1-107), his measurements *really* pertain to solutions of

$$\rho_l\, \partial^2\phi/\partial t^2 + \zeta\, \partial\phi/\partial t - T\, \partial^2\phi/\partial x^2 = \ldots\, \delta(x-x_s) \tag{1.136}$$

Any attempt to use Equation 1-134 for u(x,t) would be incorrect; Joe Engineer's convolution is restricted to the variable $\phi(x,t)$ only ... *and no other.*

Additional models. Equations 1-99 and 1-136 for u(x,t) and $\phi(x,t)$ are not the only formulations available for transverse strings. Alternative formulations are easily created to treat new applications. Suppose Joe Engineer invents a hypothetical bow that creates internal discontinuities in $\partial^2u/\partial x^2$.

How would such excitations be modeled mathematically? In this case, one can differentiate $\rho_l\, \partial^2u/\partial t^2 + \zeta\, \partial u/\partial t - T\, \partial^2u/\partial x^2 + \rho_l g = 0$ with respect to x, and add $G(t)\delta(x-x_s)$ to the right side to obtain

$$\rho_l\, \partial^2(\partial u/\partial x)/\partial t^2 + \zeta\, \partial(\partial u/\partial x)/\partial t - T\, \partial^2(\partial u/\partial x)/\partial x^2 = G(t)\delta(x-x_s) \tag{1.137}$$

Since $G(t)\delta(x-x_s)$ creates a jump in the first spatial derivative of *the dependent variable* $\partial u/\partial x$, it is responsible for a discontinuity in the required *second derivative* $\partial^2u/\partial x^2$.

In wave equation modeling, two questions arise if delta function superposition methods are used. What *physical quantity* is discontinuous through the source point? Which *dependent variable* should be used (remember, *its* first space derivative represents that physical quantity)?

Examples of these subtleties arise in MWD pulser valve modeling. *Positive pressure valves* create *pressure* discontinuities at the source, while negative pressure valves create *velocity* discontinuities; velocity jumps do not exist for the former, while pressure jumps are not found in the latter.

These valves require completely different mathematical formulations; as we will show in Chapter 10, general MWD valve models lacking pure symmetries or antisymmetries can be constructed from linear superpositions of positive and negative valve models.

While our discussion has focused on second-order equations in the context of transverse vibrations, the ideas are universal, applying to general linear systems. In each case, the exact meaning of impulse is crucial; what discontinuities they induce must be explored and made relevant to the physics, and all relevant superposition integrals must be derived.

Other delta function properties. We will list some properties of delta functions for reference, since they are not usually obvious nor available. Again, we have the best known property

$$\int_{-\infty}^{\infty} \delta(x-x_s)\, \phi(x)\, dx = \phi(x_s) \tag{1.138}$$

so that, for example,

$$\int_{-\infty}^{\infty} \delta(x)\, \phi(x)\, dx = \phi(0) \tag{1.139}$$

The limits are shown over $(-\infty,\infty)$ for brevity, but the integration may be carried out over any domain containing "x_s". Also, it can be shown that

$$\delta(-x) = \delta(x) \tag{1.140}$$

$$\delta(ax) = \delta(x)/|a| \tag{1.141}$$

$$\delta(at-bx) = \delta(t - bx/a)/|a| \tag{1.142}$$

$$x\delta(x) = 0 \tag{1.143}$$

$$\delta(x^2 - a^2) = \{\delta(x+a) + \delta(x-a)\} / \{2|a|\} \tag{1.144}$$

$$\delta(x) = -x\, \delta'(x) \tag{1.145}$$

The following derivative property

$$d\, \{\text{sgn } t\}/dt = 2\, \delta(t) \tag{1.146}$$

is also applicable, where $\text{sgn}(t) = +1,\ t > 0;\ -1,\ t < 0$. Important integral properties involving the delta function include

$$\int_{-\infty}^{\infty} \delta(x-a)\, \delta(x-b)\, dx = \delta(a-b) \tag{1.147}$$

$$\int_{-\infty}^{\infty} e^{\,i\omega(a-x)}\, d\omega = 2\pi\, \delta(x-a) \tag{1.148}$$

$$\int_{-\infty}^{\infty} \delta'(t-\tau)\, \phi(t)\, dt = -\int_{-\infty}^{\infty} \delta(t-\tau)\, \phi'(t)\, dt = -\phi'(\tau) \tag{1.149}$$

Also, if x_0 is defined by $u(x_0) = 0$, with u(x) monotonically increasing or decreasing, it can be shown that

$$\int_{-\infty}^{\infty} \delta\{u(x)\}\, \phi(x)\, dx = \phi(x_0) / |u'(x_0)| \tag{1.150}$$

If $\eta(t) = 1,\ t > 0$; and $0,\ t < 0$, it is possible to show that

$$\int_{0}^{t} \delta(\xi-\tau)\, d\xi = \eta(t-\tau) \tag{1.151}$$

Finally,

$$\int_{0}^{\infty} e^{-st}\, \delta'(t-\tau)\, dt = s\, e^{-s\tau}, \text{ where } s > 0 \tag{1.152}$$

In particular, when $\tau = 0$,

$$\int_{0}^{\infty} e^{-st}\, \delta'(t)\, dt = s, \text{ where } s > 0 \tag{1.153}$$

so that we have the inverse

$$L^{-1}\{s\} = \delta'(t) \tag{1.154}$$

Note that Equation 1-153 *alone* is not a valid transform, since it *does not* vanish as $s \to \infty$. The dipole forcing function $\delta'(t)$, though, generally appears on the right-sides of differential equations; the quotient formed by s and the Laplace transform of the differential operator, representing the transform of the quantity of physical interest, *does* vanish as s increases.

At the risk of being repetitious, we also give several relations which are derivable from the foregoing equations, but which are of significance in themselves in Fourier transform analysis.

$$\int_{-\infty}^{\infty} e^{\,i\omega x}\, \delta(x)\, dx = 1 \tag{1.155}$$

$$\int_{-\infty}^{\infty} e^{\,i\omega x}\, \delta'(x)\, dx \;=\; -\,i\omega \qquad (1.156)$$

$$\int_{-\infty}^{\infty} e^{\,-i\omega x}\, d\omega \;=\; 2\pi\, \delta(x) \qquad (1.157)$$

POINT FORCE AND DIPOLE WAVE EXCITATION

We have discussed the physical meaning and analytical modeling of point forces and dipoles in the context of the wave equation; we have also given Laplace transform solutions, and introduced δ and δ'. We now combine these ideas to solve for the transverse oscillations of a string excited by a point load at $x = x^*$ with a varying acceleration $f(t)$. That is, we consider

$$\partial^2 u/\partial t^2 \;-\; c^2\, \partial^2 u/\partial x^2 \;=\; f(t)\, \delta(x-x^*) \qquad (1.158)$$

Example 1-15: Finite string excited by a time-varying concentrated point force. We solve Equation 1-158 subject to the initial flat elevation $u(x,0) = 0$ and the static condition $\partial u(x,0)/\partial t = 0$ in the interval $0 < x < L$. The string is rigidly pinned at its end points with $u(0,t) = 0$ and $u(L,t) = 0$.

We multiply each side of Equation 1-158 by e^{-st} and time integrate the result from $t = 0$ to ∞, allowing the solution process to proceed exactly as before. Integration by parts shows that $\int e^{-st}\, \partial^2 u/\partial t^2\, dt = s^2 \int e^{-st}\, u(x,t)\, dt - su(x,0) - \partial u(x,0)/\partial t$, but we do encounter a new $f(t)\, \delta(x-x^*)$ term.

It is clear that $\int e^{-st}\, f(t)\, \delta(x-x^*)\, dt$ equals $F(s)\, \delta(x-x^*)$, where the function of x moves across the integral, and $F(s)$ represents the transform of $f(t)$ with respect to time. These steps reduce Equation 1-158 to the ordinary differential equation $s^2 U(x,s) - c^2 U''(x,s) = F(s)\, \delta(x-x^*)$, where intermediate results were simplified using initial conditions. This new equation contains $\delta(x-x^*)$, responsible for a slope discontinuity in the transverse string.

In elementary courses, this necessitates a break-up of the problem into two parts, and then, the matching of two solutions assuming continuity of displacement, plus jump conditions based on momentum considerations.

Laplace transforms conveniently circumvent these steps, in the process minimizing the possibility of algebraic error. We merely solve our ordinary differential equation *in turn* with Laplace transforms. If we analogously define

$$L\{U(x,s)\} = \mathbf{U}(\beta,s) = \int_0^\infty e^{-\beta x}\, U(x,s)\, dx,\ \beta > 0 \tag{1.159}$$

it is clear that $s^2U(x,s) - c^2U''(x,s) = F(s)\,\delta(x-x^*)$ transforms into $s^2\mathbf{U}(\beta,s) - c^2\{\beta^2\mathbf{U}(\beta,s) - \beta\, U(0,s) - U_x(0,s)\} = F(s) \int e^{-\beta x}\,\delta(x-x^*)\, dx = F(s)e^{-\beta x^*}$, where we have again used the *sifting property* of delta functions.

So far, we have defined a Laplace transform with respect to time, the conventional one used by electrical engineers, and then a second one with respect to space. Thus, we obtain the double Laplace transform

$$\begin{aligned}\mathbf{U}(\beta,s) = {} & F(s)e^{-\beta x^*}/(s^2 - c^2\beta^2) \\ & - c^2\beta\, U(0,s)/(s^2 - c^2\beta^2) \ \ - c^2U_x(0,s)/(s^2 - c^2\beta^2)\end{aligned} \tag{1.160}$$

Now, $U(0,s)$ vanishes by our assumption of a pinned *left* end (this is obtained by transforming the boundary condition $u(0,t)$ over s). The presence of *$U_x(0,s)$, also for $x = 0$,* might suggest that we have incorrectly applied an *initial value problem* solver, assuming a time-like variable x, to a *boundary value problem*.

This is not so. Although we assumed a pinned *right* end $u(L,t) = 0$, thus formulating a two-point *boundary value* problem, it is perfectly valid to define a Laplace transform over $(0,\infty)$ even though our domain is $0 < x < L$.

This use is less common, but the idea is simple: in applying spatial Laplace transforms to finite domains, we in fact assume that a semi-infinite space holds, although we ultimately confine our attention to the finite space between $x = 0$ *and* the position $x = L$ hosting the constraint $u(L,t) = 0$.

While $U_x(0,s)$ is not prescribed as part of the formulation, it can be regarded as an unknown "constant" that is fully determined by the condition $u(L,t) = 0$ at the opposite boundary. Continuing with our solution of Equation 1-160, the simplification $U(0,s) = 0$ leads to

$$\mathbf{U}(\beta,s) = U_x(0,s)/(\beta^2 - s^2/c^2) + F(s)e^{-\beta x^*}/(s^2 - c^2\beta^2) \tag{1.161}$$

for the double Laplace transform of the function $u(x,t)$.

We perform the inversion with respect to the parameter β. For the first term shown above, table look-up readily shows that the function corresponding to $1/(\beta^2 - s^2/c^2)$ is just $(c/s) \sinh \{(s/c)x\}$.

Laplace transform tables also show that the inverse of $e^{-\beta x^*}/(s^2 - c^2\beta^2)$ or $(-1/c^2)e^{-\beta x^*}/(\beta^2 - s^2/c^2)$ is $(-1/c^2)(c/s)\, \eta(x-x^*) \sinh \{(s/c)(x-x^*)\}$, where $\eta(x-x^*) = 1$ for $x > x^*$, and 0 for $x < x^*$ (an *exponential shift* property has been exploited). Therefore, we have

$$U(x,s) = F(s)(-1/c^2)(c/s)\ \eta(x-x^*)\ \sinh\{(s/c)(x-x^*)\} + U_x(0,s)(c/s)\ \sinh\{(s/c)x\} \quad (1.162)$$

Now, since $u(L,t) = 0$, it follows from taking time transforms that $U(L,s) = 0$; and because $L > x^*$, we obtain

$$0 = F(s)(-1/c^2)(c/s)\ \sinh\{(s/c)(L-x^*)\} + U_x(0,s)(c/s)\ \sinh\{(s/c)L\} \quad (1.163)$$

or

$$U_x(0,s) = F(s)(1/c^2)\ \sinh\{(s/c)(L-x^*)\}/\{\sinh\{(s/c)L\}\} \quad (1.164)$$

Back substitution into Equation 1-162 completely determines the transform $U(x,s)$ as

$$U(x,s) = F(s)(-1/c^2)(c/s)\ \eta(x-x^*)\ \sinh\{(s/c)(x-x^*)\} + F(s)(1/c^2)\ \sinh\{(s/c)(L-x^*)\}(c/s)\ \sinh\{(s/c)x\}/\{\sinh\{(s/c)L\}\} \quad (1.165)$$

Thus, the timewise solution $u(x,t)$ is in principle known for any excitation $f(t)$. The required inversion, not performed here, leads to complicated infinite series solutions in the sine eigenfunction used earlier. Because damping was ignored, the effects of each movement $f(t)$ remains in our solution for all time.

Example 1-16: Finite string excited by a time-varying point dipole (i.e., a force couple). For simplicity, we solve Equation 1-166 subject to the initial flat elevation $u(x,0) = 0$ and the static condition $\partial u(x,0)/\partial t = 0$ in $0 < x < L$. The string is rigidly pinned at the two end points, with $u(0,t) = 0$ and $u(L,t) = 0$.

$$\partial^2 u/\partial t^2 - c^2\ \partial^2 u/\partial x^2 = f(t)\ \delta'(x-x^*) \quad (1.166)$$

As before, we take Laplace transforms in time. Following exactly the same steps, we now obtain the differential equation

$$s^2 U(x,s) - c^2 U''(x,s) = F(s)\ \delta'(x-x^*) \quad (1.167)$$

where we have simplified intermediate results using initial conditions. Notice, though, that Equation 1-167 contains $\delta'(x-x^*)$ and not $\delta(x-x^*)$. In order to apply Laplace transforms in the x direction, we use Equation 1-152 in the form

$$\int_0^\infty e^{-\beta x}\ \delta'(x-x^*)\ dx = \beta\ e^{-\beta x^*}, \text{ where } \beta > 0 \quad (1.168)$$

This leads to $s^2\mathbf{U}(\beta,s) - c^2\{\beta^2\mathbf{U}(\beta,s) - \beta U(0,s) - U_x(0,s)\} = F(s)\ \beta\ e^{-\beta x^*}$, and the only difference from Example 1-15 is the extra β factor appearing on the right. The inversion procedure continues in the same manner.

Example 1-17: Splitting of an applied initial disturbance. Here, we consider an *infinite* pipe containing an MWD pulser that creates a "delta-p" signal at the valve $x = 0$. Obviously, half of the signal will propagate in one direction, with the remaining half in the opposite direction.

This splitting also applies in *finite* conduits initially, since the effects of boundaries are not immediately felt. When terminations such as drill bits and mudpumps are present, reflections will complicate the information contained in the "delta-p" signal. In Chapters 9 and 10, we show how a wave equation

$$\partial^2 u/\partial t^2 - c^2\, \partial^2 u/\partial x^2 = 0 \qquad (1.169)$$

for the fluid displacement $u(x,t)$ applies, where c is the sound speed. Here, the acoustic pressure satisfies $p = - B\, \partial u/\partial x$, where B is the bulk modulus.

Let us consider disturbances periodic in time with a frequency ω. For such excitations, we introduce the separation of variables

$$u_1(x,t) = X_1(x)\, e^{i\omega t},\ x < 0 \qquad (1.170a)$$

$$u_2(x,t) = X_2(x)\, e^{i\omega t},\ x > 0 \qquad (1.170b)$$

Substitution in Equation 1-169 leads to the unidirectional *Helmholtz equations*

$$d^2X_1(x)/dx^2 + (\omega^2/c^2)\, X_1(x) = 0 \qquad (1.171a)$$

$$d^2X_2(x)/dx^2 + (\omega^2/c^2)\, X_2(x) = 0 \qquad (1.171a)$$

Their solutions may be taken in the complex exponential forms

$$X_1(x) = C_1\, e^{i\omega x/c} + C_2\, e^{-i\omega x/c} \qquad (1.172a)$$

$$X_2(x) = C_3\, e^{i\omega x/c} + C_4\, e^{-i\omega x/c} \qquad (1.172b)$$

Thus,

$$u_1(x,t) = C_1\, e^{i(\omega x/c + \omega t)} + C_2\, e^{i(-\omega x/c + \omega t)} \qquad (1.173a)$$

$$u_2(x,t) = C_3\, e^{i(\omega x/c + \omega t)} + C_4\, e^{i(-\omega x/c + \omega t)} \qquad (1.173b)$$

The exponentials were chosen to facilitate the use of *directional filtering* or *radiation conditions.* Since the $x < 0$ and $x > 0$ solutions for $u(x,t)$ must be left and right-going, respectively, it is clear that $C_2 = C_3 = 0$. In determining the remaining C_1 and C_4, we require that the fluid remain continuous at $x = 0$,

$$u_1(0,t) = u_2(0,t) \qquad (1.174)$$

Now, the mud pulser applies its "delta-p" *signal strength* P_s, with

$$B\, \partial u_1(0,t)/\partial x - B\, \partial u_2(0,t)/\partial x = P_s\, e^{i\omega t} \qquad (1.175)$$

Substitution of Equations 1-173a,b into Equations 1-174 and 1-175 leads to

$$u_1(x,t) = -(icP_s/2\omega B)\, e^{i\omega(t+x/c)} \qquad (1.176)$$

$$u_2(x,t) = -(icP_s/2\omega B)\, e^{i\omega(t-x/c)} \qquad (1.177)$$

and the consequences

$$p_1(x,t) = -B\, \partial u_1(x,t)/\partial x = -(P_s/2)\, e^{i\omega(t+x/c)} \qquad (1.178)$$

$$p_2(x,t) = -B\, \partial u_2(x,t)/\partial x = +(P_s/2)\, e^{i\omega(t-x/c)} \qquad (1.179)$$

Thus, we have proved that half of the P_s signal propagates to the left, with the remaining half propagating to the right.

Note that we had dropped our delta function formalism, and used an "undergraduate approach" based on Equation 1-175 instead, which equivalently prescribes a nonzero jump in the first spatial derivative of u. Two functions u_1 and u_2 were required to carry out the matching in Equations 1-174 and 1-175, and the elegance provided by delta functions and transform methods was lost.

FIRST ORDER PARTIAL DIFFERENTIAL EQUATIONS

In problems governed by high-order equations, solutions are sometimes obtained by solving equivalent sets of first-order differential equations. Here we develop recipes useful to solving first-order *partial* differential equations. These ideas are covered more fully in Courant and Hilbert (1989).

In order to avoid any confusion, index notation and summation conventions are avoided, in favor of x, y, and z terminology. This section is only relevant to 3D kinematic wave theory and geophysical applications.

The solution for $\theta(x,y,z,t)$ satisfying specific equations, for instance, $\partial\theta/\partial t + a\, \partial\theta/\partial x + b\, \partial\theta/\partial y + c\, \partial\theta/\partial z = x + y$, can be handled straightforwardly. But in general, we need not focus on any one particular equation, since it is just as convenient to consider the representation

$$H(\theta, \partial\theta/\partial x, \partial\theta/\partial y, \partial\theta/\partial z, \partial\theta/\partial t, x, y, z, t) = 0 \qquad (1.180)$$

Using subscript notation for our partial derivatives, we have

$$H(\theta, \theta_x, \theta_y, \theta_z, \theta_t, x, y, z, t) = 0 \qquad (1.181)$$

in which case our example becomes $H = \theta_t + a\,\theta_x + b\,\theta_y + c\,\theta_z - x - y$.

Equation 1-181 may be *linear* or *nonlinear* in θ and its derivatives (e.g., $\theta_z - \theta_t = x$ is *linear*, while $(\theta_z)^2 - \theta_t = x$ is *nonlinear*). The explicit presence of x, y, z, and t in the argument of H indicates the existence of variable coefficients. Now, the total differential for *any* function $\theta(x,y,z,t)$ satisfies

$$d\theta = \theta_x\, dx + \theta_y\, dy + \theta_z\, dz + \theta_t\, dt \tag{1.182}$$

irrespective of H.

The idea behind *parameterization* is simple (e.g., the circle $x^2 + y^2 = 1$ can be expressed as $x = \sin\lambda$, $y = \cos\lambda$, where $0 < \lambda < 2\pi$); it will be useful to develop the general case in which θ is parametrically represented. Thus, the change in θ as a function of *any* parameter λ along a spatial trajectory is

$$d\theta/d\lambda = \theta_x\, dx/d\lambda + \theta_y\, dy/d\lambda + \theta_z\, dz/d\lambda + \theta_t\, dt/d\lambda \tag{1.183}$$

This applies to any choice of $x = x(\lambda)$, $y = y(\lambda)$, $z = z(\lambda)$, and $t = t(\lambda)$, where λ is a convenient variable. Similarly, we can show that θ_x, θ_y, θ_z and θ_t satisfy

$$d\theta_x/d\lambda = \theta_{xx}\, dx/d\lambda + \theta_{xy}\, dy/d\lambda + \theta_{xz}\, dz/d\lambda + \theta_{xt}\, dt/d\lambda \tag{1.184}$$
$$d\theta_y/d\lambda = \theta_{yx}\, dx/d\lambda + \theta_{yy}\, dy/d\lambda + \theta_{yz}\, dz/d\lambda + \theta_{yt}\, dt/d\lambda \tag{1.185}$$
$$d\theta_z/d\lambda = \theta_{zx}\, dx/d\lambda + \theta_{zy}\, dy/d\lambda + \theta_{zz}\, dz/d\lambda + \theta_{zt}\, dt/d\lambda \tag{1.186}$$
$$d\theta_t/d\lambda = \theta_{tx}\, dx/d\lambda + \theta_{ty}\, dy/d\lambda + \theta_{tz}\, dz/d\lambda + \theta_{tt}\, dt/d\lambda \tag{1.187}$$

Now, let us take partial derivatives of $H(\theta,\theta_x,\theta_y,\theta_z,\theta_t,x,y,z,t) = 0$ with respect to x, y, z, and t. For example, taking x derivatives, using the chain rule, and treating each argument as an independent variable, leads to $H_\theta\theta_x + H_{\theta x}\theta_{xx} + H_{\theta y}\theta_{yx} + H_{\theta z}\theta_{zx} + H_{\theta t}\theta_{tx} + H_x = 0$. In general, we obtain

$$H_\theta\theta_x + H_{\theta x}\theta_{xx} + H_{\theta y}\theta_{yx} + H_{\theta z}\theta_{zx} + H_{\theta t}\theta_{tx} + H_x = 0 \tag{1.188}$$
$$H_\theta\theta_y + H_{\theta x}\theta_{xy} + H_{\theta y}\theta_{yy} + H_{\theta z}\theta_{zy} + H_{\theta t}\theta_{ty} + H_y = 0 \tag{1.189}$$
$$H_\theta\theta_z + H_{\theta x}\theta_{xz} + H_{\theta y}\theta_{yz} + H_{\theta z}\theta_{zz} + H_{\theta t}\theta_{tz} + H_z = 0 \tag{1.190}$$
$$H_\theta\theta_t + H_{\theta x}\theta_{xt} + H_{\theta y}\theta_{yt} + H_{\theta z}\theta_{zt} + H_{\theta t}\theta_{tt} + H_t = 0 \tag{1.191}$$

Rearranging terms, we have

$$-H_x - H_\theta\theta_x = \theta_{xx}H_{\theta x} + \theta_{yx}H_{\theta y} + \theta_{zx}H_{\theta z} + \theta_{tx}H_{\theta t} \tag{1.192}$$
$$-H_y - H_\theta\theta_y = \theta_{xy}H_{\theta x} + \theta_{yy}H_{\theta y} + \theta_{zy}H_{\theta z} + \theta_{ty}H_{\theta t} \tag{1.193}$$
$$-H_z - H_\theta\theta_z = \theta_{xz}H_{\theta x} + \theta_{yz}H_{\theta y} + \theta_{zz}H_{\theta z} + \theta_{tz}H_{\theta t} \tag{1.194}$$
$$-H_t - H_\theta\theta_t = \theta_{xt}H_{\theta x} + \theta_{yt}H_{\theta y} + \theta_{zt}H_{\theta z} + \theta_{tt}H_{\theta t} \tag{1.195}$$

If we next compare the *left sides* of Equations 1-184 to 1-187 with those of Equations 1-192 to 1-195, we can identify

$$d\theta_x/d\lambda = -H_x - \theta_x H_\theta \tag{1.196}$$
$$d\theta_y/d\lambda = -H_y - \theta_y H_\theta \tag{1.197}$$
$$d\theta_z/d\lambda = -H_z - \theta_z H_\theta \tag{1.198}$$
$$d\theta_t/d\lambda = -H_t - \theta_t H_\theta \tag{1.199}$$

provided we consistently identify, on the respective *right sides*,

$$dx/d\lambda = H_{\theta x} \tag{1.200}$$
$$dy/d\lambda = H_{\theta y} \tag{1.201}$$
$$dz/d\lambda = H_{\theta z} \tag{1.202}$$
$$dt/d\lambda = H_{\theta t} \tag{1.203}$$

Also, Equation 1-183, using Equations 1-200 to 1-203, can be rewritten as

$$\begin{aligned} d\theta/d\lambda &= \theta_x\, dx/d\lambda + \theta_y\, dy/d\lambda + \theta_z\, dz/d\lambda + \theta_t\, dt/d\lambda \\ &= \theta_x H_{\theta x} + \theta_y H_{\theta y} + \theta_z H_{\theta z} + \theta_t H_{\theta t} \end{aligned} \tag{1.204}$$

Recapitulation. Key implications may have well been buried behind the facade of subscripts. Thus, it is convenient to introduce the simplified notation

$$p = \theta_x \tag{1.205}$$
$$q = \theta_y \tag{1.206}$$
$$r = \theta_z \tag{1.207}$$
$$s = \theta_t \tag{1.208}$$

so that Equation 1-181 becomes

$$H(\theta,p,q,r,s,x,y,z,t) = 0 \tag{1.209}$$

The implied recipe for solving the *partial* differential equation in Equation 1-181 calls for the equivalent solution of a system of coupled first-order *ordinary* differential equations in the nine "independent" unknowns shown in the argument of Equation 1-209, namely,

$$dx/d\lambda = H_p \tag{1.210}$$
$$dy/d\lambda = H_q \tag{1.211}$$
$$dz/d\lambda = H_r \tag{1.212}$$
$$dt/d\lambda = H_s \tag{1.213}$$
$$dp/d\lambda = -H_x - p H_\theta \tag{1.214}$$
$$dq/d\lambda = -H_y - q H_\theta \tag{1.215}$$
$$dr/d\lambda = -H_z - r H_\theta \tag{1.216}$$
$$ds/d\lambda = -H_t - s H_\theta \tag{1.217}$$
$$d\theta/d\lambda = p H_p + q H_q + r H_r + s H_s \tag{1.218}$$

where Equation 1-218 is obtained by simplifying Equation 1-183 with the expressions in Equations 1-210 to 1-213.

Again, the p, q, r, s, x, y, z, t and θ derivatives of $H(\theta,p,q,r,s,x,y,z,t)$ are regarded as *independent* partial derivatives. Since H is known, the expressions for the right sides of Equations 1-210 to 1-218 are explicitly available.

When initial values for θ, p, q, r, s, x, y, z, and t are prescribed, the solution can be obtained by the simultaneous integration of *nine* first-order ordinary differential equations in the *time-like* coordinate λ (simpler H functions, of course, will reduce the number of coupled equations).

In Chapter 11, this integration is carried out for the *eikonal equation* $(\partial\tau/\partial x)^2 + (\partial\tau/\partial y)^2 + (\partial\tau/\partial z)^2 = 1/c(x,y,z)^2$ corresponding to the 3D geophysical model $\partial^2P/\partial t^2 - c^2 (\partial^2P/\partial x^2 + \partial^2P/\partial y^2 + \partial^2P/\partial z^2) = 0$. For this application, the reader should now review Equations 11-1 to 11-15.

REFERENCES

Bleistein, N., *Mathematical Methods for Wave Phenomena*, Academic Press, Orlando, 1984.

Carrier, G.F., Krook, M., and Pearson, C.E., *Functions of a Complex Variable*, McGraw-Hill, New York, 1966.

Chin, W.C., *Modern Reservoir Flow and Well Transient Analysis*, Gulf Publishing, Houston, 1993.

Churchill, R.V., *Fourier Series and Boundary Value Problems*, McGraw-Hill, New York, 1941.

Churchill, R.V., *Operational Mathematics*, McGraw-Hill, New York, 1958.

Claerbout, J.F., *Fundamentals of Geophysical Data Processing*, Blackwell Scientific Publishers, Oxford, 1985.

Claerbout, J.F., *Imaging the Earth's Interior*, Blackwell Scientific Publishers, Oxford, 1985.

Courant, R., and Hilbert, D., *Methods of Mathematical Physics, Volumes I and II*, John Wiley and Sons, New York, 1989.

Graff, K.F., *Wave Motion in Elastic Solids*, Ohio State University Press, 1975.

Hildebrand, F.B., *Advanced Calculus for Applications*, Prentice-Hall, Englewood Cliffs, New Jersey, 1948.

Hudson, J.A., *The Excitation and Propagation of Elastic Waves*, Cambridge University Press, Cambridge, 1980.

Kanwal, R.P., *Generalized Functions: Theory and Technique*, Academic Press, New York, 1983.

Leighton, W., *Ordinary Differential Equations*, Wadsworth Publishing Company, Belmont, California, 1967.

Lighthill, M.J., *An Introduction to Fourier Analysis and Generalized Functions*, Cambridge University Press, Cambridge, 1959.

Morse, P.M., and Feshbach, H., *Methods of Theoretical Physics, Parts I and II*, McGraw-Hill, New York, 1953.

Tychonov, A.N., and Samarski, A.A., *Partial Differential Equations of Mathematical Physics, Vol. I*, Holden-Day, San Francisco, 1964.

2
Kinematic Wave Theory

The most important wave innovation in recent times is undoubtedly the *kinematic wave theory* (or simply, "KWT") formalism proposed by G. B. Whitham at the California Institute of Technology in the 1970s. This work is described in *Linear and Nonlinear Waves* (Whitham, 1974).

In two short decades, KWT has been applied to water waves, optics, acoustics, and many areas of engineering with success. Whitham, incidentally, also created years earlier the nonlinear theory used by aerodynamicists to model sonic booms at the earth's surface.

That theory recognized that *small local changes can lead to large cumulative farfield errors for long ray paths*: characteristics that are parallel on a linear basis converge in the farfield. Since properties remain constant along characteristics, intersections imply multi-valued, shock-like behavior.

This early sonic boom work would motivate rapid developments in *perturbation methods* to appear in the 1970s at Caltech, M.I.T., and Stanford, in particular, *multiple scaling methods, singular perturbation analysis, advanced WKB schemes,* and other *asymptotic* techniques (e.g., see van Dyke, 1964; Cole, 1968; Nayfeh, 1973).

Importantly, the new KWT formalism provides us with powerful modeling tools, but detailed recourse to differential equations is unnecessary. This requires a different outlook on mathematical physics, and here, we develop the basic ideas for use in drillstring vibrations and geophysics.

Casual observations. Numerous examples show that well-characterized waves can arise from complicated localized phenomena. Consider the irregular motions due to a pebble tossed into a lake; at some distance from impact, a circular wave pattern that propagates radially outward always emerges.

The 3D details related to cutting, shearing, and rocking motions at the rock-bit interface, for example, always vanish at some distance from the tricone or PDC bit; at the surface, axial and torsional disturbances having distinct wave-like character are always observed.

The point explosions set off by geophysical sources enter the earth as strong 3D disturbances; they ultimately undergo reflections and transmissions as well-defined plane waves. These observations suggest that the end result of transient localized disturbances may be wavetrains that vary slowly in space.

WHITHAM'S THEORY IN NONDISSIPATIVE MEDIA

Early KWT methods applied to *nondissipative* waves only, although the formalism *did* include uniform and nonuniform media, as well as linear, weakly nonlinear, and fully nonlinear motions. The general KWT equations were low-order in that high-order spatial derivative terms did not appear.

It is known that *singularities* (infinite amplitudes) obtained using low-order theory may not exist in reality, since neglected high-order terms, which become important as amplitudes increase, may alter the instability structure (e.g., see Chin (1993)). Low and high-order effects of dissipation *can* introduce changes not anticipated from the simpler model.

Chin (1976, 1980) extended KWT to include high-order wave dispersion and dissipation, showing how several contrasting models really appeared as special limits of a broader unifying theory. That work covered fully nonlinear waves, however, the complete subject falls beyond the scope of this book.

In this section, we discuss KWT for linear, nondissipative systems only, first for uniform, and then for nonuniform media, using Whitham's original nomenclature. In the next section, more complete equations are given; there, we develop new mathematical techniques that generalize early results.

Note to the reader. In introducing KWT to a new audience, an intuitive style likely to convey mathematical principles naturally is used. The physical consequences of several theorems, initially stated without proof, are developed first, to set the stage for broader discussion. Their inadequacies are explored, and only when the need for improved models is demonstrated, are more formal derivations given that rigorously extend the initial claims.

Uniform media. Here, we introduce the terminology used in KWT and the physical ideas behind kinematic wave modeling. For one-dimensional, linear wave propagation in *uniform* media (equations without variable coefficients), *elementary solutions* take the sinusoidal form

$$\alpha \cos(\kappa x - \omega t) \tag{2.1}$$

where the *dispersion relation*

$$\omega = \omega(\kappa) \tag{2.2}$$

is *real* in the absence of dissipation (see Chapter 1 for a brief review).

Equation 2-1 describes a cosine function traveling "to the right", with an amplitude α, an arbitrary wavenumber κ, and a frequency $\omega(\kappa)$. The general solution is given by Fourier superposition integrals of the form

$$\phi(x,t) = \int_0^\infty F(\kappa) \cos(\kappa x - \omega t) \, d\kappa \tag{2.3}$$

where $F(\kappa)$ is chosen to satisfy initial and boundary conditions.

Geometrically, the *wavenumber* gives the number of wave crests per unit distance, while the *frequency* describes the number of oscillations per unit time. Any dependence of the *phase velocity* ω/κ on κ leads to the *dispersion* of wave components having different wavelengths; hence, $\omega = \omega(\kappa)$ is called the *dispersion relation*.

Example 2-1: Transverse beam vibrations. Consider the lateral vibrations of a uniform beam without axial or torsional loads. If ρ is the mass density per unit volume, A the cross-sectional area, E the Young's modulus, and I the moment of inertia, the transverse displacement v(x,t) satisfies $\partial^2 v/\partial t^2 + (EI/\rho A)\, \partial^4 v/\partial x^4 = 0$.

The *uniform wave* substitution $v(x,t) = e^{i(kx-\omega t)}$ leads to $\omega^2 = (EI/\rho A)\, k^4$ or $\omega = \sqrt{(EI/\rho A)}\; k^2$. Thus, bending waves are dispersive since the phase velocity $\omega/k = \sqrt{(EI/\rho A)}\; k$ depends on k. An initial disturbance containing different length components will disperse: it will not be recognizable at large distances.

Example 2-2: Simple longitudinal oscillations. In contrast, the axial vibrations of a uniform bar follow $\partial^2 u/\partial t^2 - c^2\, \partial^2 u/\partial x^2 = 0$, where the sound speed $c = \sqrt{(E/\rho)}$. The substitution $u(x,t) = e^{i(kx-\omega t)}$ leads to $\omega = \sqrt{(E/\rho)}\; k$; since $\omega/k = \sqrt{(E/\rho)}$ is independent of k, longitudinal waves are *nondispersive*.

At least for ideal uniform bars, axial waves can be used in downhole-to-surface MWD communications; in reality, periodic *joints* in the drillstring limit the practicality of this transmission mode. In this example and in Example 2-1, we emphasize that the medium is uniform and nondissipative.

Example 2-3: Asymptotic stationary phase expansion. Although Equation 2-3 is exact, it is nearly always a complicated function of x and t, which does not render the main features and physical consequences immediately obvious.

Thus, we often obtain *asymptotic expansions* of Equation 2-3 for large x and t (e.g., see Carrier, Krook, and Pearson (1966)) to search for wave-like solutions, as suggested by the examples given early in this chapter. The *method of stationary phase* or the *saddle point method* leads to

$$\phi(x,t) \cong F(k) \sqrt{\{(2\pi)/(t\,|\omega''(k)|)\}}\cos(kx - \omega t - \pi \operatorname{sgn} \omega''/4) \tag{2.4}$$

where k(x,t) is the solution of

$$x = \omega\,'(k)t \tag{2.5}$$

In Equations 2-4 and 2-5, primes denote derivatives of the frequency with respect to the real wavenumber (e.g., $\omega'(k) = 2\sqrt{(EI/\rho A)}\,k$ in Example 2-1).

Since Equation 2-4 assumes that $\omega''(k)$ does not vanish, the results of this chapter do not apply to axial or torsional waves, which satisfy the classical wave equation; but they will figure importantly later in lateral vibrations.

We emphasize that we have not restricted our discussion to any particular area of engineering mechanics or physics. Our results will apply generally because our most severe restriction is the large time evaluation of the linear superposition integral.

Equation 2-4 *is* an oscillatory wave, however, unlike the elementary wave $\alpha \cos(\kappa x - \omega t)$, it is not uniform since k depends on x and t. But, when x and t are large relative to typical wavelengths and periods, Equation 2-5 shows how the change of k within a few wavelengths or periods is small. For example,

$$k_x/k = \{\omega\,'(k)/(k\omega''(k))\}(1/x) = O(1/x) \tag{2.6}$$

provided $\omega''(k)$ is nonzero.

Thus, apart from a simple phase change that is not relevant here, the solution given by Equation 2-3 ultimately takes the form

$$\phi(x,t) = a \cos(kx - \omega t) \tag{2.7}$$

where a, k and $\omega = \omega(k)$ are *slowly varying functions* of x and t in the sense of Equation 2-6. Importantly, *the asymptotic slowly varying solution assumes a form identical to that of the uniform wave* satisfying Equation 2-1.

It is also interesting to observe that when Equation 2-7 is rewritten in the form

$$\phi(x,t) = a \cos \Theta(x,t) \tag{2.8}$$

where the *phase function* $\Theta(x,t)$ satisfies

$$\theta(x,t) = kx - \omega t \tag{2.9}$$

the interchangeability of the differentiation $\theta_{xt}(x,t) = \theta_{tx}(x,t)$ leads to a statement describing *wave crest conservation*. That is, $\partial\theta_x/\partial t = \partial\theta_t/\partial x$ implies $\partial k/\partial t = -\partial\omega/\partial x$, or

$$\partial k/\partial t + \partial\omega/\partial x = 0 \tag{2.10}$$

In fluid mechanics, the mass conservation law $\partial\rho/\partial t + \partial(\rho q)/\partial x = 0$ states that timewise changes in density ρ must be balanced by differences in the mass flux ρq, where q is the transport velocity. Similarly, Equation 2-10 states that *timewise changes in the number of waves must be balanced by the flux of frequency* through the control length.

In 3D problems, the dispersion relation *may* depend on three wavenumbers, with $\omega = \omega(k_1,k_2,k_3,x,t) = \omega(\mathbf{k},x,t)$. For such problems, the vector description for wave crest conservation is $\partial\mathbf{k}/\partial t + \nabla\omega = 0$, and a special *irrotationality* condition for **k** applies; we consider 3D problems later, dealing with acoustic waveguide and cross-well geophysical applications.

It turns out that Equation 2-10 appears as an asymptotic result in many linear and nonlinear problems in inhomogeneous and uniform media. In KWT, we take this *kinematic* consequence as *axiomatic* and examine its implications. We will demonstrate that this assumption is consistent with conventional analysis and physical reality.

Simple consequences of kinematic wave theory. In order to illustrate the power behind KWT, we now draw upon a low-order corollary of a general amplitude theorem we derive later (e.g., see Equation 2-95). As in Equations 2-4 and 2-5, we later show *without* invoking specific engineering equations that

$$\partial a^2/\partial t + \partial(C(k)a^2)/\partial x = 0 \tag{2.11}$$

that is,

$$a_t + C(k)\, a_x + 1/2\, a\, C'(k)\, k_x = 0 \tag{2.12}$$

where

$$C(k) = d\omega(k)/dk \tag{2.13}$$

is the *group velocity*. What are the consequences of Equation 2-11?

To understand the implications, we apply the chain rule to Equation 2-10, and rewrite it as $\partial k/\partial t + d\omega(k)/dk\ \partial k/\partial x = 0$, or

$$k_t + C(k)\, k_x = 0 \tag{2.14}$$

The total differential dk is $dk = \partial k/\partial t\ dt + \partial k/\partial x\ dx$. If we divide by dt, we have $dk/dt = \partial k/\partial t + dx/dt\ \partial k/\partial x$. Comparison with Equation 2-14 shows that

$$dk/dt = 0 \tag{2.15}$$

provided that

$$dx/dt = C(k) = d\omega(k)/dk \tag{2.16}$$

Equation 2-15 states that the wavenumber k(x,t) does not change along the *characteristic* or *ray* dx/dt = C(k) = dω(k)/dk defined by Equation 2-16. In other words, *k(x,t) propagates with the group velocity* dω(k)/dk.

We turn our attention to the wave amplitude in Equation 2-12. The total differential $da = \partial a/\partial t \; dt + \partial a/\partial x \; dx$, on division by dt, gives $da/dt = \partial a/\partial t + dx/dt \; \partial a/\partial x = \partial a/\partial t + C(k) \; \partial a/\partial x$. Thus, Equation 2-12 is equivalently

$$da/dt = -1/2 \; C'(k) \; k_x \; a \tag{2.17}$$

along the ray dx/dt = C(k).

If an initial disturbance is limited to a finite interval around x = 0, then for sufficiently large x and t, we have x/t = C(k); upon partial differentiation with respect to x, we find that $1/t = C'(k) \; k_x$. Substitution in Equation 2-17 shows that da/dt = -a/2t. The simplified amplitude law

$$da/dt = -a/(2t) \tag{2.18}$$

easily integrates to

$$a = a_0(k)/t^{1/2} \tag{2.19}$$

where a_0 is an integration constant related to the initial amplitude. Equation 2-19 shows that amplitude a(t) is inversely proportional to $\sqrt{t}$, a fact consistent with Equations 2-4 and 2-5 obtained using classical methods.

Also note that the group velocity dω(k)/dk appears naturally in Equations 2-15 and 2-16 for k(x,t), and in the derivation of Equation 2-17 for a(x,t); it also follows as a consequence of classical stationary phase theory, as is evident from Equation 2-5.

Now, if Δx is the distance between two neighboring characteristics, differentiation shows that $d(a^2\Delta x)/dt = 2a(da/dt)\Delta x + a^2\Delta C = \{2a(da/dt) + a^2 C'(k)k_x\}\Delta x + O(\Delta x)^2$. Since Equation 2-17 requires $da/dt = -1/2 \; C'(k) \; k_x \; a$, it is clear that the curly bracketed { } quantity must vanish, so that

$$d(a^2\Delta x)/dt = 0 \tag{2.20}$$

Thus, *the total energy $a^2\Delta x$ between two consecutive rays defined by the group velocity is constant in a homogeneous, nondissipative medium*. All of our KWT results so far are consistent with classical physical phenomena.

Nonuniform media. Now consider wave motions in *nonuniform*, *inhomogeneous*, or *heterogeneous* media, for example, the transverse vibrations of a guitar string whose lineal mass density ρ_l varies slowly with x, or whose tension T varies slowly with time, or both, in which case $\omega = \sqrt{\{T(t)/\rho_l(x)\}}\ k$.

The restriction to *slow* is natural, since *fast* variations in properties may not support wave-like motions. We now determine what wave crest conservation implies, when applied to dispersion relations of the form

$$\omega = \omega(k,x,t) \tag{2.21}$$

where the explicit dependence on x and t is weak. Hence, we consider

$$\partial k/\partial t + \partial \omega(k,x,t)/\partial x = 0 \tag{2.22}$$

Thus, it follows from the chain rule of calculus that

$$\partial k/\partial t + \omega_k(k,x,t)\ \partial k/\partial x = -\ \omega_x(k,x,t) \tag{2.23}$$

As before, the total derivative dk satisfies $dk/dt = \partial k/\partial t + dx/dt\ \partial k/\partial x$. Hence,

$$dk/dt = -\ \omega_x(k,x,t) \tag{2.24}$$

provided that

$$dx/dt = \omega_k(k,x,t) \tag{2.25}$$

If the dispersion relation $\omega = \omega(k,x,t)$ is known, Equations 2-24 and 2-25 provide two coupled nonlinear ordinary differential equations that completely determine the wave kinematics once initial conditions on k and x are assigned.

Equations 2-24 and 2-25 show that k(x,t) still propagates at group velocity, but now it varies slowly along a ray; thus, trajectories in the x-t plane defined by Equation 2-25 will be curved and no longer straight.

Example 2-4: Numerical integration. The above equations are amenable to numerical integration. If *initial values* k_i and x_i for k and x are specified at time t_i, the propagated values at the end of a time step Δt may be determined from the *forward-differenced* forms of Equations 2-24 and 2-25, that is,

$$k_f = k_i - \omega_x(k_i,x_i,t_i)\ \Delta t \tag{2.26}$$

$$x_f = x_i + \omega_k(k_i,x_i,t_i)\ \Delta t \tag{2.27}$$

Equations 2-26 and 2-27, or their analogues in sophisticated schemes, can be computed repeatedly in order to update k and x along a ray in time.

Ease of use. In wave studies, authors investigate the consequences of partial differential equations and their auxiliary conditions. Clearly, equation-based methods can produce elegant closed form dispersion relations.

Often in research, the governing equations and boundary conditions behind a new physical phenomenon may be unknown, particularly in experimental work. Investigators focus on two objectives, the *first* of which determines formulations that replicate observed results, using equation-based modeling techniques.

For example, suppose we empirically determined that *gravity* waves in water satisfied $\Omega^r(k) = (gk \tanh kh)^{1/2}$, where g is gravitational acceleration and h is depth (Lamb, 1945). It would be difficult to guess the formulation that produces this result; in fact, it arises as the solution of *Laplace's equation* subject to free surface constraints (see Chapter 3).

The *second* objective focuses on *applying* measured results, since analytical models tend to be complicated. This is true for seismic waves in oil bearing rocks; many theoretical models have been proposed to describe rock matrix and pore fluid interaction, without definitive conclusions.

KWT provides an alternative. All that is required in order to use Equations 2-26 and 2-27 is a frequency function $\omega = \omega(k,x,t)$. The dependence of ω on k, x, and t may be obtained analytically using differential equations, numerically using computational methods, or empirically, from field or laboratory observations. The latter opens up the greatest application potential, but before discussing examples, we need to develop our ideas on attenuation modeling.

SIMPLE ATTENUATION MODELING

Wave motions in real media survive over many wavelengths or periods before damping significantly. Thus, damping is sometimes ignored during the early stages of modeling. Over larger space and time scales, damping must be accounted for since it affects wave amplitude and kinematics.

The terms *attenuation, damping, dissipation, internal friction,* and *nonconservative effects* are used interchangeably, and we follow this custom. There are two approaches that have been used to describe attenuation, and we present several unifying ideas in this section.

We will reconcile the *Q-model* popular in geophysics, with the complex frequency approaches used in vibrations and hydrodynamic stability; then, we integrate these ideas with recent developments in high-order KWT. First, let us define the Q-model and relate it to simple propagating and standing waves, before adapting it for general use in kinematic wave theory.

The Q-model. In geophysical problems, a *dimensionless measure of attenuation Q* is commonly used. If a volume of material is cycled in stress at a fixed frequency, Q can be defined as

$$1/Q = -\Delta E/2\pi E \qquad (2.28)$$

where E is the *peak strain energy* stored in the volume, and $-\Delta E$ is the energy lost *per cycle* due to material imperfections (Aki and Richards, 1980; Toksoz and Johnston, 1981; Kennett, 1983; Bourbie, Coussy, and Zinszner, 1987).

This definition is rarely used: it is difficult to drive a material element with stress waves having constant amplitude and period. More commonly, we measure the timewise decay of amplitude in standing waves at fixed wavenumber, or the spatial decay in a propagating waves at fixed frequency.

In either case, for a medium with a *linear stress-strain relationship*, the wave amplitude A is proportional to $E^{1/2}$. Hence, the foregoing definition for Q can be written in the form

$$1/Q = -\Delta A/\pi A \tag{2.29}$$

or equivalently as

$$\Delta A/A = -\pi/Q = -(2\pi)/2Q \tag{2.30}$$

Relating Q to amplitude decay in space. We wish to relate Q and A in Equation 2-30 to wave amplitude decay in space. In modeling wave propagation, recall that we had introduced the *complex frequency* in Chapter 1,

$$\Omega = \Omega^r + i\,\Omega^i \tag{2.31}$$

Equation 2-31, used in "$\exp\{i(kx-\Omega t)\}$", led to an amplitude factor "$\exp(\Omega^i t)$" that affected the propagating part "$\exp\{i(kx-\Omega^r t)\}$". What is the same amplitude factor in terms of Q?

First, let us establish *differential* (versus *per cycle*) laws describing attenuation. In the limit of infinitesimal changes, $\Delta A/A$ tends to dA/A, whereas the 2π term *per cycle* in Equation 2-30 becomes $\Omega^r\,dt$. We therefore replace Equation 2-30 with the more general statement

$$dA/A = -\Omega^r\,dt/2Q \tag{2.32}$$

Equation 2-32 can be rewritten as

$$dA/A = -(\Omega^r/2Q)\,(dt/dx)\,dx = -(\Omega^r/2cQ)\,dx \tag{2.33}$$

where we have not yet defined the speed c in $dx/dt = c$ (this is addressed later). Integration yields $\log_e A = -\Omega^r x/2cQ$ to within a constant, so that we may write

$$A(x) = A_0 \exp(-\Omega^r x/2cQ) \tag{2.34}$$

where A_0 is the initial wave amplitude at $x = 0$.

Summary. Equation 2-34 describes, as required, the spatial amplitude decay of a wave, when A_0, Ω^r, c, *and* Q are given; it is consistent with results obtained in earthquake seismology (Aki and Richards, 1980).

Relating Q to standing wave decay. We might have modeled the decay of a standing wave by introducing, as we did in Chapter 1, the imaginary part Ω^i of the complex frequency given in Equation 2-31, setting

$$A(t) = A_0 \exp(\Omega^i t) \tag{2.35}$$

If we equate timewise attenuation with its equivalent in space, Equations 2-34 and 2-35 suggest that we set $\Omega^i t = -\Omega^r x/2cQ$. But since $x = ct$, the expression simplifies to $\Omega^i = -\Omega^r/2Q$; Equation 2-29 is, alternatively,

$$1/Q = -2\,\Omega^i/\Omega^r \tag{2.36}$$

where we must have $|\Omega^i/\Omega^r| << 1$, in order to rule out critically damped waves.

Summary. When the frequency Ω^r of the standing wave *and* Q are known, the Ω^i used in more general KWT analysis is simply $\Omega^i = -\Omega^r/2Q$.

Kinematic wave generalization. We return to Equation 2-11 (that is, $\partial a^2/\partial t + \partial(C(k)a^2)/\partial x = 0$, where C(k) is the *group velocity* $\Omega^r{}_k(k)$), and examine the effect of adding $2\Omega^i a^2$ to the right side,

$$\partial a^2/\partial t + \partial(C(k)a^2)/\partial x = 2\Omega^i a^2 \tag{2.37}$$

Equation 2-37 was postulated by Landahl (1972), who derived it intuitively, so that it satisfies the simple propagating and standing wave limits introduced earlier. Chin (1976, 1980) formally obtained this result using *multiple-scaling* methods, carrying out asymptotic expansions to higher order.

Note that Landahl's equation appears in *conservation form*. It is interpreted analogously to the mass conservation equation $\partial\rho/\partial t + \partial(\rho q)/\partial x = 0$ used in fluid mechanics, which balances timewise changes in density ρ with differences in the mass flux ρq, where q is the transport velocity.

Equation 2-37 states that timewise changes to an *energy-like* a^2 must be balanced by differences in the energy flux $C(k)a^2$, where the relevant transport velocity, consistently with classical physics, is the group velocity. The right side of Equation 2-37 is *sink-like*, representing *distributed attenuation*.

Let us consider two limits of Equation 2-37. In the *propagation limit* with $\partial/\partial t = 0$, the equation $\partial(C(k)a^2)/\partial x = 2\Omega^i a^2$ on division by C(k) leads to

$$da^2/a^2 = 2\Omega^i/C(k)\,dx = -\{\Omega^r/C(k)Q\}dx \tag{2.38}$$

where we have used $1/Q = -2\,\Omega^i/\Omega^r$. Direct integration yields the result $\log_e a^2 = -\{\Omega^r/C(k)Q\}x$, or $a^2 \propto \exp\{-(\Omega^r/C(k)Q)x\}$. On taking square roots,

$$a(x) \propto \exp\{-\Omega^r x/2C(k)Q\} \tag{2.39}$$

This is consistent with Equation 2-34, but the use of Equation 2-37 importantly identifies the "c" in Equation 2-34 as the group velocity $\Omega^r{}_k(k)$.

Now we consider the *standing wave limit*, where $\partial/\partial x = 0$. Here, Equation 2-37 becomes $\partial a^2/\partial t = 2\Omega^i a^2$, and we obtain $a^2 \propto \exp(2\Omega^i t)$. Thus,

$$a(t) \propto \exp(\Omega^i t) \quad (2.40)$$

which is consistent with Equation 2-35.

Summary. In using Equation 2-37, we can replace the complex frequency on the right with $\Omega^i = -\Omega^r/2Q$, noting that the dimensionless function Q is empirically obtained. The speed "c" used earlier should be the group velocity $C(k) = d\Omega^r(k)/dk$, where $k(x,t)$ is obtained by solving Equations 2-24 and 2-25.

Recapitulation. We have introduced the Q-model used in geophysics, interpreted its meaning in terms of propagating and standing waves, and embedded these ideas within a group velocity based wave description for continuum systems which allows both changes in space and time.

Simple recipes follow. When a *propagation* experiment is conducted with the excitation frequency fixed, direct measurements can be made for the wavenumber k; the frequency function $\Omega^r(k)$ can be obtained from repeated experiments. Simultaneous attenuation observations yield $\Omega^i(k)$; then, using Equation 2-36, Q is in principle known.

Sometimes $\Omega^r(k)$ and Q are available from *standing wave* measurements. For such problems, Equation 2-36 can be used to find $\Omega^i(k)$. Substitution in Equation 2-37 allows standing wave "Q" measurements to be used in general propagation simulation.

KINEMATIC WAVE THEORY IN HOMOGENEOUS AND DISSIPATIVE MEDIA

Having introduced the essential ideas, we formally develop KWT assuming *uniform* media for now, but allowing *general frequency-dependent wave dissipation*. By dissipation, we include *instabilities* as well, since we need not restrict ourselves to purely decaying or growing waves; in fact, $\Omega^i(k)$ may generally alternate in sign.

In Chapter 1, we saw that weak dissipation (that is, $|\Omega^i(k)/\Omega^r(k)| \ll 1$) leaves the kinematics unaffected to leading order. The imaginary frequency $\Omega^i(k)$ primarily affects amplitude, leaving phase unchanged except at the second order. In this limit, our axiomatic conservation law for wave crests need *not* involve the imaginary frequency, and we again write

$$\partial k/\partial t + \partial \omega/\partial x = 0 \quad (2.41)$$

in the form

$$\partial k/\partial t + \partial\Omega^r(k)/\partial x = 0 \tag{2.42}$$

where we have replaced ω by $\Omega^r(k)$, and $\Omega^r(k)$ is any prescribed function. If we carry out the differentiation using the chain rule, we obtain

$$\partial k/\partial t + \Omega^r{}_k(k)\ \partial k/\partial x = 0 \tag{2.43}$$

In wave propagation, we often start with localized initial disturbances containing many wavelength components. The goal is to determine how wavenumber and amplitude fields evolve in space and time. Knowing how the disturbance (or information) packet disperses is vital to reassembling it, a process important to transmission line applications in communications.

Example 2-5: General initial value problem. From calculus, the total differential dk satisfies $dk = \partial k/\partial t\ dt + \partial k/\partial x\ dx$, which upon division by dt yields $dk/dt = \partial k/\partial t + dx/dt\ \partial k/\partial x$. Thus, on comparison with Equation 2-43, we have

$$dk/dt = 0 \tag{2.44}$$

provided that

$$dx/dt = \Omega^r{}_k(k) \tag{2.45}$$

Equation 2-44 indicates that k is constant along a ray. Equation 2-45, taking advantage of this fact, states that the group velocity is likewise constant; this implies straight line trajectories in the characteristic x-t plane.

The latter result, applicable to homogeneous media only, allows us to write down the integral of Equation 2-45 immediately as

$$x - \Omega^r{}_k(k)\ t = \text{constant} \tag{2.46}$$

The fact that k(x,t) is constant along rays can, therefore, be concisely expressed through the functional statement

$$k(x,t) = f\{x - \Omega^r{}_k(k)\ t\} \tag{2.47}$$

Equation 2-47 *is* the *general solution* to Equation 2-42, a nonlinear first-order partial differential equation governing k(x,t), for the initial value problem $k(x,0) = f(x)$. To see why, we simply set $t = 0$ in Equation 2-47 and find that

$$k(x,0) = f(x) \tag{2.48}$$

Thus, when the starting wavenumber distribution $k(x,0) = f(x)$ is specified, and the function $\Omega^r(k)$ is known, Equation 2-47 is the general solution to the initial value problem solving Equations 2-42 and 2-48.

We have shown that characteristics remain straight in homogeneous media, since each initial group velocity along a ray remaining unaffected and constant during the propagation. But this is not to say that all of the initial velocities $\Omega^r{}_k(k)$ equal the same constant.

This raises the possibility that our characteristics *can* intersect within a finite time (in Equation 1-1, the characteristics within each wave family *always* remain parallel). Since we have demonstrated above that k(x,t) must remain constant along a ray, the intersection of rays therefore implies that more than one value of k exists at a point, thereby forming *kinematic shocks*.

Singularities of the kinematic field. To demonstrate that *singularities* (or, *infinities*) of the general wave solution may exist, which would suggest a breakdown in the assumptions underlying Equation 2-42, we take partial derivatives of Equation 2-47 with respect to time, leading to

$$\partial k/\partial t = f'[x - \Omega^r{}_k(k)\, t]\{- \Omega^r{}_k(k) - t\, \Omega^r{}_{kk}(k)\partial k/\partial t\} \tag{2.49}$$

Some rearrangement gives

$$\partial k/\partial t + t\, \Omega^r{}_{kk}(k)\partial k/\partial t\, f'[x - \Omega^r{}_k(k)\, t] = - \Omega^r{}_k(k)\, f'[x - \Omega^r{}_k(k)\, t] \tag{2.50}$$

so that

$$\partial k/\partial t = - \Omega^r{}_k(k)\, f'[x - \Omega^r{}_k(k)\, t]/\{1 + t\, \Omega^r{}_{kk}(k)\, f'[x - \Omega^r{}_k(k)\, t]\} \tag{2.51}$$

Hence, the number of waves increases temporally without bound if the denominator vanishes at the finite time

$$t = -1/\{\Omega^r{}_{kk}(k)\, f'[x - \Omega^r{}_k(k)\, t]\} \tag{2.52}$$

provided that $t > 0$.

We can similarly differentiate Equation 2-47 with respect to x, thus yielding the result

$$\partial k/\partial x = f'[x - \Omega^r{}_k(k)\, t]\{1 - t\, \Omega^r{}_{kk}(k)\partial k/\partial x\} \tag{2.53}$$

Some rearrangement leads to

$$\partial k/\partial x = f'[x - \Omega^r{}_k(k)\, t]/\{1 + t\, \Omega^r{}_{kk}(k)\, f'[x - \Omega^r{}_k(k)\, t]\} \tag{2.54}$$

Again, a wave singularity is encountered in a finite time, dictated by Equation 2-52. However, this second derivation shows that the number of waves also increases without bound in space.

The energy singularity. What happens to the energy density as the wavenumber field becomes singular? To answer this question, we return to Equation 2-37, and rewrite it in the form

$$\partial a^2/\partial t + \partial\{\Omega^r{}_k(k)a^2\}/\partial x = 2\, \Omega^i\, a^2 \tag{2.55}$$

We carry out the partial differentiation in the second term to obtain

$$\partial a^2/\partial t + \Omega^r_k(k)\ \partial a^2/\partial x\ + \Omega^r_{kk}\partial k/\partial x\ \ a^2 = \ 2\ \Omega^i\ a^2 \tag{2.56}$$

and consequently

$$\partial a^2/\partial t + \Omega^r_k(k)\ \partial a^2/\partial x\ \ = \ \{2\ \Omega^i - \Omega^r_{kk}\partial k/\partial x\}a^2 \tag{2.57}$$

When we combine Equations 2-54 and 2-57, we find that

$$\partial a^2/\partial t + \Omega^r_k(k)\ \partial a^2/\partial x\ \ = \{2\ \Omega^i - \Omega^r_{kk}f\ '[x - \Omega^r_k(k)\ t]/\{1 + t\ \Omega^r_{kk}(k)\ f\ '[x - \Omega^r_k(k)\ t]\}\}a^2 \tag{2.58}$$

The $2\Omega^i$ term describes amplitude changes due to *damping* or *instability*, while the second term in the { } brackets describes *geometric changes* due to *ray convergence* or *divergence*. When kinematic shocks form, the above denominator vanishes, and the right side of Equation 2-58 becomes infinite. Thus, singularities in k(x,t) are accompanied by singularities in the energy.

If shocks do not form, we can obtain a closed form solution describing the variation of a^2 along a ray. From calculus, the total differential da^2 satisfies $da^2 = \partial a^2/\partial t\ dt + \partial a^2/\partial x\ dx$. Upon division by dt, we obtain $da^2/dt = \partial a^2/\partial t\ + dx/dt\ \partial a^2/\partial x$. If we compare this result with Equation 2-58, we find that

$$da^2/dt = \{2\ \Omega^i - \Omega^r_{kk}f\ '[x - \Omega^r_k(k)\ t]/\{1 + t\Omega^r_{kk}(k)\ f\ '[x - \Omega^r_k(k)\ t]\}\}a^2 \tag{2.59}$$

provided that we follow the rays defined by Equations 2-44 and 2-45.

Again, recall that along each such ray, k(x,t) remains a constant equal to its initial t = 0 value. Consequently, since the right side of Equation 2-59 is constant, it follows that

$$a^2(t) = a^2_0\ \exp\{2\Omega^i - \Omega^r_{kk}f\ '[x - \Omega^r_k(k)t]/\{1 + t\Omega^r_{kk}(k)f\ '[x - \Omega^r_k(k)t]\}\}t \tag{2.60}$$

Unlike the amplitude laws given in the previous section, where either $\partial/\partial t = 0$ or $\partial/\partial x = 0$, the generalized results derived here apply to wave motions that vary unsteadily in both t and x.

Example 2-6: Modeling dynamically steady motions. Analytical solutions can be obtained for another broad class of problems, those in which a wavemaker excites the system at constant frequency ω_0. For such problems, the wavenumber field k(x) is uniquely determined by the nonlinear equation

$$\omega_0 = \Omega^r(k) \tag{2.61}$$

where the real dispersion function $\Omega^r(k)$ is prescribed.

Setting $\partial a^2/\partial t = 0$ in Equation 2-55, we obtain $d\{\Omega^r_k(k)a^2\}/dx = 2\ \Omega^i\ a^2$. If we now carry out the *ordinary* differentiation, we have, successively,

$$\Omega^r_k(k)\ da^2/dx + a^2\Omega^r_{kk}(k)dk/dx = 2\ \Omega^i\ a^2 \tag{2.62}$$

$$\Omega^r{}_k(k)\, da^2/dx = (2\,\Omega^i - \Omega^r{}_{kk}(k)dk/dx)\, a^2 \qquad (2.63)$$

$$da^2/a^2 = 2\,\Omega^i\,(dx/\Omega^r{}_k(k)) - \Omega^r{}_{kk}(k)dk/dx\;\; (dx/\Omega^r{}_k(k)) \qquad (2.64)$$

$$da^2/a^2 = 2\,\Omega^i\, dt - \Omega^r{}_{kk}(k)/\Omega^r{}_k(k)\; dk \qquad (2.65)$$

$$\int da^2/a^2 = 2\int \Omega^i\, dt - \int \Omega^r{}_{kk}(k)/\Omega^r{}_k(k)\; dk \qquad (2.66)$$

$$\log_e a^2 = \;\; 2\int \Omega^i\, dt \;\; - \log_e \Omega^r{}_k(k) \qquad (2.67)$$

$$\log_e a^2\, \Omega^r{}_k(k) = \;\; 2\int \Omega^i\, dt \qquad (2.68)$$

and finally, the desired results

$$a^2 \propto \{\Omega^r{}_k(k)\}^{-1} \exp\,\{2\int \Omega^i\, dt\} \qquad (2.69)$$

$$a \propto \{\Omega^r{}_k(k)\}^{-1/2} \exp\,\{\int \Omega^i\, dt\} \qquad (2.70)$$

These results should be contrasted with the solutions obtained earlier, namely, $A(x) = A_0 \exp(-\Omega^r x/2cQ)$ in Equation 2-34, and $A(t) = A_0 \exp(\Omega^i t)$ in Equation 2-35, for our two simpler problems.

HIGH-ORDER KINEMATIC WAVE THEORY

So far, we have developed KWT *assuming* that Equation 2-42 (that is, $\partial k/\partial t + \partial\Omega^r(k)/\partial x = 0$) applies, and *assuming* that $\partial a^2/\partial t + \partial(C(k)a^2)/\partial x = 2\Omega^i a^2$ given by Equation 2-37 holds. Since the former is independent of a(x,t), these assumptions imply that we can solve for k(x,t) and a(x,t) sequentially.

This is not absolutely true: from Chapter 1, wave phase (and hence, k(x,t)) *is* affected by dissipation *through amplitude* at the second order. The limitations behind our *ad hoc* assertions need to be established, ideally by examining how they arise as limits of a more complete theory.

In this section, we formally derive both of the above low-order results. In fact, we give their extensions to *two additional orders of magnitude*, and demonstrate the consequences in truncating the new *phase* and *amplitude* expansions.

In low-order wave theory, it was sufficient to consider the kinematics of the motion using the dispersion relation for the uniform plane wave only, which was completely independent of amplitude. *This is not the case with higher order corrections, even when nonlinearities are completely neglected.*

The results obtained here are extremely important to a number of physical problems, ranging from the study of kinematic shocks, to ray tracing through dissipative formations in crosswell tomography, to Kirchoff wave migration in seismics, and to lateral drillstring vibrations deep in the wellbore.

Basic assumptions. We start with a homogeneous uniform medium, but again allow general frequency-dependent attenuation or instability. We assume that the complex dispersion relation corresponding to the uniform sinusoidal wavetrain is available, obtained, say, using the methods of Chapter 1.

Now let us examine a general linear superposition of *monochromatic wave* components, each behaving like exp $i(kx-\omega t)$ and satisfying the complex dispersion relation

$$\omega = \Omega(k) = \Omega^r(k) + i\,\Omega^i(k) \tag{2.71}$$

We assume that

$$\Omega^i(k)/\Omega^r(k) \approx O(\varepsilon) \tag{2.72}$$

where $\varepsilon << 1$ is a small dimensionless number (e.g., one proportional to Q^{-1} in Equation 2-36).

This flexibility allows nontrivial wave motions that are not highly or critically damped. Here, k is a real wavenumber, and $\Omega^r(k)$ and $\Omega^i(k)$ are the real and imaginary parts of the complex frequency $\Omega(k)$ respectively. The general Fourier superposition integral can be written as

$$F(x,t) = \int B(k)\, e^{i(kx-\omega t)}\, dk \tag{2.73}$$

where the limits $(-\infty,\infty)$ of integration are omitted for clarity. Equation 2-73 extends the simpler integral of Equation 2-3 in order to model dissipative effects.

In any particular problem, the function B(k) would be proportional to the Fourier transform of the initial condition; this function exists, of course, when the initial disturbances are suitably confined.

The general $\psi(x,t)$ equation. If we now focus our attention at the neighborhood of a (real) *center wavenumber* $k = k_0$, with a corresponding (real) *center frequency* $\Omega^r{}_0$, we can certainly write

$$F(x,t) = \psi(x,t) \exp i(k_0x - \Omega^r{}_0 t) \tag{2.74}$$

Equation 2-74 represents the first step of many that may not be obvious to the reader. Its strategic role was clear only after intensive thought, upon which Equations 2-94 and 2-95 would emerge. *The reader who wishes to proceed directly to those results may do so without break in continuity.*

In this explicit (and still exact) representation, F(x,t) contains a purely periodic part, together with a complex amplitude function $\psi(x,t)$ that incorporates dissipative effects. The latter function identically satisfies

$$\psi(x,t) = \int B(k) \exp i[(k-k_0)x - (\Omega - \Omega^r_0)t]\, dk = \int G(k)\, dk \tag{2.75}$$

where G(k) has been introduced to simplify our notation.

Now, the complex dispersion relation given above holds for every component wave solution. Taylor expansion about centered properties gives

$$\omega = \Sigma\, \Omega_{nk}(k_0)(k-k_0)^n/n! \tag{2.76}$$

where the summation Σ is taken from $n = 0$ to ∞. Applying the operator

$$\int \ldots B(k) \exp i[(k-k_0)x - (\Omega - \Omega^r_0)t]\, dk \tag{2.77}$$

to both sides, and noting that

$$i\, \partial\psi/\partial t = \int (\omega - \Omega^r_0)\, G\, dk \tag{2.78}$$

$$(-i)^n\, \partial^n\psi/\partial x^n = \int (k - k_0)^n\, G\, dk \tag{2.79}$$

leads to an important identity that serves as the *exact* partial differential equation satisfied by $\psi(x,t)$,

$$i\, \partial\psi/\partial t = i\, \Omega^i{}_0\psi + \Sigma\, \{(-i)^n/n!\}\, \Omega_{nk,0}\, \partial^n\psi/\partial x^n \tag{2.80}$$

Equation 2-80 will serve as the focal point of our subsequent discussion, one which determines its asymptotic wave properties.

Method of multiple scales. In many physical problems, multiple time scales exist. For example, the solution to the damped wave equation given by Equation 1-44 is characterized by a period defined by the undamped system, a second time scale associated with amplitude decay, and a third which introduces dissipation-dependent changes to wave phase.

Or consider the oscillations of a mass-spring-damper system. The obvious time scale is the period, which depends for the most part on mass and spring stiffness. The damping term affects amplitude over larger time scales, and again, after longer times, induces changes to the period itself.

In the *method of multiple scales*, this *a priori* knowledge of the physical system is explicitly exploited (Cole, 1968; Nayfeh, 1973). We specifically seek separable, asymptotic, wave-like (a.k.a., "WKB") solutions of the form

$$\psi(x,t) = a(X,T)\, e^{i\,\theta(x,t)} \tag{2.81}$$

where $\theta(x,t)$, a rapidly varying disturbance phase function, is motivated by the discussion leading to Equations 2-8 and 2-9. Its space and time derivatives are related to wavenumber and frequency as before, but following Equation 2-6, we insist that the latter vary slowly in space and time.

To develop the theory formally, we introduce new uppercase wavenumber and frequency functions K and ω. We require that they vary slowly with respect to x and t, and accomplish this by having them depend explicitly on the large-scale variables X and T where

$$X = \varepsilon x \tag{2.82}$$

$$T = \varepsilon t \tag{2.83}$$

Again, ε is the small parameter inferred from Equation 2-72, which states that $\Omega^i(k)/\Omega^r(k) \approx O(\varepsilon)$. That is, K and ω may vary significantly, but only over large space and time scales for which X and T are O(1), or equivalently, for which x and t are $O(1/\varepsilon)$. Thus, we formally assume that

$$\theta_x = k - k_0 = K(X,T) \tag{2.84}$$

and

$$-\theta_t = (\omega - \Omega^r{}_0) = \omega(X,T) \tag{2.85}$$

The amplitude a(X,T) in Equation 2-81, following Equations 2-4 and 2-5, is likewise assumed to vary slowly with x and t, changing significantly only over X and T scales of O(1).

Interestingly, partial differentiation of Equation 2-84 with respect to time, of Equation 2-85 with respect to space, and use of the identity $\partial\theta_x/\partial t = \partial\theta_t/\partial x$ lead to the *law for wave crest conservation*

$$\partial K/\partial T + \partial\omega(X,T)/\partial X = 0 \tag{2.86}$$

previously taken as axiomatic.

Now, let us digress and consider *any* function u(x,t) that possesses slowly varying wavelike solutions of the form

$$u(x,t) = U(\theta,X,T) \tag{2.87}$$

Using the chain rule of calculus, we deduce the property

$$\partial u(x,t)/\partial x = U_X X_x + U_\theta \theta_x = \varepsilon U_X + K U_\theta \tag{2.88}$$

In Equation 2-88 and in the subsequent analysis, subscripts are used to represent partial derivatives. With some effort, we can show by *mathematical induction* (Chin, 1976) that the n*th* partial derivative of $U(\theta,X,T)$ with respect to x satisfies

$$u_{nx}(x,t) = K^n U_{n\theta} \tag{2.89}$$
$$+ \varepsilon \ [nK^{n-1}U_{(n-1)\theta,X} + 1/2\ n(n-1)K^{n-2}K_X U_{(n-1)\theta}]$$
$$+ \varepsilon^2 \ [1/2\ n(n-1)K^{n-2}U_{(n-2)\theta,XX}$$
$$+ \quad 1/2\ n(n-1)(n-2)K^{n-3}K_X U_{(n-2)\theta,X}$$
$$+ \quad 1/8\ n(n-1)(n-2)(n-3)K^{n-4}K_X^2 U_{(n-2)\theta}$$
$$+ \quad 1/6\ n(n-1)(n-2)K^{n-3}K_{XX}U_{(n-2)\theta}\ ]$$
$$+ \varepsilon^3 \ [1/12\ n(n-1)(n-2)(n-3)(n-4)K^{n-5}K_X K_{XX} U_{(n-3)\theta}$$
$$+ \quad 1/6\ n(n-1)(n-2)(n-3)K^{n-4}K_{XX}U_{(n-3)\theta,X}$$
$$+ \quad 1/4\ n(n-1)(n-2)(n-3)K^{n-4}K_X U_{(n-3)\theta,XX}$$
$$+ \quad 1/6\ n(n-1)(n-2)K^{n-3}U_{(n-3)\theta,XXX}$$
$$+ \quad 1/24\ n(n-1)(n-2)(n-3)K^{n-4}K_{XXX}U_{(n-3)\theta}$$
$$+ \quad 1/8\ n(n-1)(n-2)(n-3)(n-4)K^{n-5}K_X^2 U_{(n-3)\theta,X}$$
$$+ \quad 1/48\ n(n-1)(n-2)(n-3)(n-4)(n-5)K^{n-6}K_X^3 U_{(n-3)\theta}\ ]$$
$$+ \ldots$$
$$+ \ \varepsilon^n\ U_{nX}$$

Generalized wave results. We now substitute Equation 2-81 into Equation 2-80, and apply the derivative operations defined by Equation 2-89 throughout. After considerable algebra, we obtain a complicated equation, whose real and imaginary parts lead to

$$a\,\omega = \varepsilon\, \Sigma\, (\Omega^i_{nk,0}/n!)\ [nK^{n-1}a_X + 1/2\ n(n-1)K^{n-2}K_X\, a] \tag{2.90}$$
$$+ \Sigma\, (\Omega^r_{nk,0}/n!)\ [aK^n - \varepsilon^2\ \{1/2\ n(n-1)K^{n-2}a_{XX}$$
$$+1/2\ n(n-1)(n-2)\ K^{n-3}K_X\, a_X + 1/8\ n(n-1)(n-2)(n-3)K^{n-4}K_X{}^2\, a$$
$$+\ 1/6\ n(n-1)(n-2)K^{n-3}K_{XX}\, a\}]$$

$$\varepsilon a_T = \Omega^i{}_0\, a - \varepsilon\, \Sigma\, (\Omega^r_{nk,0}/n!)\ [nK^{n-1}a_X + 1/2\ n(n-1)K^{n-2}K_X\, a] \tag{2.91}$$
$$+ \Sigma\, (\Omega^i_{nk,0}/n!)\ [aK^n - \varepsilon^2\ \{1/2\ n(n-1)K^{n-2}a_{XX}$$
$$+1/2\ n(n-1)(n-2)\ K^{n-3}K_X\, a_X + 1/8\ n(n-1)(n-2)(n-3)K^{n-4}K_X{}^2\, a$$
$$+\ 1/6\ n(n-1)(n-2)K^{n-3}K_{XX}\, a\}]$$
$$+ \varepsilon^3\, \Sigma\, (\Omega^r_{nk,0}/n!)\ [1/12\ n(n-1)(n-2)(n-3)(n-4)K^{n-5}K_X K_{XX} a$$
$$+\ 1/6\ n(n-1)(n-2)(n-3)K^{n-4}K_{XX}\, a_X$$
$$+\ 1/6\ n(n-1)(n-2)K^{n-3}\ a_{XXX}$$
$$+\ 1/4\ n(n-1)(n-2)(n-3)K^{n-4}K_X a_{XX}$$
$$+\ 1/24\ n(n-1)(n-2)(n-3)K^{n-4}K_{XXX} a$$
$$+\ 1/8\ n(n-1)(n-2)(n-3)(n-4)K^{n-5}K_X^2 a_X$$
$$+\ 1/48\ n(n-1)(n-2)(n-3)(n-4)(n-5)K^{n-6}K_X^3 a\ ]$$

Here, the summations are taken from n=1 to infinity. Equation 2-90 describes the *perturbation phase*, while Equation 2-91 describes the *amplitude* of the *slowly varying wave*. Both of these equations are cumbersome, though, since they contain infinite numbers of Taylor series terms.

But our use of *centered wavenumber expansions*, while the end objective itself in many stability analyses, in our study serves only an intermediate purpose. It functions as a mathematical artifice, allowing us to obtain Equations 2-90 and 2-91, which we *will not* retain as our final results.

Since these two equations *must* apply independently of any specific centered values, we use binomial formulas to sum them analytically. We had assumed that Taylor expansions in k were legitimate: having carried our derivations using these expansions, we now to sum the series analytically to obtain more compact results.

This is analogous to our viewing the Taylor series $1 + x + x^2 + x^3 + \ldots$, which is convergent only for $|x| < 1$, as implying the more general function $1/(1-x)$, which is defined for all x, except $x = 1$.

This *analytical continuation* eliminates the subscript 0s, and allows us to return Equations 2-90 and 2-91 to functional forms having complete k's. The generalized *phase* (or *eikonal*) relation that results is

$$\omega = \Omega^r(k) + \varepsilon a_x \Omega^i{}_k(k)/a + 1/2\, \varepsilon k_x \Omega^i{}_{kk} - 1/2\, \varepsilon^2 a_{xx} \Omega^r{}_{kk}/a$$

$$- 1/2\, \varepsilon^2 a_x k_x \Omega^r{}_{kkk}/a - 1/8\, \varepsilon^2 k_x{}^2 \Omega^r{}_{kkkk} - 1/6\, \varepsilon^2 k_{xx} \Omega^r{}_{kkk} \qquad (2.92)$$

Since Ω^i is small, the terms on the first and second lines of Equation 2-92 other than $\Omega^r(k)$ are both $O(\varepsilon^2)$. Those on the first line represent high-order effects due to dissipation; those on the second are high-order effects due to dispersion. The corresponding equation governing a^2 takes the form

$$\partial a^2/\partial T + \partial(\Omega^r{}_k a^2)/\partial X = 2\Omega^i(k)a^2/\varepsilon \qquad (2.93)$$

$$- \varepsilon^2 (a a_{xx} \Omega^i{}_{kk}/\varepsilon + a a_x k_x \Omega^i{}_{kkk}/\varepsilon$$

$$+ 1/4\, a^2 k_x{}^2 \Omega^i{}_{kkkk}/\varepsilon + 1/3\, a^2 k_{xx} \Omega^i{}_{kkk}/\varepsilon)$$

$$+ \varepsilon^2 (1/3\, a a_{xxx} \Omega^r{}_{kkk} + 1/2\, a a_{xx} k_x \Omega^r{}_{kkkk}$$

$$+ 1/3\, a a_x k_{xx} \Omega^r{}_{kkkk} + 1/12\, a^2 k_{xxx} \Omega^r{}_{kkkk}$$

$$+ 1/6\, a^2 k_x k_{xx} \Omega^r{}_{kkkkk} + 1/4\, a a_x k_x{}^2 \Omega^r{}_{kkkkk}$$

$$+ 1/24\, a^2 k_x{}^3 \Omega^r{}_{kkkkkk})$$

The left side of Equation 2-93 appears in *conservation form*. The right "sink-like" side indicates the expected "$2\Omega^i(k)a^2/\varepsilon$" damping, but higher order terms related to wave dissipation and dispersion are both present.

Equations 2-92 and 2-93 appear in slow (X,T) variables. Returning to physical coordinates (x,t), we have, to two higher orders beyond classical ray theory (with $\varepsilon \approx O(\Omega^i(k)/\Omega^r(k)) << 1$), the phase relationship

$$\omega = \Omega^r(k) + a_x\Omega^i{}_k(k)/a + 1/2\, k_x\Omega^i{}_{kk} - 1/2\, a_{xx}\Omega^r{}_{kk}/a - 1/2\, a_x k_x\Omega^r{}_{kkk}/a - 1/8\, k_x{}^2\Omega^r{}_{kkkk} - 1/6\, k_{xx}\Omega^r{}_{kkk} \quad (2.94)$$

for the *slowly varying* wave, to be contrasted with the dispersion relationship $\omega = \Omega(k) = \Omega^r(k) + i\,\Omega^i(k)$ given by Equation 2-71, which only applies to the *uniform plane wave*. We also obtain the generalized amplitude equation

$$\begin{aligned}\partial a^2/\partial t + \partial(\Omega^r{}_k a^2)/\partial x = \;& 2\Omega^i(k)a^2 \\ & - (aa_{xx}\Omega^i{}_{kk} + aa_x k_x\Omega^i{}_{kkk} \\ & + 1/4\, a^2 k_x{}^2\Omega^i{}_{kkkk} + 1/3\, a^2 k_{xx}\Omega^i{}_{kkk}) \\ & + (1/3\, aa_{xxx}\Omega^r{}_{kkk} + 1/2\, aa_{xx}k_x\Omega^r{}_{kkkk} \\ & + 1/3\, aa_x k_{xx}\Omega^r{}_{kkkk} + 1/12\, a^2 k_{xxx}\Omega^r{}_{kkkk} \\ & + 1/6\, a^2 k_x k_{xx}\Omega^r{}_{kkkkk} + 1/4\, aa_x k_x{}^2\Omega^r{}_{kkkkk} \\ & + 1/24\, a^2 k_x{}^3\Omega^r{}_{kkkkkk})\end{aligned} \quad (2.95)$$

which shows, quite naturally, how group velocity emerges as the physically relevant propagation velocity. Equations 2-94 and 2-95, again, are not independent. They are mathematically coupled through Equation 2-86, which we now rewrite in the form

$$\partial k/\partial t + \partial\omega/\partial x = 0 \quad (2.96)$$

Substitution of Equation 2-94 in Equation 2-96, as we had shown earlier in a simpler capacity, reveals the group velocity $\Omega^r{}_k$ as the relevant transport speed for the wavenumber k(x,t).

The low-order limit. We are finally in a position to derive, with mathematical rigor, the low-order limit arising out of Equations 2-94 to 2-95. If in Equation 2-94, we retain only the term in "$\omega = \Omega^r(k) + \ldots$" shown, substitution in Equation 2-96 leads to

$$\partial k/\partial t + \partial\Omega^r(k)/\partial x = 0 \quad (2.97)$$

Deletion of the high-order terms in Equation 2-95 leads to

$$\partial a^2/\partial t + \partial(\Omega^r{}_k a^2)/\partial x = 2\Omega^i(k)a^2 \quad (2.98)$$

which are *exactly* Equations 2-42 and 2-55 assumed earlier.

Closing remarks. Despite their apparent generality, our results are limited to linear systems because linear assumptions were invoked, e.g., Fourier superposition, sinusoidal solutions, exponential damping, and so on. Also, the derivation disallows explicit heterogeneities (e.g., a vibrating string with varying density $\rho_l(\varepsilon x)$ or tension $T(\varepsilon t)$). But the *structure* of the high-order terms will apply to weakly heterogeneous systems, to weakly nonlinear systems, or both, where Ω^r and Ω^i are based on the primary wave or harmonic. These high order terms will figure importantly in our geophysical applications.

EFFECT OF LOW-ORDER NONUNIFORMITIES

Regardless of discipline, whether we deal with acoustics, electromagnetic waves, or capillary waves on water, all linear waves are marked by a single characteristic: *they are completely characterized by the dispersion relation connecting the complex frequency ω to a real wavenumber k.*

This relationship, which may depend explicitly on k, x and t, takes the general form

$$\omega(k,x,t) = \Omega^r(k,x,t) + i\,\Omega^i(k,x,t) \tag{2.99}$$

and may be obtained analytically or experimentally.

Again, any dependence on x and t must necessarily be weak, for if not, slowly varying motions would not result, and the consequent motions would invalidate the assumptions behind our wave model.

The presence of heterogeneities, of course, does not affect the fundamental ideas behind wave crest conservation. When nonconservative effects are weak, waves are still conserved, and we can still write

$$\partial k/\partial t + \partial\Omega^r(k,x,t)/\partial x = 0 \tag{2.100}$$

Detailed formal analysis. We *introduced* waves in *nondissipative*, nonuniform media earlier in Equations 2-21 to 2-27. We now develop the theory for dissipative, nonuniform media, restricting ourselves to low-order effects, in much greater detail; quantities similar to the high-order terms in Equations 2-94 and 2-95 are not considered.

Again, Equation 2-100 can be simplified and solved in characteristic form. Using the chain rule of calculus, we have

$$\partial k/\partial t + \Omega^r{}_k(k,x,t)\,\partial k/\partial x = -\,\Omega^r{}_x(k,x,t) \tag{2.101}$$

The total derivative of k(x,t) is just $dk = \partial k/\partial t\ dt + \partial k/\partial x\ dx$. Dividing through by dt, we obtain $dk/dt = \partial k/\partial t + \partial k/\partial x\ dx/dt$. On comparison with Equation 2-101, it is clear that

$$dk/dt = -\,\Omega^r{}_x(k,x,t) \tag{2.102}$$

along the ray path

$$dx/dt = \Omega^r{}_k(k,x,t) \tag{2.103}$$

In the Equations 2-102 and 2-103, the *total derivative* d/dt denotes changes *following the wave* (a.k.a., the *convective, material,* or *substantive* derivative). The speed of interest is the *group velocity* $C_g = \Omega^r{}_k(k,x,t)$ obtained as the partial derivative of $\Omega^r{}_k(k,x,t)$ with respect to k, holding x and t fixed.

Since the $\omega(k,x,t)$ is available, from analytical modeling or empirical measurement, these first-order ordinary differential equations in time are easily integrated to give x and k along a wave trajectory; Equations 2-26 and 2-27 apply exactly as before.

When the dispersion function does not depend explicitly on x and t, the result dk/dt = 0 indicates that k is a constant equal to its initial value; the corresponding equation $dx/dt = \Omega^r{}_k(k)$ = constant shows that the characteristic must be a straight line. On the other hand, when a dependence on x is allowed, k will vary along the ray, and the ray will no longer be straight.

In Chapter 6, we show that ray convergence due to nonuniformities is responsible for *wave trapping* and high amplitude lateral vibrations in drillstrings. The (mathematical) inhomogeneities, as far as the beam equation is concerned, arise from axial stress variations along the drillstring. *The results given here will be significant when we discuss vibration modeling.*

Wave energy and momentum. In our earlier work dealing with waves in *uniform* media, we found that $\partial a^2/\partial t + \partial\{\Omega^r{}_k(k)a^2\}/\partial x = 2\,\Omega^i\,a^2$ describes amplitude (see Equation 2-55). But when *heterogeneities* are present, *work interaction* with the underlying medium in general exists.

In mechanics books, concepts such as kinetic, potential strain, or potential gravitational energy are developed from differential equation models for specific distributed systems, but the subject can be treated more generally.

Noether's Theorem (Courant and Hilbert, 1989; Gelfand and Fomin, 1963) describes energy and momentum concepts from a theoretical physics point of view. Essentially, *energy is associated with a variational principle that is invariant with respect to translations in time*, and similarly, *momentum is associated with translations in space*.

Then, using *multiple-scaling* arguments such as thosed introduced in Equation 2-87, it is possible to show that the *wave energy density* E(x,t) and the *wave momentum density* M(x,t) satisfy

$$\partial E/\partial t + \partial(\Omega^r{}_k(k,x,t)E)/\partial x = E\{2\Omega^i + \Omega^r{}_t(k,x,t)/\Omega^r\} \tag{2.104}$$

$$\partial M/\partial t + \partial(\Omega^r{}_k(k,x,t)M)/\partial x = M\{2\Omega^i - \Omega^r{}_x(k,x,t)/k\} \tag{2.105}$$

following, again, our convention where $\omega(k) = \Omega^r(k,x,t) + i\,\Omega^i(k,x,t)$ describes the uniform wave (simple dispersion relations are given in Chapter 1-6).

Let us consider Equation 2-104, where *E(x,t) contains both kinetic and potential energy contributions.* The left side appears in *conservation form.* If the right side were zero, then $\partial E/\partial t + \partial(\Omega^r{}_k(k,x,t)E)/\partial x = 0$ shows that timewise increases in wave energy density are balanced by differences in energy flux, with group velocity $\Omega^r{}_k(k,x,t)$ being the energy transport velocity.

Since the simplified equation for E in the above paragraph takes the same form as Equation 2-11, the result analogous to Equation 2-20 states that the total energy between two group velocity rays is constant.

The right side of Equation 2-104, which is proportional to E, represents a distributed energy source or sink. It models damping or instability through the imaginary frequency Ω^i, but the presence of $\Omega^r{}_t(k,x,t)/\Omega_r$ indicates that *time-dependent* heterogeneities will affect energy in or loss from the system.

Note that Ω^i and $\Omega^r{}_t(k,x,t)/\Omega_r$ both contain k(x,t) in their arguments; thus, the use of Equation 2-104 requires the solution of Equations 2-102 and 2-103 beforehand.

Similar comments apply to Equation 2-105 for wave momentum density, the important emphasis being the $\Omega^r{}_x(k,x,t)/k$ at the right. Momentum sources arise from dissipative effects, and also from the effects of *spatial* heterogeneities (e.g., the variable mass on a guitar string, or in water waves, the gentle slope of an underlying beach).

Changes to energy and momentum density arise from three effects: heterogeneities in the propagation medium, dissipation, and *geometric spreading*, all of which are included in Equations 2-104 and 2-105.

Finally, in nonuniform media, E and M are different from the a^2 used for waves propagating in uniform homogeneous media. Both E and M are proportional to a^2, and all may be used as qualitative measures of stability; however, E, M and a^2 are physically different, behave differently, and have different units.

Example 2-7: String with variable properties. We consider weakly damped transverse vibrations of a string with *constant* lineal mass density ρ_l and tension T. In Chapter 1, we examined the model $\rho_l\ \partial^2u/\partial t^2 + \gamma\ \partial u/\partial t - T\ \partial^2u/\partial x^2 = 0$, and determined that $\rho_l\ (\Omega^{r2} - \Omega^{i2}) = Tk^2 + \gamma\ \Omega^i$ and $\Omega^i = -\gamma/(2\rho_l)$.

For small dissipations satisfying $|\ \Omega^i/\Omega^r\ | << 1$, the first equation is approximated by $\rho_l\Omega^{r2} = Tk^2$. We also indicated that this dispersion relation holds to leading order for slow variations when $\rho_l = \rho_l(x)$ and $T = T(t)$, provided the dependences on x and t are weak. This *fact* arises formally using *asymptotic WKB methods* (Nayfeh, 1973). For our purposes,

$$\Omega^r(k,x,t) = \{T(t)/\rho_l(x)\}^{1/2}\ k \tag{2.106}$$

$$\Omega^i(k,x,t) = -\gamma/2\rho_l(x) \tag{2.107}$$

To evaluate the right sides of Equations 2-104 and 2-105, we take the derivative $\Omega^r_t(k,x,t) = 1/2\ \{T(t)/\rho_l(x)\}^{-1/2}\ \{\partial T(t)/\partial t/\rho_l(x)\}k$, and then, $\Omega^r_x(k,x,t) = 1/2\ \{T(t)/\rho_l(x)\}^{-1/2}\ \{T(t)\}^{1/2}\ (-1/2)\{\rho_l(x)\}^{-3/2}\{\partial \rho_l(x)/\partial x\}k$. The group velocity is simply $\Omega^r_x(k,x,t) = \{T(t)/\rho_l(x)\}^{1/2}$.

Physically, Equation 2-104 indicates that energy changes to the system are expected on a guitar string that is being stretched or relieved in tension in time. Equation 2-105 indicates that momentum changes are expected whenever variations in spatial mass density distribution exist.

Computational solution. Integration of Equations 2-104 and 2-105, in general, must proceed numerically. For E(x,t), we have the sequence of steps,

$$\partial E/\partial t + \partial(\Omega^r_k(k,x,t)E)/\partial x = E\{2\Omega^i + \Omega^r_t(k,x,t)/\Omega^r\} \tag{2.108}$$

$$\partial E/\partial t + \Omega^r_k(k,x,t)\ \partial E/\partial x + E\ \partial\Omega^r_k(k,x,t)/\partial x = E\{2\Omega^i + \Omega^r_t(k,x,t)/\Omega^r\} \tag{2.109}$$

$$\partial E/\partial t + \Omega^r_k(k,x,t)\ \partial E/\partial x + E\ \partial\Omega^r_{kk}(k,x,t)\partial k/\partial x + E\ \partial\Omega^r_{kx}(k,x,t) = E\{2\Omega^i + \Omega^r_t(k,x,t)/\Omega^r\} \tag{2.110}$$

$$\partial E/\partial t + \Omega^r_k(k,x,t)\ \partial E/\partial x = E\{2\Omega^i + \Omega^r_t(k,x,t)/\Omega^r - \partial\Omega^r_{kx}(k,x,t) - \partial\Omega^r_{kk}(k,x,t)\partial k/\partial x\} \tag{2.111}$$

Drawing upon earlier work, we immediately write the required characteristic equations for wavenumber and position,

$$dk/dt = -\Omega^r_x(k,x,t) \tag{2.112}$$

along the ray path

$$dx/dt = \Omega^r_k(k,x,t) \tag{2.113}$$

Since $dE/dt = \partial E/\partial t + \Omega^r_k(k,x,t)\ \partial E/\partial x$, Equation 2-111 becomes

$$dE/dt = E\{2\Omega^i + \Omega^r_t(k,x,t)/\Omega^r - \partial\Omega^r_{kx}(k,x,t) - \partial\Omega^r_{kk}(k,x,t)\partial k/\partial x\} \tag{2.114}$$

The analysis for wave momentum, using Equation 2-105, proceeds along similar lines. Since $dM/dt = \partial M/\partial t + \Omega^r_k(k,x,t)\ \partial M/\partial x$, we analogously have

$$dM/dt = M\{2\Omega^i - \Omega^r_x(k,x,t)/k - \partial\Omega^r_{kx}(k,x,t) - \partial\Omega^r_{kk}(k,x,t)\partial k/\partial x\} \tag{2.115}$$

In Equations 2-114 and 2-115, the $\partial k/\partial x$ term requires *numerical differentiation across rays*, using values computed by *integration along rays*. The latter integrations can be performed using the previously suggested scheme in Equations 2-26 and 2-27, namely,

$$k_f = k_i - \omega_x(k_i,x_i,t_i)\ \Delta t \tag{2.116}$$

$$x_f = x_i + \omega_k(k_i,x_i,t_i)\,\Delta t \tag{2.117}$$

For example, energy and momentum integrations proceed by writing

$$(E_f - E_i)/\Delta t = E_i\{2\Omega^i(k_i,x_i,t) + \Omega^r{}_t(k_i,x_i,t)/\Omega^r(k_i,x_i,t) - \partial\Omega^r{}_{kx}(k_i,x_i,t) - \partial\Omega^r{}_{kk}(k_i,x_i,t)\partial k_i/\partial x\} \tag{2.118}$$

$$(M_f - M_i)/\Delta t = M_i\{2\Omega^i(k_i,x_i,t) - \Omega^r{}_x(k_i,x_i,t)/k_i - \partial\Omega^r{}_{kx}(k_i,x_i,t) - \partial\Omega^r{}_{kk}(k_i,x_i,t)\partial k_i/\partial x\} \tag{2.119}$$

and solving for E_f and M_f individually as in Equations 2-116 and 2-117 above. More accurate numerical schemes are available in the literature, but Equations 2-118 and 2-119 illustrate the basic ideas.

Dynamically steady problems. Often, constant frequency sources are used to create dynamically steady wave patterns. This constancy provides an immediate integral of Equation 2-100; since $\partial k/\partial t = 0$, we find that

$$\Omega^r(k,x) = \omega_0 \tag{2.120}$$

for the complex dispersion relation

$$\omega(k,x) = \Omega^r(k,x) + i\,\Omega^i(k,x) \tag{2.121}$$

For Equation 2-120 to apply, the excitation frequency ω_0 *must* be constant; *there must be no explicit dependence of $\omega(k,x)$ on time*. Note that Equation 2-120 may be an *algebraic* equation, but it may also be *transcendental*. It may be solved analytically or numerically for k(x).

Once the wavenumber distribution k(x) is available, the energy or momentum density can be obtained by integrating the following steady forms,

$$\partial(\Omega^r{}_k(k,x)E)/\partial x = E\{2\Omega^i + \Omega^r{}_t(k,x)/\Omega_r\} \tag{2.122}$$

$$\partial(\Omega^r{}_k(k,x)M)/\partial x = M\{2\Omega^i - \Omega^r{}_x(k,x)/k\} \tag{2.123}$$

The derivatives in Equations 2-122 and 2-123 are *ordinary* derivatives, and standard numerical integration procedures may be used.

In Chapter 6, KWT is used to model bending waves propagating on drillstrings in the presence of nonuniform axial strain. E(x) and M(x) are shown to vary inversely with group velocity $\Omega^r{}_k(k,x)$, as expected from Equations 2-122 and 2-123. Interestingly, all solutions to Equation 2-120 for k(x) admit small values of $\Omega^r{}_k$ near the *neutral point*, thus demonstrating the natural tendency toward lateral instability.

Waves in nonuniform moving media. We defer our discussion of KWT in moving media (e.g., ocean currents, steady winds, and gusts) until Chapter 12, where applications to water waves are given. For now, we summarize the principal results. If $\Omega^r(k,x,t)$ and $\Omega^i(k,x,t)$ describe the wave propagation problem *without* any flow, then the kinematic equations for the problem *with* a mean flow U(x,t) are

$$dk/dt = \Omega^r_x(k,x,t) + U_x k \quad (2.124)$$

$$dx/dt = U(x,t) + \Omega^r_k(k,x,t) \quad (2.125)$$

$$\partial E/\partial t + \partial((U(x,t) + \Omega^r_k(k,x,t))E)/\partial x = \quad (2.126)$$
$$E\{2\Omega^i + \{\Omega^r_t(k,x,t) + U\Omega^r_x(k,x,t) - k\Omega^r_k(k,x,t)\, U_x\}/\Omega^r\}$$

$$\partial M/\partial t + \partial((U(x,t) + \Omega^r_k(k,x,t))M)/\partial x = \quad (2.127)$$
$$M\{2\Omega^i - \Omega^r_x(k,x,t)/k - U_x\}$$

Average Lagrangian formalism. In this book, only the simplest *recipes* obtained in kinematic wave theory appear; in order to understand the complete theory and the full power behind KWT, training is required in *variational calculus* (Gelfand and Fomin, 1963). We give a *flavor* of these methods only; the reader may ignore this work without loss of continuity.

Example 2-8: Wave action conservation. We consider the transverse vibrations of a string, which satisfy $\rho_l\, \partial^2 u/\partial t^2 - T\, \partial^2 u/\partial x^2 = 0$, allowing $\rho_l = \rho_l(x)$ and $T = T(t)$. The kinetic energy density is $1/2\ \rho_l\, (\partial u/\partial t)^2$, while the potential strain energy density is $1/2\ T\, (\partial u/\partial x)^2$. The *Lagrangian* L of the dynamical system equals their difference, $L = 1/2\ \rho_l\, (\partial u/\partial t)^2 - 1/2\ T\, (\partial u/\partial x)^2$.

The uniform plane wave solution is $u(x,t) = a \sin(kx-\omega t)$, for which $\partial u/\partial t = - a\omega \cos(kx-\omega t)$ and $\partial u/\partial x = ak \cos(kx-\omega t)$. If we substitute these in L, and take an average over the *rapidly varying phase* $\theta = kx-\omega t$, we obtain an *average Lagrangian* $L = 1/4\ a^2\, (\rho_l\omega^2 - Tk^2)$.

Taking *variations* in *a* leads to the *Euler equation* $L_a = 0$, which is just the phase relation of the *slowly varying wave*,

$$\rho_l\omega^2 - Tk^2 = 0 \quad (2.128)$$

but taking *variations* in θ leads to $\partial L_\omega/\partial t - \partial L_k/\partial x = 0$ (since $\omega = - \partial\theta/\partial t$ and $k = \partial\theta/\partial x$, noting that θ disappears because averages over it are taken), where $L_\omega = 1/2\ a^2\rho_l\omega$ and $L_k = - 1/2\ a^2 Tk$. Thus,

$$\partial(a^2\rho_l\omega)/\partial t + \partial(a^2 Tk)/\partial x = 0 \quad (2.129)$$

Equation 2-129 reduces to our amplitude equation for a^2 when the properties ρ_l and T are *constant* (see Equation 2-11).

Importantly, for *variable* properties, algebraic manipulation shows that *Equation 2-129 can be rewritten as either Equations 2-104 or 2-105 for wave energy or momentum density!* Equation 2-129 is clearly the more "universal" conservation law, and the quantity L_ω is known as the *wave action density*.

Equally significant is the applicability of KWT to *fully nonlinear waves*, where sine solutions and Fourier superposition no longer apply; here, *average Lagrangian formulations* provide elegant solutions to otherwise intractable problems. Whitham (1974) gives an exposition of the early ideas; high-order linear and nonlinear effects are discussed by Chin (1980).

THREE-DIMENSIONAL KINEMATIC WAVE THEORY

Here we develop KWT in three dimensions. Our derivation will be restricted to the low-order terms; high-order diffusive and dispersive effects, such as those in Equations 2-94 and 2-95, are not modeled, but the required modifications will be apparent to the skilled reader.

We will avoid index notation in order to eliminate any unnecessary confusion. Following our approach for one-dimensional waves, we begin with the complex dispersion relation for the uniform wave,

$$\omega = \Omega^r(k_x,k_y,k_z,x,y,z,t) + i\,\Omega^i(k_x,k_y,k_z,x,y,z,t) \qquad (2.130)$$

Wave crest conservation requires that

$$\partial \mathbf{k}/\partial t + \nabla\Omega^r = 0 \qquad (2.131)$$

or, in component Cartesian form,

$$\partial k_x/\partial t + \partial\Omega^r/\partial x = 0 \qquad (2.132)$$

$$\partial k_y/\partial t + \partial\Omega^r/\partial y = 0 \qquad (2.133)$$

$$\partial k_z/\partial t + \partial\Omega^r/\partial z = 0 \qquad (2.134)$$

Let us expand Equations 2-132 to 2-134 using the chain rule of calculus, so that

$$\partial k_x/\partial t + \Omega^r_{kx}\,\partial k_x/\partial x + \Omega^r_{ky}\,\partial k_y/\partial x + \Omega^r_{kz}\,\partial k_z/\partial x = -\Omega^r_x \qquad (2.135)$$

$$\partial k_y/\partial t + \Omega^r_{kx}\,\partial k_x/\partial y + \Omega^r_{ky}\,\partial k_y/\partial y + \Omega^r_{kz}\,\partial k_z/\partial y = -\Omega^r_y \qquad (2.136)$$

$$\partial k_z/\partial t + \Omega^r_{kx}\,\partial k_x/\partial z + \Omega^r_{ky}\,\partial k_y/\partial z + \Omega^r_{kz}\,\partial k_z/\partial z = -\Omega^r_z \qquad (2.137)$$

Wave irrotationality. Recall our use of the *phase function* in Equations 2-84 and 2-85. In one dimension, its motivation was furnished by solutions taking the form sin $(kx - \omega t)$ or *sin* θ, where $\omega = -\partial\theta/\partial t$ and $k = \partial\theta/\partial x$ (see Equations 2-8 and 2-9).

For slowly varying waves, we *assume* the existence of $\theta(x,y,z,t)$. Since differentiation order can be interchanged, e.g., $\partial\theta_y(x,y,z,t)/\partial x = \partial\theta_x(x,y,z,t)/\partial y$, we have, setting $k_x = \theta_x(x,y,z,t)$ and $k_y = \theta_y(x,y,z,t)$, the result $\partial k_y/\partial x = \partial k_x/\partial y$. In general, the *wave irrotationality* conditions

$$\partial k_x/\partial y = \partial k_y/\partial x \tag{2.138}$$

$$\partial k_x/\partial z = \partial k_z/\partial x \tag{2.139}$$

$$\partial k_y/\partial z = \partial k_z/\partial y \tag{2.140}$$

apply, using terminology adopted from fluid mechanics. Therefore, Equations 2-135 to 2-137 can be rewritten in the form

$$\partial k_x/\partial t + \Omega^r_{kx}\,\partial k_x/\partial x + \Omega^r_{ky}\,\partial k_x/\partial y + \Omega^r_{kz}\,\partial k_x/\partial z = -\Omega^r_x \tag{2.141}$$

$$\partial k_y/\partial t + \Omega^r_{kx}\,\partial k_y/\partial x + \Omega^r_{ky}\,\partial k_y/\partial y + \Omega^r_{kz}\,\partial k_y/\partial z = -\Omega^r_y \tag{2.142}$$

$$\partial k_z/\partial t + \Omega^r_{kx}\,\partial k_z/\partial x + \Omega^r_{ky}\,\partial k_z/\partial y + \Omega^r_{kz}\,\partial k_z/\partial z = -\Omega^r_z \tag{2.143}$$

Since the $k_x(x,y,z,t)$ depends on four variables, namely, t, x, y, and z, its total derivative is known from calculus as $dk_x = \partial k_x/\partial t\ dt + \partial k_x/\partial x\ dx + \partial k_x/\partial y\ dy + \partial k_x/\partial z\ dz$. Division by dt shows that

$$dk_x/dt = \partial k_x/\partial t + dx/dt\ \partial k_x/\partial x + dy/dt\ \partial k_x/\partial y + dz/dt\ \partial k_x/\partial z \tag{2.144}$$

Similarly,

$$dk_y/dt = \partial k_y/\partial t + dx/dt\ \partial k_y/\partial x + dy/dt\ \partial k_y/\partial y + dz/dt\ \partial k_y/\partial z \tag{2.145}$$

$$dk_z/dt = \partial k_z/\partial t + dx/dt\ \partial k_z/\partial x + dy/dt\ \partial k_z/\partial y + dz/dt\ \partial k_z/\partial z \tag{2.146}$$

The ray equations. On comparing Equations 2-144 to 2-146 with Equations 2-141 to 2-143, we obtain the coupled set of nonlinear ordinary differential equations

$$dk_x/dt = -\Omega^r_x \tag{2.147}$$

$$dk_y/dt = -\Omega^r_y \tag{2.148}$$

$$dk_z/dt = -\Omega^r_z \tag{2.149}$$

$$dx/dt = \Omega^r_{kx} \tag{2.150}$$

$$dy/dt = \Omega^r_{ky} \tag{2.151}$$

$$dz/dt = \Omega^r_{kz} \tag{2.152}$$

Equations 2-147 to 2-152 are solved by numerical integration using standard computational schemes for ordinary differential equations.

Only when the right sides of Equations 2-147 to 2-149 identically vanish (that is, *no* explicit dependences on x, y, and z) do the wavenumbers k_x, k_y, and k_z remain unchanged along rays. Then, since the right-side arguments in

Equations 2-150 to 2-152 depend *at most* on k_x, k_y, and k_z, and are thus constant, the rays therefore remain straight.

Frequency variation. We now determine how frequency varies along rays in inhomogeneous media. Since the function $\Omega^r(k_x,k_y,k_z,x,y,z,t)$ depends on *seven* independent variables, namely, k_x, k_y, k_z, x, y, z, and t, its total differential $d\Omega^r$ is known from calculus as

$$d\Omega^r = \Omega^r_{kx}\, dk_x + \Omega^r_{ky}\, dk_y + \Omega^r_{kz}\, dk_z + \Omega^r_x\, dx + \Omega^r_y\, dy + \Omega^r_z\, dz + \Omega^r_t\, dt \tag{2.153}$$

Division by dt gives

$$d\Omega^r/dt = \Omega^r_{kx}\, dk_x/dt + \Omega^r_{ky}\, dk_y/dt + \Omega^r_{kz}\, dk_z/dt + \Omega^r_x\, dx/dt + \Omega^r_y\, dy/dt + \Omega^r_z\, dz/dt + \Omega^r_t \tag{2.154}$$

The terms in Equation 2-154 can be regrouped in the form

$$d\Omega^r/dt = (\Omega^r_{kx}\, dk_x/dt + \Omega^r_x\, dx/dt) + (\Omega^r_{ky}\, dk_y/dt + \Omega^r_y\, dy/dt) + (\Omega^r_{kz}\, dk_z/dt + \Omega^r_z\, dz/dt) + \Omega^r_t \tag{2.155}$$

Equation 2-155 can be simplified using Equations 2-147 to 2-152, so that

$$d\Omega^r/dt = (-\Omega^r_{kx}\,\Omega^r_x + \Omega^r_x\,\Omega^r_{kx}) + (-\Omega^r_{ky}\,\Omega^r_y + \Omega^r_y\,\Omega^r_{ky}) + (-\Omega^r_{kz}\,\Omega^r_z + \Omega^r_z\,\Omega^r_{kz}) + \Omega^r_t \tag{2.156}$$

$$= \Omega^r_t \tag{2.157}$$

which shows that frequency varies along rays only when the medium varies in time (e.g., as tension increases, the pitch on a guitar string increases).

Energy variation. Equation 2-130 assumes that the uniform wave dispersion relation $\omega = \Omega^r(k_x,k_y,k_z,x,y,z,t) + i\,\Omega^i(k_x,k_y,k_z,x,y,z,t)$ is available. The 3D energy law corresponding to Equation 2-104 is

$$\partial E/\partial t + \partial(\Omega^r_{kx}(k_x,k_y,k_z,,x,y,z,t)E)/\partial x + \partial(\Omega^r_{ky}(k_x,k_y,k_z,,x,y,z,t)E)/\partial y + \partial(\Omega^r_{kz}(k_x,k_y,k_z,,x,y,z,t)E)/\partial z = E\{2\Omega^i(k_x,k_y,k_z,,x,y,z,t) + \Omega^r_t(k_x,k_y,k_z,,x,y,z,t)/\Omega^r\} \tag{2.158}$$

Equation 2-158 can be written along the rays of Equations 2-150 to 2-152, by straightforwardly extending the steps taken in Equations 2-108 to 2-112.

Changes to energy density arise from three effects: time-dependent heterogeneities, dissipation, and geometric spreading, all of which are included in Equation 2-158.

Ray topology. Earlier we showed how rays remain straight in homogeneous media, while in heterogeneous media, they are typically curved. Now we demonstrate that they need *not* in general remain orthogonal to their wave fronts. Again, wave fronts are surfaces are defined by the constancy of the phase function θ,

$$\Theta(x,y,z,t) = \theta(x,y,z,t) - \text{constant} = 0 \qquad (2.159)$$

From vector calculus, the unit normal to this surface is given by

$$\mathbf{n} = (\theta_x\mathbf{i} + \theta_y\mathbf{j} + \theta_z\mathbf{k})/\sqrt{(\theta_x^2 + \theta_y^2 + \theta_z^2)} \qquad (2.160a)$$

where the subscripts represent partial derivatives, and **i**, **j**, and **k** are unit vectors in the x, y, and z directions. Since the wavenumber components satisfy $k_x = \partial\theta/\partial x$, $k_y = \partial\theta/\partial y$ and $k_z = \partial\theta/\partial z$, we can write Equation 2-160a as

$$\mathbf{n} = (k_x\mathbf{i} + k_y\mathbf{j} + k_z\mathbf{k})/\sqrt{(k_x^2 + k_y^2 + k_z^2)} \qquad (2.160b)$$

In general, rays are obtained as characteristics of the *eikonal* equation, that is, the trajectories in $dx/dt = \Omega^r_{kx}$, $dy/dt = \Omega^r_{ky}$, and $dz/dt = \Omega^r_{kz}$ (see Equations 2-150 to 2-152).

Consider the dispersion relation in Equation 3-46, describing three-dimensional acoustic waves in *isotropic* homogeneous media,

$$\omega = \Omega^r = c\,(k_x^2 + k_y^2 + k_z^2)^{1/2} \qquad (2.161)$$

We easily see that the velocities in Equations 2-150 to 2-152 satisfy

$$\begin{aligned} dx/dt = \Omega^r_{kx} &= 1/2\ c\,(k_x^2 + k_y^2 + k_z^2)^{-1/2}\,2k_x \\ &= c\,k_x\,(k_x^2 + k_y^2 + k_z^2)^{-1/2} \end{aligned} \qquad (2.162)$$

$$\begin{aligned} dy/dt = \Omega^r_{ky} &= 1/2\ c\,(k_x^2 + k_y^2 + k_z^2)^{-1/2}\,2k_y \\ &= c\,k_y\,(k_x^2 + k_y^2 + k_z^2)^{-1/2} \end{aligned} \qquad (2.163)$$

$$\begin{aligned} dz/dt = \Omega^r_{kz} &= 1/2\ c\,(k_x^2 + k_y^2 + k_z^2)^{-1/2}\,2k_z \\ &= c\,k_z\,(k_x^2 + k_y^2 + k_z^2)^{-1/2} \end{aligned} \qquad (2.164)$$

The right sides of Equations 2-162 to 2-164 are proportional to the components in Equation 2-160b; thus, the rays corresponding to the isotropic dispersion relation in Equation 2-161 are parallel to the normals to the wave front.

Now, consider the *anisotropic* problem, with different sound speeds c_x, c_y, and c_z in the x, y and z directions, that is

$$\omega = \Omega^r = (c_x^2k_x^2 + c_y^2k_y^2 + c_z^2k_z^2)^{1/2} \qquad (2.165)$$

The new velocity components are

$$dx/dt = \Omega^r_{kx} = c_x^2k_x\,(c_x^2k_x^2 + c_y^2k_y^2 + c_z^2k_z^2)^{-1/2} \qquad (2.166)$$

$$dy/dt = \Omega r_{ky} = c_y{}^2k_y\,(c_x{}^2k_x{}^2 + c_y{}^2k_y{}^2 + c_z{}^2k_z{}^2)^{-1/2} \tag{2.167}$$

$$dz/dt = \Omega r_{kz} = c_z{}^2k_z\,(c_x{}^2k_x{}^2 + c_y{}^2k_y{}^2 + c_z{}^2k_z{}^2)^{-1/2} \tag{2.168}$$

The vector associated with Equations 2-166 to 2-168 is clearly *not* parallel to the normal $\mathbf{n} = (k_x\mathbf{i} + k_y\mathbf{j} + k_z\mathbf{k})/\sqrt{(k_x{}^2 + k_y{}^2 + k_z{}^2)}$ in Equation 2-160b.

Straightness and wave front orthogonality are only true of specialized motions satisfying the three-dimensional wave equation in homogeneous isotropic media. The normal in Equations 2-160a,b is perpendicular to the wave front, but it is related to the phase velocity, and not the ray velocity associated with energy propagation.

REFERENCES

Bourbie, T., Coussy, O., and Zinszner, B., *Acoustics of Porous Media*, Gulf Publishing Company, Houston, 1987.

Chin, W.C., *Physics of Slowly Varying Wavetrains in Continuum Systems, Ph.D. Thesis*, Massachusetts Institute of Technology, Cambridge, Massachusetts, 1976.

Chin, W.C., "Effect of Dissipation and Dispersion on Slowly Varying Wavetrains," AIAA Journal, Feb. 1980, pp. 149-158.

Chin, W.C., *Modern Reservoir Flow and Well Transient Analysis*, Gulf Publishing, Houston, 1993.

Cole, J.D., *Perturbation Methods in Applied Mathematics*, Blaisdell Publishing, Waltham, Massachusetts, 1968.

Courant, R., and Hilbert, D., *Methods of Mathematical Physics, Volumes I and II*, John Wiley and Sons, New York, 1989.

Gelfand, I.M., and Fomin, S.V., *Calculus of Variations*, Prentice-Hall, Englewood Cliffs, New Jersey, 1963.

Landahl, M.T., "Wave Mechanics of Breakdown," Journal of Fluid Mechanics, Vol. 56, Part 4, 1972, pp. 775-802.

Nayfeh, A., *Perturbation Methods*, John Wiley and Sons, New York, 1973.

van Dyke, M., *Perturbation Methods in Fluid Mechanics*, Academic Press, New York, 1964.

Whitham, G.B., *Linear and Nonlinear Waves*, John Wiley and Sons, New York, 1974, pp. 363-402.

3
Elementary Examples from Classical Mechanics

We will introduce a range of wave propagation problems showing the diversity of model equations taking forms *other* than $\partial^2u/\partial t^2 - c^2\, \partial^2u/\partial x^2 = 0$. We also illustrate the concepts discussed in Chapters 1 and 2 with examples found in structural, fluid-dynamic, and acoustical applications.

EXAMPLE 3-1: LATERAL VIBRATION OF SIMPLE BEAMS

Consider the lateral bending vibrations of a uniform bar with mass density ρ, cross-sectional area A, Young's modulus E, and moment of inertia I (see Figure 3-1). The transverse displacement v(x,t) satisfies

$$\rho A\, \partial^2 v/\partial t^2 + EI\, \partial^4 v/\partial x^4 = 0 \qquad (3.1)$$

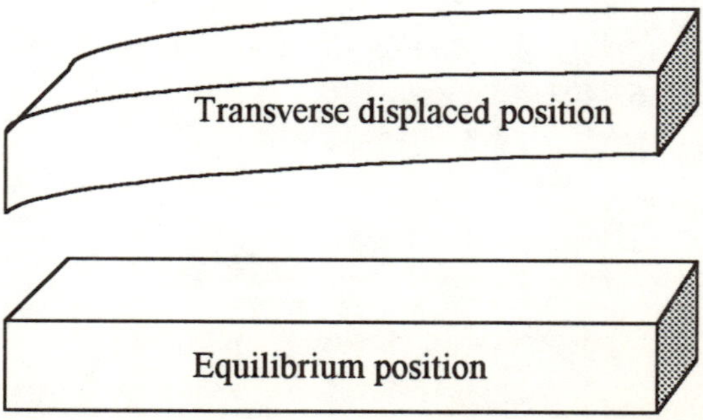

Figure 3-1. A simple beam.

Numerous textbooks and papers on beam theory are available. Among the better known are Timoshenko and Goodier (1934), Love (1944), Clark and Reissner (1951), Den Hartog (1952), and Sechler (1952); also, Timoshenko (1958), Abramson, Plass, and Ripperger (1958), Timoshenko and Gere (1961), Clough and Penzien (1975), and Graff (1975).

These references provide a range of analytical and numerical solution techniques that have proven useful in engineering calculations. In this example, we wish to introduce basic analysis concepts and discuss difficulties associated with different approaches.

Let us *attempt* a traveling wave solution of the form v(x,t) = f(x-ct), where the speed c is constant for all waves, as we did for the classical wave equation. We find that $\rho A c^2 f'' + EI f'''' = 0$, which is obviously not true universally; thus, different waves are likely to propagate with different speeds.

Now, the substitution $v(x,t) = \sin(kx - \omega t)$ leads to $-\rho A\omega^2 + EI k^4 = 0$ or $\omega = \pm \sqrt{(EI/\rho A)}\, k^2$. Since the phase velocity $\omega/k = \pm \sqrt{(EI/\rho A)}\, k$ depends on the wavenumber k, beam disturbances are dispersive, with the exact propagation velocities depending on the form of the initial disturbance. As an alternative to "f(x-ct)", we now separate variables, taking

$$v(x,t) = q(t)r(x) \tag{3.2}$$

Substitution in Equation 3-1 and division by $\rho A q(t) r(x)$ leads to $q''(t)/q(t) = -(EI/\rho A) r''''(x)/r(x)$. Since the left side is a function of time, and the right depends only on space, each must equal the same constant. Thus,

$$q''(t)/q(t) = -(EI/\rho A) r''''(x)/r(x) = -\omega^2 \tag{3.3}$$

where the right side is a negative constant (positive or zero values lead to trivial solutions). Hence, we have the two ordinary differential equations

$$r''''(x) - \beta^4 r(x) = 0 \tag{3.4}$$

$$q''(t) + \omega^2 q(t) = 0 \tag{3.5}$$

where

$$\beta^4 = \rho A \omega^2/EI \tag{3.6a}$$

$$\omega = \beta^2 \sqrt{(EI/\rho A)} \tag{3.6b}$$

The general solutions to Equations 3-4 and 3-5 are

$$r(x) = A \sin \beta x + B \cos \beta x + C \sinh \beta x + D \cosh \beta x \tag{3.7}$$

$$q(t) = E \sin \omega t + F \cos \omega t \tag{3.8}$$

In order to explore the consequences of Equation 3-1 more fully, we consider a finite length beam in $0 < x < L$, and two boundary condition models.

Example 3-1a: Hinged ends. The boundary conditions r(0) = 0, r"(0) = 0, r(L) = 0, and r"(L) = 0 assume the same zero displacement and moment at each end. Following the detailed analysis described in Example 3-1b, we find that trivial solutions are obtained unless sin βL = 0.

This requires βL= π, 2π, 3π, ... , that is, βL = nπ where n = 1, 2, 3, ... or β = nπ/L. Equation 3-7, which reduces to r(x) = sin βx here, implies the eigenfunction $r_n(x) = \sin n\pi x/L$, for which the corresponding time amplitude takes the form $(E_n \sin \{(n\pi/L)^2\sqrt{(EI/\rho A)}\}t + F_n \cos \{(n\pi/L)^2\sqrt{(EI/\rho A)}\}t)$. Hence, the general solution is

$$v(x,t) = \Sigma(E_n \sin \{(n\pi/L)^2\sqrt{(EI/\rho A)}\}t + F_n \cos \{(n\pi/L)^2\sqrt{(EI/\rho A)}\}t) \sin n\pi x/L \tag{3.9}$$

where the summation index n varies from 1 to ∞.

The coefficients E_n and F_n are still undetermined, but following the procedure in Example 1-3, they are fixed once the initial transverse displacement and speed v(x,0) and $v_t(x,0)$ are specified. Separation of variables, in this and other elementary examples, leads to "trouble-free" results.

Example 3-1b: Clamped end, other end free. Here we introduce difficulties that highlight the weaknesses of analytical methods. The conditions r(0) = 0 and r'(0) = 0 imply zero displacement and slope, while r"(L) = 0 and r'"(L) = 0 call for zero shear force and moment. From the first two conditions, we have B + D = 0 and A + C = 0, and from the last two, we obtain

$$C \{\sin \beta L + \sinh \beta L\} + D\{\cos \beta L + \cosh \beta L\} = 0 \tag{3.10}$$

$$C \{\cos \beta L + \cosh \beta L\} + D\{\sinh \beta L - \sin \beta L\} = 0 \tag{3.11}$$

For nontrivial solutions, the determinant formed by the curly-bracketed coefficients of C and D must vanish, leading to the *eigenvalue relation*

$$\cos \beta L \cosh \beta L = -1 \tag{3.12}$$

This *transcendental equation* has an infinite number of *eigenvalue* solutions, each of which require intensive numerical solution. In fact, the first six for β_1, β_2, ..., β_6 are determined from

$$\beta L = 1.875, 4.694, 7.855, 10.996, 14.137, 17.279 \tag{3.13}$$

The corresponding *eigenfunction* is no longer the "$r_n(x) = \sin n\pi x/L$" found for Example 3-1a, but the unwieldy trigonometric-hyperbolic function

$$r_\beta(x) = \sinh \beta x - \sin \beta x - (\sinh \beta L + \sin \beta L)(\cosh \beta x - \cos \beta x)/(\cosh \beta L + \cos \beta L) \tag{3.14}$$

The general solution takes the form

$$v(x,t) = \Sigma(E_\beta \sin \beta^2\sqrt{(EI/\rho A)}\, t + F_\beta \cos \beta^2\sqrt{(EI/\rho A)}\, t)\, r_\beta(x) \qquad (3.15)$$

where the summation is taken over the eigenvalue β.

Unlike Example 3-1a, the coefficients E_β and F_β are *not* easily obtained in terms of the initial transverse displacement and speed $v(x,0)$ and $v_t(x,0)$. The difficulty arises from the fact that these initial conditions must be expanded *not* in sine or cosine series, but in terms of the eigenfunctions expressed by Equation 3-14.

In principle, the simple boundary conditions used here do show that the above eigenfunctions are *complete*, so that procedures similar to those used in deriving formulas for Fourier coefficients will lead to solutions. If the initial conditions vary rapidly with space, then more terms in β must be summed, and Equation 3-12 must be correspondingly solved.

For simple boundary condition combinations, e.g., clamped-free, pinned-pinned, and deflection-torsional spring, it is possible to establish the *orthogonality* of the resulting eigenfunctions, so that representation using series such as that obtained in Equation 3-15 is guaranteed.

However, time-dependent boundary conditions resulting from mass or dashpot contributions lead to difficulties, since orthogonality is not achieved, and alternative solutions are required. Of course, even for the present example, the "exact" solution to Equation 3-15 may be less accurate and more cumbersome than a direct numerical integration of Equation 3-1. Finite difference solutions are developed in Chapter 6.

EXAMPLE 3-2: ACOUSTIC WAVES IN WAVEGUIDES

In Chapter 1, we studied the classical equation $\partial^2u/\partial t^2 - c^2\, \partial^2u/\partial x^2 = 0$, for which the nondispersive speed characterizing *all* disturbances is the *constant* c. In our simple beam example, we learned that the fourth-order beam equation $\rho A\, \partial^2v/\partial t^2 + EI\, \partial^4v/\partial x^4 = 0$ also possesses traveling wave solutions. However, the waves are dispersive, with different component wavelengths propagating at different speeds.

Here, we will demonstrate how geometrical constraints cause some disturbances to propagate, while damping out others. We will show how $\partial^2p/\partial t^2 - c^2\, (\partial^2p/\partial x^2 + \partial^2p/\partial y^2 + \partial^2p/\partial z^2) = 0$ (a natural extension to $\partial^2u/\partial t^2 - c^2\, \partial^2u/\partial x^2 = 0$) admits wave motions with speeds *other* than c, and also, how group velocity concepts arise for 3D waveguide motions.

A *waveguide* is just a medium that transmits wave motion. A vibrating string, for example, represents a 1D waveguide, but the term is used to describe wave motions with one and two *cross-space* or *modal* dimensions. Waves in waveguides are called *modal waves.*

In anticipation of Part III, which deals with MWD mud pulse telemetry in drillpipes, and swab-surge in borehole annuli, we consider a model waveguide problem for fluid flow acoustics. To keep the analysis simple, we study sound propagation in the 3D rectangular duct shown in Figure 3-2.

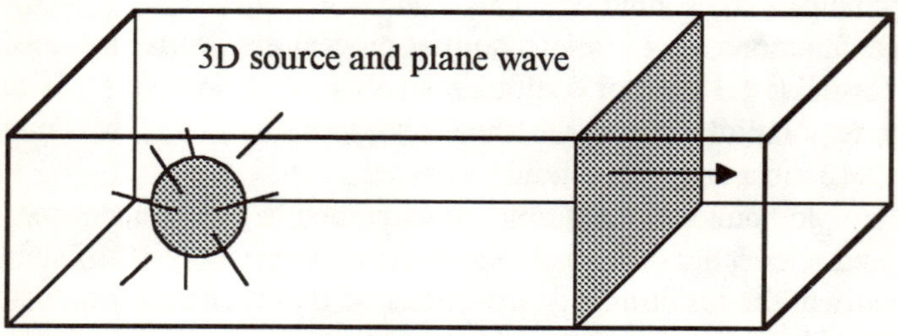

Figure 3-2. Simple rectangular waveguide.

Many excellent references to classical acoustics are available, e.g., Mason (1942), Kinsler and Frey (1950), Morse and Ingard (1968), and Pierce (1981). In this example, we hope to cover the fundamental ideas behind physical acoustics and the mathematical techniques underlying waveguide analysis.

In the simplest case, the *momentum laws* governing fluid flow are the *Navier-Stokes equations* for constant viscosity flow,

$$\rho\,(\partial u/\partial t + u\,\partial u/\partial x + v\,\partial u/\partial y + w\,\partial u/\partial z) = -\,\partial p/\partial x + \mu\,(\partial^2 u/\partial x^2 + \partial^2 u/\partial y^2 + \partial^2 u/\partial z^2) \qquad (3.16)$$

$$\rho\,(\partial v/\partial t + u\,\partial v/\partial x + v\,\partial v/\partial y + w\,\partial v/\partial z) = -\,\partial p/\partial y + \mu\,(\partial^2 v/\partial x^2 + \partial^2 v/\partial y^2 + \partial^2 v/\partial z^2) \qquad (3.17)$$

$$\rho\,(\partial w/\partial t + u\,\partial w/\partial x + v\,\partial w/\partial y + w\,\partial w/\partial z) = -\,\partial p/\partial z + \mu\,(\partial^2 w/\partial x^2 + \partial^2 w/\partial y^2 + \partial^2 w/\partial z^2) \qquad (3.18)$$

Here u(x,y,z,t, v(x,y,z,t), and w(x,y,z,t) are *Eulerian velocities* at the fixed point (x,y,z), in the directions of the x, y, and z axes, respectively, and t is time. Also, p(x,y,z,t) and ρ(x,y,z,t) represent *pressure* and *mass density per unit volume*. In transient compressible flows, density and velocity fields are coupled by *mass conservation*,

$$\partial\rho/\partial t + \partial(\rho u)/\partial x + \partial(\rho v)/\partial y + \partial(\rho w)/\partial z = 0 \qquad (3.19)$$

Despite the apparent generality of the above equations, the right sides in Equations 3-16 to 3-18 apply to *Newtonian* flows only, and then, specifically to fluids having constant viscosity.

Newtonian fluids have linear *stress-strain* relationships, but *drilling muds*, typically *non-Newtonian*, are characterized by nonlinear laws. Their analogous right sides are considerably more complicated, the form depending on the *rheological model* used, e.g., Bingham plastic, power law, viscoelastic, etc.

Simple hydraulic flows. When hydraulic (principally *incompressible*) *flowing* effects are to be modeled, the convective left sides of Equations 3-16 to 3-18, which describe acceleration, must be retained. These terms sometimes vanish, e.g., for straight ducts with pressure constant across the cross-section.

The details of the stress tensor must also be preserved, and the complete nonlinear equations must be solved together with *no-slip* velocity boundary conditions. An exposition of the state-of-the-art in non-Newtonian flow modeling is given in a recent textbook by this author (Chin, 1992a,b).

Acoustic simplifications. For acoustic (or sound) motions, two simplifications are possible. We assume zero mean background velocity, so that the quadratic velocity products on the left sides of Equations 3-16 to 3-18 remain small during the passage of disturbance waves; nonzero velocities are considered in Chapters 9 and 10.

If the resultant wave motions are not critically damped, the rheology of the fluid can also be neglected, in a crude first approximation. Rheological effects, as they affect attenuation, are typically reintroduced *after-the-fact* using imaginary frequencies, as indicated in Chapter 2.

More exact procedures and solutions dealing with the effects of flow velocity, pipe diameter, and fluid type on attenuation, are given in the literature. Basic analysis techniques are described in the classic work of Lamb (1945); these specialized results, though, fall beyond the scope of this book.

When fluid rheology is neglected, the right side terms of Equations 3-16 to 3-18 (or their counterparts in non-Newtonian flow) vanish. If fluid speeds are small, the nonlinear convective terms on the left sides can also be ignored as discussed, leaving

$$\rho_m \, \partial u/\partial t = - \, \partial p/\partial x \tag{3.20}$$

$$\rho_m \, \partial v/\partial t = - \, \partial p/\partial y \tag{3.21}$$

$$\rho_m \, \partial w/\partial t = - \, \partial p/\partial z \tag{3.22}$$

where ρ_m is a mean density (formally, we set $\rho = \rho_m + \rho'$ where ρ' is the disturbance part). Similarly, Equation 3-19 can be linearized to give

$$\partial \rho'/\partial t + \rho_m (\partial u/\partial x + \partial v/\partial y + \partial w/\partial z) = 0 \tag{3.23}$$

We can write Equations 3-20 to 3-23 in vector form as

$$\rho_m \, \partial \mathbf{q}/\partial t = -\nabla p \tag{3.24}$$

$$\partial \rho'/\partial t + \rho_m \nabla \circ \mathbf{q} = 0 \tag{3.25}$$

where we have introduced the velocity vector $\mathbf{q} = u\mathbf{i} + v\mathbf{j} + w\mathbf{k}$ using standard Cartesian notation.

To simplify the discussion, we employ some terminology from *vector calculus* (e.g., see Hildebrand, 1948). Now, we will take the *divergence* of Equation 3-24 so that we obtain $\rho_m \, \partial \nabla \circ \mathbf{q}/\partial t = -\nabla^2 \mathbf{p}\}$, and substitute the result $\nabla \circ \mathbf{q} = -(1/\rho_m) \, \partial \rho'/\partial t$ from Equation 3-25. This yields

$$\partial^2 \rho'/\partial t^2 = \nabla^2 p \tag{3.26}$$

Three-dimensional wave equation. At this point, the *constitutive laws* relating ρ' and p enter the derivation. If the fluid is a *gas*, an *equation of state* relates the two, whereas a fluid analogy to *Hooke's law* $\sigma = E\varepsilon$ applies to *slightly compressible liquids* (see Chapter 9 for a detailed development of the ideas). In either case, the 3D wave equation

$$\partial^2 p/\partial t^2 - c^2(\partial^2 p/\partial x^2 + \partial^2 p/\partial y^2 + \partial^2 p/\partial z^2) = 0 \tag{3.27}$$

is obtained, where the constant c represents the speed of sound in the medium. Again, Equation 3-27 assumes zero mean flow and physical damping.

Modal solution. If we assume that our waveguide walls are infinitely rigid, Figure 3-2 suggests that wave energy must travel along the duct centerline, which we define as the z axis. Therefore, the unknowns in Equations 3-20 to 3-22 may be taken in the separable form

$$p(x,y,z,t) = P(x,y) \exp i(\omega t - k_z z) \tag{3.28}$$

$$u(x,y,z,t) = U(x,y) \exp i(\omega t - k_z z) \tag{3.29}$$

$$v(x,y,z,t) = V(x,y) \exp i(\omega t - k_z z) \tag{3.30}$$

$$w(x,y,z,t) = W(x,y) \exp i(\omega t - k_z z) \tag{3.31}$$

In each of these product solutions, the "exp $i(\omega t - k_z z)$" represents a wave propagating in the ± z direction, while the (x,y) reflects the *modal* or *cross-structure* dependence of the solution. The assumptions $\rho_m \, \partial u/\partial t = -\partial p/\partial x$, $\rho_m \, \partial v/\partial t = -\partial p/\partial y$, and $\rho_m \, \partial w/\partial t = -\partial p/\partial z$ in Equations 3-20 to 3-22 lead to

$$U(x,y) = -P_x(x,y)/i\omega\rho_m \tag{3.32}$$

$$V(x,y) = -P_y(x,y)/i\omega\rho_m \tag{3.33}$$

$$W(x,y) = P(x,y) \, k_z/\omega\rho_m \tag{3.34}$$

while Equation 3-27 becomes

$$P_{xx} + P_{yy} + (\omega^2/c^2 - k_z^2)\, P = 0 \qquad (3.35)$$

As in Chapter 1, we attempt separable solutions, but because the problem is three-dimensional, the procedure must be applied twice in sequence. We therefore take

$$P(x,y) = X(x)Y(y) \qquad (3.36)$$

Substitution in Equation 3-35 leads to

$$- X''(x)/X(x) = Y''(y)/Y(y) + (\omega^2/c^2 - k_z^2) \qquad (3.37)$$

Since the left side is a function of x (independent of y), and the right side is a function of y (independent of x), each side must separately equal the same positive constant k_x^2,

$$- X''(x)/X(x) = Y''(y)/Y(y) + (\omega^2/c^2 - k_z^2) = k_x^2 \qquad (3.38)$$

This implies two equations, namely,

$$X''(x) + k_x^2\, X(x) = 0 \qquad (3.39)$$

$$Y''(y)/Y(y) + (\omega^2/c^2 - k_z^2) = k_x^2 \qquad (3.40)$$

The solution to Equation 3-39 is $X(x) = A \sin k_x x + B \cos k_x x$. Infinitely rigid, impermeable walls are assumed at $x = 0$ and $x = L_x$. Now, from Equation 3-32, the requirement $U = 0$ translates into one for the derivative $P_x(x,y)$ and therefore $X'(x)$. The condition $X'(0) = 0$ implies that $A = 0$, while the second leads to $X'(L_x) = - k_x B \sin k_x L_x = 0$. If nontrivial solutions are sought, we must set $k_x L_x = l\pi$, where $l = 0,1,2,3,$ ... and so on. This leads to

$$X_l(x) = B_l \cos l\pi x/L_x \qquad (3.41)$$

Next, let us consider Equation 3-40, and rewrite it in the form

$$Y''(y)/Y(y) = k_x^2 - (\omega^2/c^2 - k_z^2) = - k_y^2 \qquad (3.42)$$

or

$$Y''(y) + k_y^2\, Y(y) = 0 \qquad (3.43)$$

where k_y^2 is positive. This has the solution $Y(y) = C \sin k_y y + D \cos k_y y$. Infinitely rigid, impermeable walls are also assumed at $y = 0$ and $y = L_y$. We similarly find that $V = 0$ implies a vanishing of $P_y(x,0)$. Hence $Y'(0) = 0$, so that $C = 0$. Then, $Y'(L_y) = - D k_y \sin k_y L_y = 0$ forces us to choose $k_y L_y = m\pi$, $m = 0,1,2,3,$ ... in order to obtain the nontrivial solutions

$$Y_m(y) = D_m \cos m\pi y/L_y \qquad (3.44)$$

Thus, Equations 3-36, 3-41, and 3-44 together imply that p(x,y,z,t) can be constructed from superpositions of the component solutions

$$p_{l,m}(x,y,z,t) = P_{l,m} \cos l\pi x/L_x \cos m\pi y/L_y \exp i(\omega t - k_z z) \tag{3.45}$$

The dispersion relation. The reader should repeat our analysis using separation constants with different signs, in order to show that trivial solutions are obtained. With our selection, it is interesting to observe that the resulting right side of Equation 3-42 can be written as $k_x^2 - (\omega^2/c^2 - k_z^2) = - k_y^2$ or

$$\omega^2/c^2 = k_x^2 + k_y^2 + k_z^2 \tag{3.46}$$

This is *exactly* the dispersion relation that would have resulted had we directly substituted elementary solutions of the form

$$p(x,y,z,t) = \exp i(\omega t \pm k_x x \pm k_y y \pm k_z z) \tag{3.47}$$

in Equation 3-27. Thus, the solution given by Equation 3-45 can be interpreted as the correct superposition of elementary solutions that satisfies the assumed rigid wall boundary conditions. Other models, e.g., elastic, attenuative, or porous wall constraints, are likewise modeled. Now, Equation 3-46 can be manipulated to give

$$k_z = \pm \sqrt{\{\omega^2/c^2 - k_x^2 - k_y^2\}} \tag{3.48}$$

Here the ± sign indicates the existence of two wave families propagating in opposite directions along the z axis. Consider the argument of the square root function.

When $\omega/c > \sqrt{\{k_x^2 + k_y^2\}}$, k_z is real: the wave advances in the z direction, and the *eigenfunction* in Equation 3-45 is called a *propagating mode*. The limiting frequency for which k_z is real is the *cutoff frequency* given by

$$\omega_{cut\text{-}off} = c\sqrt{\{k_x^2 + k_y^2\}} \tag{3.49}$$

for the (*l,m*) mode (the meaning of *mode* is addressed in the next section).

If the input frequency is below cutoff, the argument of the square root in $k_z = \sqrt{\{\omega^2/c^2 - k_x^2 - k_y^2\}}$ becomes negative, and k_z is pure imaginary with

$$k_z = \pm i\sqrt{\{k_x^2 + k_y^2 - \omega^2/c^2\}} \tag{3.50}$$

If our waveguide is defined on the half-space z > 0, with the excitation source at z = 0, the sign in Equation 3-50 is chosen so that p(x,y,z,t) physically tends to 0 as z approaches infinity. In this limit, Equation 3-45 becomes

$$\begin{aligned} p(x,y,z,t) = {} & P_{l,m} \cos l\pi x/L_x \cos m\pi y/L_y \\ & \times \exp i\omega t \exp(-\sqrt{\{k_x^2 + k_y^2 - \omega^2/c^2\}}\, z) \end{aligned} \tag{3.51}$$

which is a *standing wave* that decays exponentially with z.

Physical interpretation. To understand how the contrasting solutions given by Equations 3-45 and 3-51 might arise physically, consider a "lumpy" piston located at z = 0, whose face contains surface irregularities.

A theory of *double Fourier series*, analogous to that for *single Fourier series* in Chapter 1, can be used to handle product eigenfunctions such as the "cos $l\pi x/L_x$ cos $m\pi y/L_y$" encountered above.

In principle, the lumpy geometry and its axial motion can be resolved into components that are associated with the mode shapes cos $l\pi x/L_x$ cos $m\pi y/L_y$. The amplitudes $P_{l,m}$ can be determined uniquely in terms of integrals involving surface features, just as we had indicated in Chapter 1 (Hildebrand, 1948; Tychonov and Samarski, 1964, 1967). The complete solution is obtained by superposing all modal contributions.

Each of the *(l,m)* modes so constructed *either* propagates with a speed of its own (see Equations 3-55 and 3-56 below) *or* fails to propagate. In the above example where the input frequency is below cutoff, the solution in Equation 3-51 is a standing wave that decays exponentially with z. The eigenfunction is termed an *evanescent mode*, and no energy is propagated down the waveguide.

If the waveguide is excited with a frequency just below the cutoff frequency of a particular mode, then this and higher modes are evansescent, and not important at appreciable distances from the source. All modes having cutoff frequences below the driving frequency may propagate energy and be detectable at large distances.

MWD notes. In *mud pulse telemetry*, *positive pressure pulsers* and *sirens* create 3D wave modes by mechanically blocking mud flow in the drillpipe. Bypass flows, vortexes, and other local inefficiencies are created, but these small-scale disturbances do not travel to the surface; only the plane wave associated with area blockage propagates uphole. This low-order mode will, of course, be affected by dissipation.

Negative pressure pulsers introduce local three-dimensional waves by creating differential pressure signals directly, by porting drillpipe mud into the lower pressure annular flow. Again, only the plane wave component propagates to the surface.

Downhole MWD *turbines* supply energy to power sensor-telemetry electronics and hydraulic components. While *thin* turbine blades superficially resemble thick *mud siren lobes*, they do *not* significantly block the oncoming mud; thus, turbine noise does *not* propagate uphole.

Again, only downhole events associated with pressures uniform across the area impact MWD surface signal processing, since this lowest-order mode propagates most efficiently. Such events include shock sub noise, PDC drilling noise, and bit bounce. Further details appear in Chapter 10.

Phase and group velocity. Let us consider *propagating waves* in more detail. From Equation 3-46, we have $\omega^2/c^2 = k_x^2 + k_y^2 + k_z^2$ or $\omega = c\sqrt{(k_x^2 + k_y^2 + k_z^2)}$. Now, from the pressure solution in Equation 3-45, a phase velocity can be defined by $c_p = \omega/k_z$ or,

$$c_p = c\sqrt{(k_x^2 + k_y^2 + k_z^2)}\,/k_z \tag{3.52}$$

which is clearly not the "expected" c of Equation 3-27.

In order to understand Equation 3-52, we introduce (in standard Cartesian notation) a *wavenumber vector* $\mathbf{k} = k_x \boldsymbol{i} + k_y \boldsymbol{j} + k_z \boldsymbol{k}$ whose magnitude is $k = \sqrt{(k_x^2 + k_y^2 + k_z^2)}$. First, we explore its physical significance.

Consider a harmonic component $p(x,y,z,t) = \sin(\omega t - k_x x - k_y y - k_z z)$. Surfaces of constant p (or phase) are described by $f(x,y,z) = \omega t - k_x x - k_y y - k_z z$ = constant, where t is a parameter. From vector calculus, normals to surfaces are proportional to $-\nabla f$. But this gradient function is just $k_x \boldsymbol{i} + k_y \boldsymbol{j} + k_z \boldsymbol{k}$. Thus, the wavenumber vector **k** is perpendicular to the wave front.

The phase velocity. Now, it is easy to see that the propagation vector **k** for each of the traveling waves (inferred from Equation 3-47) makes an angle θ with the z axis given by

$$\cos\theta = k_z/k = k_z/\sqrt{(k_x^2 + k_y^2 + k_z^2)} \tag{3.53}$$

Since Equation 3-52 states that $c_p = ck/k_z$ or $c_p/c = k/k_z$, combination with Equation 3-53 shows that the phase velocity is related to "c" by

$$c_p = c/\cos\theta \tag{3.54}$$

The group velocity. Note that each component wave carries energy down the waveguide by the process of *continual reflection* from the walls (via suitable combinations of Equation 3-47). The energy of a wave is propagated with speed c in the direction **k**, but the speed with which energy travels *in the z direction* (defined as the group velocity c_g) must be given by the component of the plane wave velocity c along the waveguide axis, namely,

$$c_g = c\cos\theta = c\,k_z/\sqrt{(k_x^2 + k_y^2 + k_z^2)} \tag{3.55}$$

From Equation 3-46, we have $\omega = c(k_x^2 + k_y^2 + k_z^2)^{1/2}$. Note that the partial derivative of Equation 3-55 with respect to k_z is

$$\begin{aligned}\partial\omega/\partial k_z &= c\,(1/2)\,(k_x^2 + k_y^2 + k_z^2)^{-1/2}\,(2k_z) \\ &= ck_z/\sqrt{(k_x^2 + k_y^2 + k_z^2)}\end{aligned} \tag{3.56}$$

Thus, on comparing Equations 3-55 and 3-56, we find that

$$c_g = \partial\omega/\partial k_z \tag{3.57}$$

which is consistent with the general theory of Chapter 2.

The velocity potential. We take the opportunity to introduce the *velocity potential* while we are dealing with fluid mechanics. This variable is important in simulating the MWD acoustic fields produced by *negative pressure pulsers.* To facilitate the discussion, we assume some familiarity with *vector calculus* (Hildebrand, 1948).

We make use of Equation 3-24 in the form $\nabla p = - \rho_m \partial \mathbf{q}/\partial t$. Now, we take the *curl* of each side, or $\nabla \times \nabla p = - \rho_m \partial(\nabla \times \mathbf{q})/\partial t$. But $\nabla \times \nabla p$ vanishes for *any* function $p(x,y,z,t)$, thus leaving the result $\partial(\nabla \times \mathbf{q})/\partial t = 0$. Therefore, $\nabla \times \mathbf{q}$ is a function of space; if it is zero at $t = 0$, it remains zero for all $t > 0$.

That $\nabla \times \mathbf{q} = \mathbf{0}$ initially, and hence for all time, is well accepted in *inviscid* fluid mechanics (that is, in flows absent of viscosity, as was assumed in this section). This vanishing refers to the fact that (with special exceptions in stratified flows and compressible flows with shockwaves) fluid elements in inviscid media do not spin about their own axes while in translatory motion.

Note that the magnitudes of the elements of the *vorticity vector* $\zeta = \nabla \times \mathbf{q}$ are proportional to the rotation rates of spinning particles. In the absence of spin, we have *irrotational* flow; the non-rotating carts on a ferris wheel illustrate the assumptions typical in acoustic analyses. An exception is sound propagation through *shear flow*, a specialty important in jet engine design.

Again from vector calculus, any function $\mathbf{q}$ satisfying $\nabla \times \mathbf{q} = \mathbf{0}$, can be represented as the *gradient* of a *potential* function $\phi(x,y,z,t)$, that is, as $\mathbf{q} = \nabla\phi$. Returning to scalar notation, the velocity components take the form

$$u = \partial\phi/\partial x \tag{3.58a}$$

$$v = \partial\phi/\partial y \tag{3.58b}$$

$$w = \partial\phi/\partial z \tag{3.58c}$$

In summary, the result $\nabla \times \mathbf{q} = \mathbf{0}$ allows us to express our velocities in the form given by Equation 3-58a,b,c, that is, through a *single* function ϕ and not three. This is important to our deducing Equation 3-59 below.

For simplicity, consider 1D wave motions satisfying Equations 3-20 and 3-23, that is, $\rho_m \partial u/\partial t = - \partial p/\partial x$ and $\partial \rho'/\partial t + \rho_m \partial u/\partial x = 0$. The first equation, together with $u = \partial\phi/\partial x$, suggests that $\partial(p + \rho_m \partial\phi/\partial t)/\partial x = 0$, or $p + \rho_m \partial\phi/\partial t$ = function of time only. This time function is purely constant if the flow is steady far upstream; here, $p = -\rho_m \partial\phi/\partial t$ applies to within an insignificant constant.

If we introduce a *compressibility* κ defined by $\rho' = \kappa \rho_m p$, the equation $\partial \rho'/\partial t + \rho_m \partial u/\partial x = 0$ transforms into $\kappa\, \partial p/\partial t + \partial u/\partial x = 0$. Substitution of $p = -\rho_m \partial\phi/\partial t$ and $u = \partial\phi/\partial x$ yields or $\partial^2\phi/\partial t^2 - (1/\rho_m \kappa)\, \partial^2\phi/\partial x^2 = 0$ or

$$\partial^2\phi/\partial t^2 - c^2\, \partial^2\phi/\partial x^2 = 0 \tag{3.59}$$

where the constant speed of sound c satisfies $c^2 = 1/\rho_m\kappa$. Thus, the velocity potential also satisfies the wave equation.

Modeling MWD sources. *Why* should we deal with Equation 3-59 for $\phi(x,t)$, when the one-dimensional form of Equation 3-27 for p(x,t) is already available? It turns out that pressure, useful in conventional engineering analyses, has limited value in modeling positive or negative MWD pulsers!

This observation motivated our detailed discussion of wave equation sources in Chapter 1. In order to model pulser sources correctly, describing their velocity and pressure symmetries or anti-symmetries accurately, different classes of dependent variables are needed.

Many books discuss *acoustic pressure* only, which satisfy the wave equation in Equation 3-27. But other physical properties, e.g., the velocities u, v, and w, the disturbance density ρ', the fluctuating temperature, and the velocity potential, are also wave-like.

The dependent variable appropriate to a particular problem depends on specific boundary condition details. For the oscillating end piston popular in simple examples, the axial velocity may be natural; or, the Lagrangian fluid displacement itself may be used with its time derivative prescribed at the end.

In many petroleum engineering applications, especially those in Measurement-While-Drilling, the excitation source actually resides *internally* within the waveguide, and not at the very end. Mathematical modeling is more challenging in several respects.

Acoustic energy is created within the medium, and propagated in both directions; then, all subsequent reflections must be allowed to pass *transparently* through the source point. But not all sources function similarly, even on a qualitative level.

For example, as we will see in Chapter 10, the *created pressure* on either side of a *positive pressure pulser* is *antisymmetric*: when one side is over-pressured relative to ambient conditions, the other is under-pressured. For *negative pressure pulsers*, both are either over-pressured or under-pressured, and the *created pressure* field is *symmetric*. *Symmetries* and *antisymmetries* must be correctly accounted for.

We explore these subtleties by considering specific models. In Chapter 1, we indicated that sources can be introduced into wave equations by appending *delta functions*. Therefore, let us *naively* add a delta function to the right side of Equation of 3-27 for pressure,

$$\partial^2 p/\partial t^2 - c^2\, \partial^2 p/\partial x^2 = A\, \delta(x-x^*) \tag{3.60}$$

where A is a normalizing constant and $0 < x, x^* < L$.

What, exactly, does this *blind* application of delta functions model? From Chapter 1, we understand that $A\, \delta(x-x^*)$ produces an internal discontinuity in the spatial derivative $\partial p/\partial x$; the right side of Equation 3-60 is responsible for a sudden jump in the *pressure gradient* $\partial p/\partial x$.

Now, positive pressure and siren pulsers will produce internal jumps in the *pressure level* itself. Thus, it is not clear what Equation 3-60 models, and how it can be used, unless a mechanical device capable of generating *pressure gradient* discontinuities can be designed.

Next, we repeat this exercise with the velocity potential of Equation 3-59, and add a delta function to its right side. Thus, we consider

$$\partial^2\phi/\partial t^2 - c^2\, \partial^2\phi/\partial x^2 = A\,\delta(x-x^*) \qquad (3.61)$$

Equation 3-61 now models a physical system where the spatial first derivative $\partial\phi/\partial x$ is discontinuous; that is, the *axial velocity* u is discontinuous, on noting Equation 3-58a. How might this be accomplished in practice?

For instance, a "pulsating balloon" introduced into our drillstring at $x = x^*$ would produce such acoustic disturbances. The balloon surfaces at the sides of x^* *always* head away from (or toward) each other symmetrically, thus providing the required axial velocity discontinuity.

For example, if $u(x^*-) = -100$ ft/sec and $u(x^*+) = +100$ ft/sec, then the difference $u(x^*+) - u(x^*-) = +200$ ft/sec $\neq 0$ is nonzero, providing the discontinuity in axial velocity. Thus, Equation 3-61 represents one possible way to model negative pressure pulsers.

The reader might consider how pulsers that create *pressure discontinuities* should be modeled; certainly, $\phi(x,t)$ and $p(x,t)$ are inappropriate. This problem will be considered in Chapters 9 and 10.

The role of source point modeling is also important in drillstring vibrations. Conventional boundary condition models *prescribe periodic displacements at the bit* $x = 0$ in order to simulate bit motions. But periodic displacements have zero time averages; by invoking such models, non-zero rate-of-penetration and bit bounce cannot be modeled*!*

In Chapter 5, we will show how *displacement sources* (as demonstrated in the above example) can be used to model tricone bit kinematics; the additional degree of freedom that this flexibility offers *then* permits us to apply *rock-bit interactions* as boundary conditions at $x = 0$, thus opening up new classes of possible dynamical motions at the bit.

EXAMPLE 3-3: GRAVITY-CAPILLARY WAVES IN DEEP WATER

We have encountered wave phenomena hosted by different formulations, e.g., the classical equation $\partial^2u/\partial t^2 - c^2\, \partial^2u/\partial x^2 = 0$, the fourth-order beam equation $\rho A\, \partial^2v/\partial t^2 + EI\, \partial^4v/\partial x^4 = 0$, and the 3D acoustic model $\partial^2p/\partial t^2 - c^2 (\partial^2p/\partial x^2 + \partial^2p/\partial y^2 + \partial^2p/\partial z^2) = 0$. Their respective dispersion relations, i.e., $\omega = ck$, $\omega = \sqrt{(EI/\rho A)}\, k^2$, and $\omega^2/c^2 = k_x^2 + k_y^2 + k_z^2$, represent contrasting results.

Here, we will consider the propagation of gravity-capillary waves in deep water, as shown in Figure 3-3. We will demonstrate how *wave-like* solutions to *elliptic* equations like $\partial^2\phi/\partial x^2 + \partial^2\phi/\partial y^2 = 0$ arise (general solutions to such equations take the form $\phi(x,y) = f(x+iy) + g(x-iy)$). However, wave-like motions are made possible by special *free surface* boundary conditions.

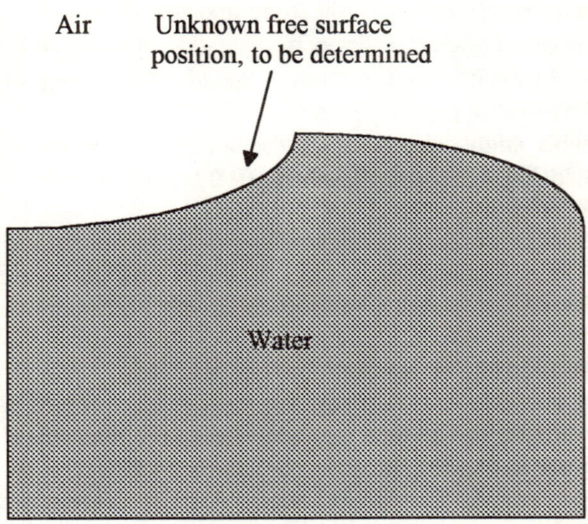

Figure 3-3. Water waves with free surface.

For readers new to hydrodynamics, *gravity waves* are those dominated by *gravity* or weight; *capillary waves* are, by contrast, controlled by *surface tension*. We introduce the basic concepts here, but a more detailed treatment is offered in Chapter 12. A classic hydrodynamics reference is Lamb (1945).

Governing Laplace equation. We consider *incompressible* wave motions in *constant density* water; these exclude sound effects, which would arise from fluid compressibility. Since the density ρ in the mass conservation law $\partial\rho/\partial t + \partial(\rho u)/\partial x + \partial(\rho v)/\partial y = 0$ is constant, we have

$$\partial u/\partial x + \partial v/\partial y = 0 \qquad (3.62)$$

Again, we assume inviscid flow with zero initial rotation. Then, the irrotationality requirement $\nabla \times \mathbf{q} = \mathbf{0}$ allows us to write the *Eulerian velocities* u(x,y,t) and v(x,y,t) as x and y derivatives of a *potential* function φ(x,y,t),

$$u = \partial\varphi/\partial x \qquad (3.63)$$

$$v = \partial\varphi/\partial y \qquad (3.64)$$

(in general, the velocity in a direction s takes the value $\partial\varphi/\partial s$). If we substitute Equations 3-63 and 3-64 into Equation 3-62, we obtain Laplace's equation

$$\partial^2\varphi/\partial x^2 + \partial^2\varphi/\partial y^2 = 0 \qquad (3.65)$$

which holds for steady and unsteady incompressible flows.

Incidentally, Figure 3-3 shows a *slope discontinuity* in the free surface. The suggested *wave breaking* does not arise in linear theory (as is considered here), but only in weakly and fully nonlinear wave models.

Boundary conditions. The boundary condition at any *rigid stationary boundary* is *zero normal flow*, that is, $\partial\varphi/\partial n = 0$ where n is the normal distance from the boundary.

Kinematic condition at interfaces. Consider an *interface* located anywhere within a flow, described by the locus of points f(x,y,z,t) = 0, and the property that fluid does not cross it. The velocity of the *fluid* normal to the interface therefore equals the velocity of the *interface* normal to itself.

The normal velocity of a surface is $-f_t/\sqrt{(f_x^2 + f_y^2 + f_z^2)}$, while the normal velocity of the fluid is $(uf_x+vf_y+wf_z)/\sqrt{(f_x^2 + f_y^2 + f_z^2)}$. The condition that these be equal results in the interface condition $f_t + uf_x + vf_y + wf_z = 0$. If we take our *free surface* in the form f(x,y,t) = y - η(x,t) = 0, or

$$y = \eta(x,t) \qquad (3.66)$$

we have $\partial(y-\eta)/\partial t + u\ \partial(y-\eta)/\partial x + v\ \partial(y-\eta)/\partial y = 0$. If no background flow exists, the complete law $-\partial\eta/\partial t - u\ \partial\eta/\partial x + v = 0$ can be replaced by $-\partial\eta/\partial t + v \approx 0$ since the nonlinear term is small. Using Equation 3-64, this becomes

$$\partial\varphi/\partial y = \partial\eta/\partial t \qquad (3.67)$$

Dynamic surface condition. At a free surface, the dynamic pressure boundary condition is $p = -T(1/R_1 + 1/R_2)$, where T is the surface tension, and R_1 and R_2 are the principal radii. For (x,y) coordinates, the linearized form is

$$p = -T\, \partial^2\eta/\partial x^2 \quad (3.68)$$

The complete unsteady *Bernoulli equation* (Lamb, 1945) is

$$p/\rho = -\partial\varphi/\partial t - gy - (u^2 + v^2)/2 + F(t) \quad (3.69)$$

where g is the acceleration due to gravity. If p is constant at the free surface, e.g., equal to atmospheric pressure, then $F(t) = 0$. Since a mean flow does not exist, the quadratic terms in Equation 3-69 are smaller than the linear ones, and we have the approximation

$$p/\rho = -\partial\varphi/\partial t - gy \quad (3.70)$$

Equations 3-66, 3-68, and 3-70 imply the combined kinematical-dynamical free surface condition

$$\partial^2\varphi/\partial t^2 = (T/\rho\, \partial^2/\partial x^2 - g)\, \partial\varphi/\partial y \quad (3.71)$$

Problem solution. We will solve Equation 3-65 for $\varphi(x,y,t)$, subject to Equation 3-71 at the *unknown* free surface $y = \eta(x,t)$, and the condition that all motions cease at infinity. As usual, we attempt separable solutions,

$$\varphi(x,y,t) = \phi(y)\, e^{i(kx-\omega t)} \quad (3.72)$$

where k is a wavenumber and ω is the frequency (waveguide motions are assumed in the x direction only).

The function $\phi(y)$ provides the *cross-structure* for the *modal* boundary value problem. Substitution of Equation 3-72 into $\partial^2\varphi/\partial x^2 + \partial^2\varphi/\partial y^2 = 0$ leads to $\phi''(y) - k^2\, \phi(y) = 0$. The general solution takes the form $\phi(y) = Ae^{-ky} + Be^{+ky}$ where A and B are constants, but only the contribution $\phi(y) = e^{+ky}$ is meaningful since it gives decaying motions at $y = -\infty$. For this solution, we obtain $\phi'(y) = ke^{+ky}$, $\phi(0) = 1$, and $\phi'(0) = k$.

Note that this modal problem, unlike our previous modal problem for acoustical waveguides, is *not* a discrete eigenvalue problem. In fact, substitution of Equation 3-72 in Equation 3-71 shows that

$$(-i\omega)^2\, \phi(y) = T/\rho\, (ik)^2\, \phi'(y) - g\phi'(y) \quad (3.73)$$

$$(-i\omega)^2\, \phi(0) = T/\rho\, (ik)^2\, \phi'(0) - g\phi'(0) \quad (3.74)$$

Consequently, we have the dispersion relation

$$\omega^2 = Tk^3/\rho + gk \quad (3.75)$$

for any wave component with wavenumber k.

Now, since $\phi(y) = e^{+ky}$ is real, we consider only the real part of Equation 3-72 and write

$$\varphi(x,y,t) = C\, e^{+ky} \cos(kx - \omega t) \tag{3.76}$$

with a *cosine* dependence. The free surface position is also an unknown, and a functional assumption for it must be made. Since Equation 3-67 requires that $\partial\varphi/\partial y = \partial\eta/\partial t$, we choose

$$\eta(x,t) = -\, a \sin(kx - \omega t) \tag{3.77}$$

with a *sine* dependence, so that the trigonometric functions cancel. In Equation 3-77, "a" represents the *wave elevation* or *amplitude*. Substitution of Equations 3-76 and 3-77 in Equation 3-67 shows that $C = a\omega/k$, so that

$$\varphi(x,y,t) = (a\omega/k)\, e^{+ky} \cos(kx - \omega t) \tag{3.78}$$

thus completing the solution.

Energy considerations. In order to demonstrate some elementary properties of waves, we drop surface tension for simplicity, so that the dispersion relation in Equation 3-75 reduces to $\omega^2 = gk$ or

$$\omega = (gk)^{1/2} \tag{3.79}$$

In this discussion, we assume the positive square root, and ignore the "±" for brevity. This result for *deep water gravity waves* has the *phase velocity*

$$c_p = \omega/k = \sqrt{(g/k)} \tag{3.80}$$

and the *group velocity*

$$c_g = \partial\omega/\partial k = (1/2)\sqrt{(g/k)} \tag{3.81}$$

so that

$$c_g = (1/2)\, c_p \tag{3.82}$$

If we refer to a mass element $\rho(\eta dx)$ and consider its potential energy increase for a wave elevation η, simple center of gravity considerations lead to

$$PE = g\rho \int_0^{\lambda/2} \eta^2\, dx = 1/2\, g\rho \int_0^{\lambda} \eta^2\, dx \tag{3.83}$$

where $\lambda = 2\pi/k$ is the *wavelength*. Substituting Equations 3-77 and 3-78 into Equation 3-83, we find that the potential energy per unit wavelength is

$$PE = 1/4\, g\rho a^2 \lambda \tag{3.84}$$

Now, the kinetic energy per unit wavelength is

$$KE = 1/2\ \rho \int_{-\infty}^{0} \int_{0}^{\lambda} \{(\partial\varphi/\partial x)^2 + (\partial\varphi/\partial y)^2\}\ dx\ dy \tag{3.85}$$

Direct evaluation using Equations 3-77 and 3-78 shows that

$$KE = 1/4\ g\rho a^2\lambda \tag{3.86}$$

Comparing Equations 3-84 and 3-86, we find that $PE = KE$. Thus, the total energy $1/2\ g\rho a^2\lambda$ is *half* potential and *half* kinetic: this *equipartition of energy* is characteristic of linear systems.

Let us now consider the rate of work dW/dt at a fixed x, performed by the fluid at the left, on the fluid at the right,

$$dW/dt = \int_{-\infty}^{0} p\ \partial\varphi/\partial x\ dy \tag{3.87}$$

Use of Equations 3-70 and 3-78 shows that

$$dW/dt = 1/2\ g\rho a^2(\omega/k)\ \sin^2 (kx - \omega t) \tag{3.88}$$

the integral of which over a period $2\pi/\omega$ is $1/4\ g\rho a^2\lambda$.

This is just one-half of the total energy $1/2\ g\rho a^2\lambda$ contained in one wavelength. Thus, the supply of energy is sufficient only to advance an isolated wave group at a speed equal to half the velocity of individual waves, in complete agreement with Equation 3-82.

EXAMPLE 3-4: FLUID-SOLID INTERACTION - WAVES ON ELASTIC MEMBRANES

We give another modal example containing free surface conditions, namely, the stability of an inviscid shear flow *and* its interaction with an elastic membrane. This requires coupled fluid-solid equations for the flow eigenfunction and for the unknown membrane position.

Unlike the previous example, we will demonstrate how *bounds* on the eigenvalues can be obtained for general flows without carrying out explicit integrations. We show how the use of *inequalities* can be exploited, in order to produce general conclusions that may not be apparent from specific examples.

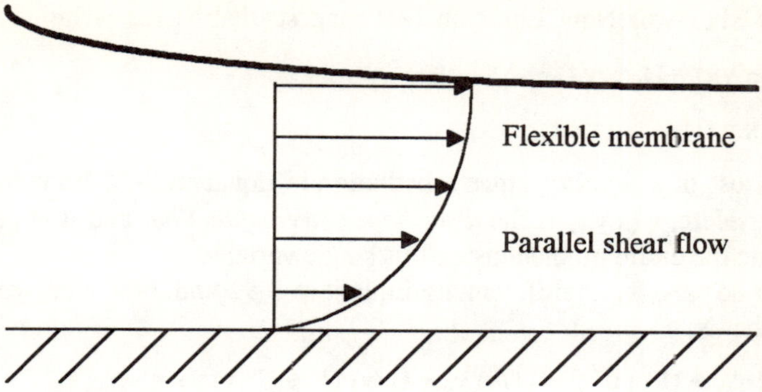

Figure 3-4. Fluid instability waves on elastic membrane.

Governing Rayleigh equation. For simplicity, we consider the Navier-Stokes equations given by Equations 3-16 to 3-18 in the constant density, zero viscosity, two-dimensional limit. Thus, we have

$$\rho\,(\partial u/\partial t + u\,\partial u/\partial x\;+ v\,\partial u/\partial y) = -\,\partial p/\partial x \tag{3.89}$$

$$\rho\,(\partial v/\partial t + u\,\partial v/\partial x\;+ v\,\partial v/\partial y) = -\,\partial p/\partial y \tag{3.90}$$

where u(x,y,t) and v(x,y,t) are *Eulerian velocities* at the fixed point (x,y), in the x and y directions, and t is time. Also, p(x,y,t) and ρ represent *pressure* and *mass density*. Our velocities conserve mass via Equation 3-19, so that

$$\partial u/\partial x\;+ \partial v/\partial y = 0 \tag{3.91}$$

We consider the stability of a *parallel shear flow*, with a velocity U(y) in the x direction, to wave-like disturbances. This *hydrodynamic stability* problem deals with *aeroelastic* applications since it combines elements of fluid dynamics and elasticity.

If we set u(x,y,t) = U(y) + u'(x,y,t), v(x,y,t) = v'(x,y,t), and p(x,y,t) = P(y) + p'(x,y,t), where U and P are mean quantities, and (primed) disturbances are smaller, substitution in Equations 3-89 to 3-91, and linearization, lead to

$$\rho\,(\partial u'/\partial t + U(y)\,\partial u'/\partial x\;+ U'(y)\,v\,) = -\,\partial p'/\partial x \tag{3.92}$$

$$\rho\,(\partial v'/\partial t + U\,\partial v'/\partial x) = -\,\partial p'/\partial y \tag{3.93}$$

$$\partial u'/\partial x\;+ \partial v'/\partial y = 0 \tag{3.94}$$

where U'(y) denotes dU(y)/dy.

We wish to simplify the analysis by dealing with one dependent variable as opposed to two. Now, Equation 3-94 suggests that we can write

$$u'(x,y,t) = \psi_y(x,y,t) \tag{3.95}$$

$$v'(x,y,t) = - \psi_x(x,y,t) \tag{3.96}$$

without loss of generality, since substitution in Equation 3-94 leads to a trivial identity. Here, $\psi(x,y,t)$ is the *disturbance streamfunction*, and it is possible to formulate the entire problem using this single variable.

To do this, let us differentiate Equations 3-92 and 3-93 with respect to y and x, that is,

$$\rho \{u'_{yt} + U(y)\, u'_{yx} + U'(y)\, v_y + U'(y)\, u'_x + U''(y)\, v\} = - p'_{yx} \tag{3.97}$$

$$\rho \{v'_{xt} + Uv'_{xx} \} = - p'_{xy} \tag{3.98}$$

Subtraction gives

$$u'_{yt} + U(y)\, u'_{yx} + U'(y)\, v'_y + U'(y)\, u'_x + U''(y)\, v' - v'_{xt} - Uv'_{xx} = 0 \tag{3.99}$$

Using Equations 3-94 to 3-96, we obtain

$$\psi_{yyt} + U(y)\, \psi_{yyx} - U''(y)\, \psi_x + \psi_{xxt} + U\psi_{xxx} = 0 \tag{3.100}$$

We simplify Equation 3-100 further by introducing the separation of variables

$$\psi(x,y,t) = \phi(y) \exp ik(x-ct) \tag{3.101}$$

Substitution in Equation 3-100 leads to the so-called *Rayleigh equation*

$$(U-c)\{\phi''(y) -k^2\phi\} - U''\phi(y) = 0 \tag{3.102}$$

where k is a specified wavenumber, and $\phi(y)$ is the modal eigenfunction; the factor exp ik(x-ct) allows wave propagation in the x direction, and $c = c_r + i\, c_i$ is the *complex eigenvalue*.

Boundary conditions for $\phi(y)$. We will assume a rigid wall placed along $y = - H < 0$, at which the normal velocity v vanishes. Thus,

$$\phi(-H) = 0 \tag{3.103}$$

Now consider a flexible membrane whose deviation from equilibrium $y = 0$ satisfies $y = \eta(x,t)$, where $\eta(x,t)$ is an unknown in addition to $\phi(y)$. The disturbance pressure acting on the membrane satisfies the force balance

$$p'(y=0) = \rho g\eta - T\eta_{xx} + s\eta + m\eta_{tt} \tag{3.104}$$

Here we assume a constant fluid density ρ, a gravitational acceleration g, a membrane tension T, a spring constant s, and a lineal mass density m. If we now take

$$p'(x,y,t) = p(y) \exp ik(x-ct) \quad (3.105)$$

$$\eta(x,t) = a \exp ik(x-ct) \quad (3.106)$$

we obtain

$$p(0) = (\rho g + Tk^2 + s - mk^2c^2)a \quad (3.107)$$

The kinematic condition

$$v = \eta_t + U \eta_x \quad (3.108)$$

leads to

$$\phi(0) = -\{U(0) - c\}a \quad (3.109)$$

Hence

$$\phi'(0)/\phi(0) = U'(0)/\{U(0) - c\} + \{\rho g + Tk^2 + s - mk^2c^2\}/\{\rho(U(0)-c)^2\} \quad (3.110)$$

Equations 3-102, 3-103, and 3-110 specify the entire *inviscid stability problem.* The reader should examine the similarities and differences between the present formulation and our previous one for water waves.

Eigenvalue bounds. For prescribed velocity profiles made up of simple straight lines and arcs, complex eigenvalue relationships connecting c (that is, $(\omega_r + i\omega_i)/k$) to g, ρ, s, m and the wavenumber k can be found, and the corresponding eigenfunctions can be calculated as in the previous example.

However, we choose instead to demonstrate a procedure available to mathematicians: we obtain *upper and lower bounds* on the eigenvalues c *without* restricting the form of the parallel flow U(y).

Let us introduce $\phi(y) = (c-U)F(y)$ and $W = U - c$, so that Equation 3-102 takes the form $(W^2F')' - k^2W^2F = 0$. Multiplication by the *complex conjugate* F^* and integration over (-H,0) leads to

$$\int W^2\{|F'|^2 + k^2|F|^2\} \, dy = F^*(W^2F')| = G|F_0|^2 \quad (3.111)$$

where all limits of integration at y = 0 and y = -H are henceforth omitted for brevity.

In Equation 3-111, we have set $\rho G = \rho(G_r + iG_i) = \rho g + Tk^2 + s - mk^2c^2$, taken Equations 3-110 and 3-103 in the form $W^2F' = GF$ and $F(-H) = 0$, and written $F_0 = F(0)$. Now, define $Q = |F'|^2 + k^2|F|^2 > 0$ so that real and imaginary parts give

$$\int [(U - c_r)^2 - c_i^2]\, Q\, dy = G_r\, |F_0|^2 \quad (3.112)$$

$$-2c_i \int (U - c_r)\, Q\, dy = G_i\, |F_0|^2 \quad (3.113)$$

Expansion of Equation 3-112 using Equation 3-113 leads to

$$\int U^2 Q\, dy = (G_r - c_r G_i / c_i)\, |F_0|^2 + (c_r^2 + c_i^2) \int Q\, dy \quad (3.114)$$

assuming $c_i > 0$. Next, let A and B be the minimum and maximum of U(y) in (-H,0). Then, the identity

$$\begin{aligned} 0 &\geq \int (U-A)(U-B)\, Q\, dy \\ &= \int U^2 Q\, dy - (A+B) \int UQ\, dy + AB \int Q\, dy \end{aligned} \quad (3.115)$$

on using Equations 3-113 and 3-114 leads to

$$\begin{aligned} 0 \geq\ & [(c_r^2 + c_i^2) - (A+B)c_r + AB][mk^2|F_0|^2/\rho + \int Q dy] \\ & +[g + s/\rho + (k^2/\rho)(T - mAB)]|F_0|^2 \end{aligned} \quad (3.116)$$

Note that each term in the bracketed quantity $[mk^2|F_0|^2/\rho + \int Q dy]$ is positive ($Q > 0$ follows from an earlier definition). If $g + s/\rho + (k^2/\rho)(T - mAB) \geq 0$, it follows that $(c_r^2 + c_i^2) - (A+B)c_r + AB \leq 0$, or

$$[c_r - 1/2\, (A+B)]^2 + c_i^2 \leq [1/2\, (A+B)]^2 - AB = [1/2\, (B-A)]^2 \quad (3.117)$$

Thus, we have the general theorem: *the complex wave velocity c for any unstable mode must lie inside the semicircle in the upper half of the complex c-plane which has the range of U for diameter.*

Quantitative results have been obtained *without* knowing U(y), and *without* recourse to calculations for the dispersion relation $c = \omega/k = f(g, \rho, s, m, k)$ or the eigenfunction $\phi(y)$.

Sufficient conditions for stability ($\omega_i < 0$), and specific calculated results are given in Chin (1979), where the model is used to study flow stability over compliant surfaces.

EXAMPLE 3-5: PROBLEMS IN HYDRODYNAMIC STABILITY

Hydrodynamic stability problems abound in nature. The best known example is the *flapping flag* in wind. Ideally, a flag "should" remain flat and undisturbed as uniform winds at either side flow by tangentially. But this never occurs: the slightest perturbations set up instabilities that do not decay.

Shear flow instability. Previously we examined the behaviour of the disturbance quantities u', v', and p' to the parallel mean flow U(y) (i.e., *shear flow*) and its pressure P(y). We substituted u(x,y,t) = U(y) + u'(x,y,t), v(x,y,t) = v'(x,y,t), and p(x,y,t) = P(y) + p'(x,y,t) into complete *inviscid* equations, and derived the governing *Rayleigh* equation.

Then we determined bounds on the complex eigenvalue. Although we did not derive any explicit eigenvalue relations, these can be obtained straightforwardly for specific velocity profiles (Chin, 1979).

A similar process applies to the complete *viscous* equation for *Newtonian* fluids; the analogous disturbances satisfy the *Orr-Sommerfeld* equation. Again, streamfunctions proportional to $\phi(y)$ exp ik(x-ct) are assumed; the objective is the eigenvalue relation $c(k) = c^r(k) + ic^i(k)$ and its eigenfunction $\phi(y)$.

Once these are available, the solution of the complete *initial value problem* for general *wave-like disturbances* to a particular flow U(y) is in principle known. If $c^i(k) > 0$, the flow is said to be *unstable*, whereas it is *stable* if $c^i(k) < 0$; otherwise, the flow is *neutrally stable*. A comprehensive survey of the Orr-Sommerfeld stability literature appears in Schlichting (1968).

Neutral stability diagrams. Calculated hydrodynamic stability results are usually presented by means of *stability diagrams*. A typical example is shown in Figure 3-5, derived for a particular velocity profile U(y).

On the horizontal axis is the dimensionless *Reynolds number* $R = \rho U\delta/\mu$ for a *boundary layer flow*, where the length scale is based on boundary layer thickness δ (ρ and μ represent density and viscosity). On the vertical axis is the dimensionless product $k\delta$, where k is the wavenumber of the disturbance.

The *neutral stability curve* shown, determined from detailed eigenvalue analysis, delineates stable and unstable regions of $(k\delta, \rho U\delta/\mu)$ space. Stability may be desirable in delaying the onset of laminar-to-turbulent flow transition, while instability may be useful in promoting rapid mixing. The *critical Reynolds number* is that number below which all disturbances remain stable.

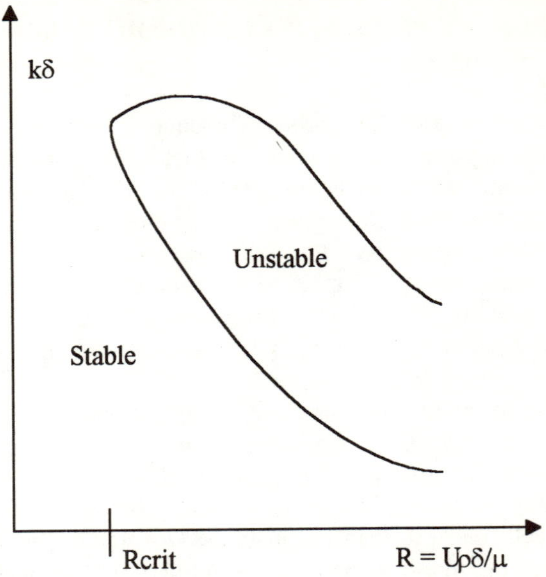

Figure 3-5. Neutral stability diagram.

Borehole flow stability. The laminar stability of drilling mud and cement flows in the borehole annulus, important to drilling and completions, should be studied using hydrodynamic stability methods. Instead, the critical Reynolds number of 2,300 is often invoked by practitioners, but *this applies to parabolic laminar profiles of Newtonian fluids in circular pipes only!*

Following Schlichting (1968), the velocity profile of interest must first be obtained. For any eccentric annulus, this is easily accomplished using finite difference methods applied to boundary conforming grid systems; for example, refer to the author's recent book *Borehole Flow Modeling in Horizontal, Deviated, and Vertical Wells* (e.g., see Chin, 1992a,b) for numerical algorithms and computed results.

This calculation will have assumed a particular rheological model, e.g., Bingham plastic, power law fluid, or Herschel-Bulkley flow. Then, the viscous disturbance equations corresponding to the constitutive model used should be derived and solved on the same boundary-conforming mesh, with waves allowed to propagate in the cross-plane and axial directions.

Typical *steady flow* results are given in Figures 3-6a,b, where axial velocities in *in/sec* are labelled on the annular cross-section. These show that velocity profiles are far from parabolic. Chapter 9 gives detailed results for different fluid types and annuli of practical importance.

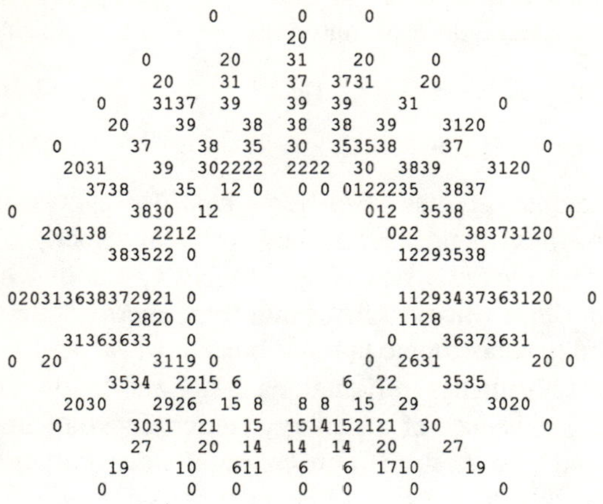

Figure 3-6a. Power law model, concentric flow with cuttings bed.

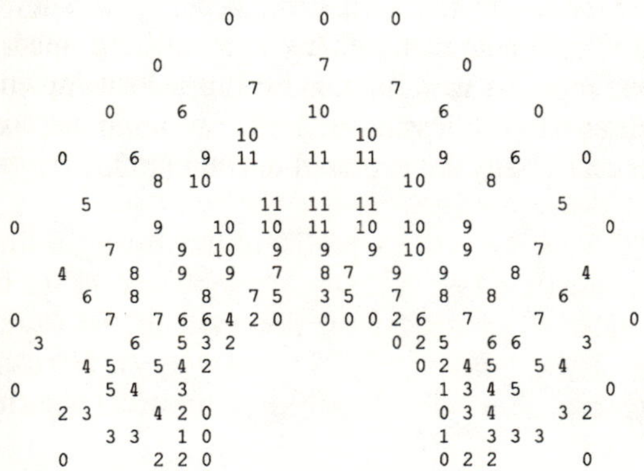

Figure 3-6b. Power law fluid, highly eccentric circular geometries.

Stability of irrotational flows. Very often, the parallel flow velocity U may *not* depend on the cross-coordinate y; instead, it may vary in the direction of the flow only, and additionally with time, so that U = U(x,t). For such problems, other stability criteria can be defined, based on

$$\partial E/\partial t + \partial(\Omega^r_k(k,x,t)E)/\partial x = E\{2\Omega^i + \Omega^r_t(k,x,t)/\Omega^r\} \quad (2.104)$$

$$\partial M/\partial t + \partial(\Omega^r_k(k,x,t)M)/\partial x = M\{2\Omega^i - \Omega^r_x(k,x,t)/k\} \quad (2.105)$$

Landahl (1972), for example, studies *shear flow stability* over *slowly growing boundary layers*, allowing dependences of the horizontal velocity U on x, t, *and* y. His classic paper uncovered a new wave instability, one in which heterogeneities in the medium cause waves to focus and trap; energy "pile-up" and unstable amplification lead to catastrophic laminar breakdown.

Consider the limit where Ω^i vanishes in Equations 2-104 and 2-105. The remaining terms show that the existence of an energy source does not imply that of a momentum source, and vice versa. If nonzero right sides are used to define instability, one must specify whether E or M is under consideration.

Even so, subtleties still remain. For example, it is possible for the *total* energy (or momentum) between two rays to be constant, while *energy (or momentum) density versus position within the wave group* changes with time, in such a way that *integrable* amplitude singularities appear.

Interestingly, this is the case for the drillstring bending instabilities considered in Chapter 6. Finally we note that, while wavenumber, amplitude, energy, momentum and action propagate with the same group velocity in linear theory, the propagation velocities will differ with each other in nonlinear wave flows. These conceptual ideas on stability are explored in Chin (1976).

REFERENCES

Abramson, H.N., Plass, H.J., and Ripperger, E.A., "Stress Wave Propagation in Rods and Beams", in *Advances in Applied Mechanics, Vol. V*, edited by Hugh Dryden and Theordore von Karman, Academic Press, New York, 1958.

Chin, W.C., *Physics of Slowly Varying Wavetrains in Continuum Systems, Ph.D. Thesis*, Massachusetts Institute of Technology, Cambridge, Massachusetts, 1976.

Chin, W.C., "Stability of Inviscid Shear Flow Over Flexible Membranes," AIAA Journal, June 1979.

Chin, W.C., *Borehole Flow Modeling in Horizontal, Deviated, and Vertical Wells*, Gulf Publishing, Houston, 1992.

Chin, W.C., *Petrocalc 14: Horizontal and Vertical Annular Flow Modeling, Petroleum Engineering Software for the IBM PC and Compatibles*, Gulf Publishing, Houston, 1992.

Clark, R.A., and Reissner, E., "Bending of Curved Tubes", in *Advances in Applied Mechanics, Vol. II*, edited by Richard von Mises and Theordore von Karman, Academic Press, New York, 1951.

Clough, R.W., and Penzien, J., *Dynamics of Structures*, McGraw-Hill, New York, 1975.

Den Hartog, J.P., *Advanced Strength of Materials*, McGraw-Hill, New York, 1952.

Graff, K.F., *Wave Motion in Elastic Solids*, Ohio State University Press, 1975.

Hildebrand, F.B., *Advanced Calculus for Applications*, Prentice-Hall, Englewood Cliffs, New Jersey, 1948.

Kinsler, L.E., and Frey, A.R., *Fundamentals of Acoustics*, John Wiley and Sons, New York, 1950.

Lamb, H., *Hydrodynamics*, Dover Publications, New York, 1945.

Landahl, M.T., "Wave Mechanics of Breakdown," Journal of Fluid Mechanics, Vol. 56, Part 4, 1972, pp. 775-802.

Love, A.E.H., *A Treatise on the Mathematical Theory of Elasticity*, Dover Publications, New York, 1944.

Mason, W.P., *Electromechanical Transducers and Wave Filters*, D. Van Nostrand Company, New York, 1942.

Morse, P.M., and Ingard, K.U., *Theoretical Acoustics*, McGraw-Hill, New York, 1968.

Pierce, A.D., *Acoustics: An Introduction to Its Physical Principles and Applications*, McGraw-Hill, New York, 1981.

Schlichting, H., *Boundary Layer Theory*, McGraw-Hill, New York, 1968.

Sechler, E.E., *Elasticity in Engineering*, Dover Publications, New York, 1952.

Timoshenko, S.P., *Strength of Materials, Part II: Advanced Theory and Problems*, Van Nostrand Reinhold Company, New York, 1958.

Timoshenko, S.P., and Gere, J.M., *Theory of Elastic Stability*, McGraw-Hill, New York, 1961.

Timoshenko, S.P., and Goodier, J.N., *Theory of Elasticity*, McGraw-Hill, New York, 1934.

Tychonov, A.N., and Samarski, A.A., *Partial Differential Equations of Mathematical Physics, Vol. I*, Holden-Day, San Francisco, 1964.

Tychonov, A.N., and Samarski, A.A., *Partial Differential Equations of Mathematical Physics, Vol. II*, Holden-Day, San Francisco, 1967.

Part II
Drillstring Vibrations

4
Typical Downhole Vibration Environment

The subject of *drillstring vibrations* has attracted increasing interest among oilfield practitioners in recent years. For example, its roles in pipe fatigue, borehole instability, increased *rate of penetration* (using controlled resonances), and as a potential "see *ahead* of the drillbit" downhole imaging device (drillstring vibrations *have* been correlated with formation type), among others, have been noted in the literature.

For these reasons, *isolated* analytical models have been formulated in recent years to simulate empirically observed results, in order to improve our understanding of different aspects of the physical phenomenon.

Ultimately, *integrated* drilling simulators would predict penetration rate, fatigue hazards, violent lateral vibrations, stick-slip phenomena, given the kinematics of the drillbit, appropriate measures of formation hardness or rock-bit interaction, drillstring configuration (i.e., *bottom hole assembly* or *BHA*), and so on. Optimal well planning would then become a reality.

What is wave motion? Rigorous definitions for wave propagation can be found in physics and mathematics books. Here, we will introduce fundamental ideas and terminology using intuitive arguments and examples, as we wish only to provide beginning students with a physical feeling for the subject.

Imagine a metal bar without sudden changes (or *discontinuities*) in cross-sectional area or material properties. If this bar (better off, a stretched toy *Slinky*™) is struck at one end with a hammer parallel to the axis, energy propagates along axis through *wave motions* to be felt at the opposite end; multiple reflections may occur, depending on material attenuation and the range of excitation frequencies.

It is important to observe that the material at the struck end is not physically transported to the opposite end: there is no permanent material transport anywhere, although material elements will be displaced temporarily as the wave *propagates* through. This transient displacement is associated with local changes to material velocity, acceleration, and stress.

In fluid mechanics, this is called *sound* or *acoustic wave motion*, as opposed to *wind* or *hydraulic flow*. Wave motions arise from the *compressibility* of the medium. If the opposite end of our struck bar responded immediately, there would be no wave motion, since the *incompressible* medium will have transmitted the excitation instantaneously at infinite speed. Such materials act like infinitely rigid billiard balls: the displacement of any one immediately affects the others.

Compressibility gives rise to *wave propagation* having a distinct speed called the *sound speed*. In the case of *one-dimensional* motions, where the *wavelength* is large compared to a typical cross-sectional length scale, the sound speed $c = \sqrt{(E/\rho)}$, where E is *Young's modulus* and ρ is the *mass density* per unit volume (for acoustic waves in fluids, the *bulk modulus* B replaces E).

Many subtleties arise in wave propagation. For one, the sound speed may not be the only speed with which disturbances propagate in 3D systems (see Chapter 3). The incorrect conclusion may have been reached by those whose only exposure to waves is the classical equation $\partial^2 u/\partial t^2 - c^2 \partial^2 u/\partial x^2 = 0$.

And in physical systems that engineers may *perceive* as one-dimensional, the propagation may in fact be three-dimensional if the wavelengths in question are small compared to typical cross-dimensions. A vigorous "handshake" will excite one-dimensional long waves on a stretched jump rope; however, merely "pinching" the rope will cause little in the way of large-scale wave propagation.

At sudden changes in geometry or material properties, an *incident wave* arriving at the *discontinuity* may partially reflect and partially transmit; if so, we refer to the change as an *impedance mismatch* (such reflections and transmissions need not occur if impedances remain unchanged through the transition). The physical quantities that are *conserved* through the discontinuity depend on the physics of the particular problem.

The drillstring is a *waveguide* supporting many types of vibrations. General waveguide motions represent sums of distinct mode shapes or eigenfunctions, which are either *propagating* or *evanescent* (see Chapter 3).

Different modes contain different amplitude distributions in the cross-space; the lowest (propagating) mode is the constant amplitude plane wave mode.

We showed earlier that *waveguide effects* and *dissipation* may preclude certain waves from propagating; thus, full 3D analysis is not always required, if the end motions of interest are planar 1D waves. Hence, we need not model the *inserts* on a bit; a *boundary condition* accounting for *rock-bit interaction* and its effect on *long waves* suffices, if we are not interested in 3D events near the bit. Wave equation models may be combined with *lumped parameter models* describing small-scale physics in aggregate macroscopic form.

DRILLSTRING VIBRATION MODES

The three primary classes of drillstring motions are *axial*, *torsional*, and *lateral* vibrations. Secondary classifications are often made; for example, *stick-slip* oscillations form one subset of torsional vibrations, while *whirling* and *precessional instabilities* are subsets of lateral vibrations. *Bit bounce* is a severe instance of axial vibration.

Drillstring vibrations can be *fully transient*, or they may be *dynamically steady*. Furthermore, energy transfer between the mean and harmonic components of the strain field of a single axial, torsional, or lateral vibration mode, *and* energy transfer between these latter modes, are *both* possible. Thus, numerous degrees of freedom exist.

Axial vibrations. The vibrations of the bar described above are *axial* or *longitudinal* ones because they occur along the longitudinal axis of the rod. It is not necessary for the rod to be straight. A rod bent in the shape of a pretzel, so long as the "intersections" are restricted from contact, will produce axial vibrations if struck at the ends or excited from within. The mathematics is one-dimensional, utilizing a single coordinate "x" measured along the axis.

The drillpipe system generally contains discontinuities in properties along its length, e.g., cross-sectional area, material density, Young's modulus, sound speed, rectangular and polar moments of inertia, and so on. At these sudden changes, incident waves will partially reflect and transmit, following well-defined conservation laws that are specific to the application.

Other elements that control the vibrations are rig details at the surface, the kinematics of the drillbit used, and rock-bit interaction occurring at the formation interface. Rock-bit interactions are identical in form to *mixed boundary conditions* used by mathematicians; this means that advances in vibration modeling are possible with modest levels of technology transfer.

Transverse vibrations. Axial vibrations are to be contrasted with *transverse* vibrations, e.g., the oscillations seen on jump ropes, violin and guitar strings. Such "strings" are 1D systems that support *tension* only; without tension, they collapse and do not admit wave propagation.

Drillstrings are not "strings" in this sense: they are *beams* that support transverse vibrations even in *compression* because they possess *stiffness*. For more details, refer to the beam references cited in Chapter 3. Transverse vibrations are also known as *lateral* or *bending* vibrations.

The subtleties behind lateral bending are easily visualized in a simple *must* experiment using the long flexible rubber erasers found in draftsmen's tools. Apply an axial load with your fingers, and watch the eraser bend *in-plane*. Next, twist the eraser somewhat, and reapply the same load: now the eraser bends in *two* planes; when the eraser is long, the two bending deflections occur in mutually *orthogonal* (or, perpendicular) planes.

Thus *two coupled modes* of lateral deflection are the rule in practice. As if the physics were not complicated enough, understand that the million-pound, mile-long drillstring never completely rests on the rock bottom; if it did, the formation would crack, and the bit would never rotate with noticeable speed.

Because the drillstring is lifted from the surface, it is in tension there; since it must contact the formation in order to drill, it must be in compression at the bit. Hence, the axial load in general changes from tension at the surface to compression at the bit, the exact point where the sign of the normal stress changes being the *neutral point* (the neutral point is located in the drill collar).

Other neutral point definitions are available, but ours suffices for this book's objectives. Students with *strength of materials* experience understand that beam deflections satisfy a fourth-order differential equation. This contains a second-derivative term related to the superimposed axial load. At the neutral point, its sign changes, leading to a *turning point* problem.

This turning point, due to sign reversals in axial stress, is primarily responsible for violent lateral vibrations near the neutral point, and their failure to be observed and detected from the surface. We will provide an unequivocal proof later in Chapter 6. Understanding the physics, therefore, requires a firm appreciation of the mathematics; both, consequently, are necessary before computer models can be properly formulated or developed.

Torsional vibrations. Torsional or rotational oscillations are exactly what they imply: twists and turns. They are (somewhat) elementary to model, because like axial vibrations, both satisfy the classical wave equation.

While the partial differential *equations* governing axial and torsional vibrations do not couple, dynamic coupling is achieved downhole at the bit through boundary conditions. For example, the *dynamic weight-on-bit* due to longitudinal oscillations will affect the *dynamic torque-on-bit*, and likewise, bit bounce implies simultaneous axial and torsional free ends.

As shown in Chapters 7 and 8, transfer of energy between mean and dynamic components of the torsional strain field can also lead to *stick-slip* vibrations, nonuniform rpm's, and *torque reversals*, their time scales being determined by the dynamic weight-on-bit. Axial and torsional vibrations, through interactions at the bit, mutually affect each other's evolution in time when nonlinear back-interaction through lateral coupling is allowed.

As we demonstrated in our eraser experiment, torsion couples both static bending modes, and likewise, it must couple both lateral vibration modes. Thus, the single beam theory studied in elementary elasticity must be replaced by coupled beam equations, which generally contain variable torques and axial forces. All of the vibrations modes discussed will be coupled and transient.

Whirling vibrations. *Whirling* motions are often confused with torsional vibrations. The confusion arises because whirling vibrations, like torsional ones, cannot exist without established drillstring rotation. By contrast, a drillstring will deform axially and laterally even without rotation, e.g., a case in point is the non-rotating BHA for a *turbodrilled* well.

Whirling motions are a subset of lateral vibrations; these represent rotating displacements *about* the equilibrium axis of the drillstring. The motion of the *drill collars* may vary from simple whirling motion, like that of an unbalanced centrifuge, to the highly irregular motions induced by nonlinear effects such as fluid forces, stabilizer clearance, and borehole-wall contact.

More specialized discussions on whirl appear in Vandiver, Nicholson, and Shyu (1989), and Jansen (1992), who respectively consider axial-lateral wave coupling, and the effects of *eccentricity* of the drill collar's center of mass, fluid damping, stabilizer clearance, plus friction, on *forward* and *backward whirl*.

Coupled axial, torsional, and lateral vibrations. Beginning students often ask, "How are drillstring vibrations different from those of conventional rotating machine shafts?" Let us restate the above observations with slightly different emphasis.

First, machine shafts are *not* loaded by axial stress fields that change signs. Second, the boundary condition at the rock-bit interface introduces a strong dynamic coupling *not* addressed in the interdisciplinary literature. Third, high static torque levels, due to large weight-on-bit, will strongly couple both lateral bending modes. Unsteady torque reversals and stick-slip, in particular, introduce difficulties new to the engineering literature.

Thus, while each of the axial, torsional and lateral vibration modes *in itself* is of *academic* interest, it is the *combined effect* that is of *practical* importance. But because all of the displacement modes interact with each other, it is essential to study each separately, at least initially, in order to ensure physical integrity and to minimize the possibility of formulation errors.

Transient and dynamically steady oscillations. When drillstrings rotate at constant speed, all of the above vibration modes may be excited, leading to superpositions in axial, torsional and bending displacements. If the resultant motion is a *standing wave pattern* that does not change in time, the system is *dynamically steady*, which is *not* to say that the drillstring is stationary.

When each of the vibrations is studied separately, we find that their standing waves contain *nodes* where motion ceases. By contrast, deflections are maximum at *antinodes*, where vibration dampeners might be placed. Understanding the nodal nature of vibrations is important: a sensor for lateral vibrations detection, for example, is useless if it is mounted at a nodal point.

Good engineering skill, of course, is necessary in deciphering the many complications that are likely downhole. For instance, a collar-mounted strain gage will measure elongations due to axial, torsional, and lateral vibrations, not to mention those due to thermal effects. Positioning at nodes and antinodes will affect signal-to-noise ratio, and hence, measurement integrity.

Motions that are not *dynamically steady* are *transient*: they track irregularly with time. This is so when bits bounce, when drillstrings *make hole* too fast, and on startup of rotary drilling. Bit bounce, for example, induces surges in drillpipe mud pressure, causing MWD signal detection problems.

Transient loads may be destructive, leading to drillpipe fatigue and wear. When their accelerations are large and occur over extremely short time periods, the loads are called *shocks*. Shocks are measured in *g's* together with *milliseconds*; very often, they are also characterized by *shock frequency*. In qualifying MWD tools for field use, shock levels and frequencies must be specified together with the manner in which the tool is shock-tested, that is, its *drop axis* or *axes*.

Understanding the environment. Whether our objective is designing a better BHA for directional use, or drilling with less likelihood of fatigue failure, or making hole faster, it is important to understand what the downhole environment is and what its consequences might be.

For example, the failure of a real-time MWD logging tool due to bounce on a flatbed truck may not presage downhole catastrophe, since the failure modes may be completely different; one might ask, rightly so, *what* the roles of borehole constraints are and *how* drillbit motions excite the system.

In designing MWD tools, for instance, vibration table testing is of little value unless the actual conditions and shock loadings encountered downhole while drilling are simulated. Similar comments apply to computer models and numerical simulators.

Before we discuss the mathematics in detail, it is useful to define the most basic time and space scales. Consider a drillstring rotating at 1 rpm; its rotary speed is therefore 1 revolution per 60 seconds, or 1/60 *th* Hz. If it rotates at N_{rpm} and drills with a *tri*cone bit, the axial excitation frequency takes on the value $f = 3N_{rpm}/60$ Hz or $N_{rpm}/20$ Hz. Thus, for $N_{rpm} = 60$, we have f = 3 Hz; higher frequencies are possible depending on the construction of the bit.

The wavelength λ associated with a frequency f and a sound speed c is $\lambda = c/f$. If c = 15,000 ft/sec, and f = 3 Hz, then λ = 5,000 ft. Since real drillstring lengths range from a significant fraction to many multiples of λ, *wave equation* (as opposed to *lumped mass*) *modeling* is appropriate.

It is difficult, and even dangerous, to assign "ballpark numbers" to different modes of downhole vibration, since much depends on the BHA used, the formation being drilled, the drillbit design, and the degree of cone wear (e.g., the average human weight is 150 lbf, *but* the range covers 1- 1,500 lbf!). We therefore avoid the temptation to quote "real numbers".

LONG-STANDING VIBRATIONS ISSUES

We have described the vibration modes of drilling interest, but before we pursue the subject of modeling, it is helpful to consider *the big picture* and discuss the long-standing issues in vibrations analysis. Philosophically, one always learns and executes more effectively *if* real world problems are posed.

Example 4-1: Case of the missing waves. It is well known that far downhole, severe lateral vibrations are the predominant reason for drillstring failure; yet it is generally acknowledged that bending oscillations cannot be detected at the surface.

Lateral vibrations with bending moments *routinely* exceeding 20,000 ft-lbfs, for example, have been measured in drill collars. However, simultaneous surface measurements for axial and torsional waves, which propagate effortlessly, provide few clues as to their occurrence. These observations, true even in vertical holes where borehole wall damping is likely to be minimal, seem contradictory and mutually exclusive. Where, if any, is the fallacy?

Example 4-2: Looking for resonance in all the wrong places. In vibrations modeling, researchers assume simplified axial, torsional, and lateral mode boundary conditions, plus sinusoidal excitations where appropriate, in order to study *normal modes* and *system resonances.* Workers tend to force-fit drillstring models to classical solutions.

Yet, many drillers have *attempted* to drill at resonance, in order to make hole faster, or simply to evaluate theoretical predictions, without any success (Dareing, 1972). Recent studies (cited in later chapters) even question the applicability of classical normal mode analysis in real world drilling.

Example 4-3: Drillstrings that don't drill. The common use of sinusoidal bit displacement boundary conditions, which bears superficial semblance to general Fourier analysis, *requires* that the bit remain stationary on the average.

Thus, it does *not* make hole, it does *not* bounce, and it generally does *not* simulate reality. How can we use rock-bit interaction models obtained in the laboratory, drillbit kinematics and rotary speed, and bottomhole assembly layout, to predict rate-of-penetration, bit bounce, and real vibration instabilities?

Example 4-4: Modeling coupled vibrations. Axial, torsional, and lateral vibration modes are highly coupled and transient, yet no published model has addressed the *physics* of the interaction process, let alone develop a model for the full problem. In Chapters 5-8, we will formulate and test *each* of the axial, torsional, and lateral building blocks *individually*, and *then* integrate them in numerical fashion, following physical arguments that make physical sense.

Example 4-5: Energy transfer mechanisms. Recent workers in drillstring vibrations have recognized the need to have continual *dynamic* energy transfer among axial, torsional, and bending modes of vibration, with the complete system viewed as the entity of physical interest.

This objective is only partially complete: *energy transfer between mean and dynamic stresses must be modeled.* Consider torsional stresses: at the start of drilling, *torque-up* merely *winds* the pipe, increasing its mean strain energy. Once rotation begins, torque oscillations propagate up and down the drillstring, and if bit friction cannot maintain this *status quo*, torque reversals are likely, which significantly affect the mean strain field.

This is also the case, although less obvious, with axial vibrations. The mean *DC* field is usually ignored, but it turns out that dynamic *AC* results *can* effect changes to *DC* displacement levels and vice-versa, for certain types of rock-bit interaction. *Making hole* requires, therefore, a strong coupling between mean and dynamic stress fields, one that is traditionally ignored.

PRACTICAL APPLICATIONS

The basic objectives in vibrations research, and particularly for this book, are easily stated: explain the *why's* and *how's* behind well known drilling mysteries, couple all vibration modes through physically consistent formulations, and produce efficient simulation algorithms for operational use.

An understanding of the physics, plus appropriate explanations using sound mathematics, constitute the prime motivation for Part II. Still, why emphasize *fundamentals* in a lean industry known for its practical orientation? These anecdotal (but true) stories will illustrate the point.

Anecdotal stories. In 1988, just having uncovered the proof indicating natural tendencies for catastrophic bending fatigue at the neutral point, a failure mechanism known to drillers, this author met with researchers from a major oil company that had built at great expense a *supercomputer* simulator.

The authors, noting that computational instabilities tended to form at the neutral point, "successfully" suppressed their occurrence by reformulating the finite element program! The stable algorithm evolving out of their efforts, while it ran every time, was obviously useless to the funding company.

In another instance, a structures engineer, well versed in *running* finite element codes, was now promoted to a new position which required *writing* software. His algorithm refused to converge; and when it did, it generated mysterious solutions.

The real problem, due to incorrect boundary conditions, and hence, ill-conditioned matrixes, escaped unnoticed; instead, the errors were removed by unrealistically increasing drillstring length, a "proprietary secret" which only reduced the severity of the discrepancies.

Applications to the field. Behind each equation and boundary condition in this book stand three objectives: rigorous mathematics, stable algorithms, and uncompromised modeling of reality. In order to avoid the above pitfalls, the physics will always drive analysis, without exception.

Drillstring vibration, in the final analysis, finds no shortage of applications. We will cite some of these, although we have not attempted to compile a comprehensive directory of subjects. Among the more obvious are those described below.

Structural damage. Of first and foremost importance are costly disasters due to drillpipe breakage, twistoffs, fishing jobs, fatigue damage, lost or broken cones, PDC bit cutter damage, and casing wear. Understanding vibrations can help in drilling efficiency and safety.

Vibration affects roller cone performance and bit bearing life. Dareing and Radzimovsky (1965) note that dynamic forces play appreciable roles in bearing life, affecting sealed and unsealed bearings; large dynamic forces appear to be more damaging than static loads.

Polycrystalline diamond compact (PDC) bit damage results from backward whirling. When a bit whirls, its instantaneous center moves about the bit face, rather than around its true center as with smooth rotation. Cutter paths are not circular: they strike rock at off-design angles, leading to bit failure in the form of spalling and chipping.

Formation damage. Wave propagation figures importantly in formation damage. For example, swab-surge may cause borehole instability in the annulus, precipitated by sudden drillpipe movement. Also, formations can be fractured by improper drilling practices; high dynamic bit loads may fracture rock and result in lost circulation.

Directional drilling. Directional control and bit walk are crucial to directional drilling. In the past, drillers used static loads only, but industry consensus has taken the position that dynamic forces can play important roles. Calculations should be directed at moments, torques, and forces *at the bit*; empirical information on doglegs and rock-bit interaction should be related to boundary condition modeling as it affects directional control.

Increasing rate of penetration. Rate-of-penetration is related to drillstring dynamics, e.g., identical bits drilling identical formations will perform differently with different BHAs. Early workers noted the potential of drillstring vibrations in increasing penetration rate, studying force versus displacement phase relations at the rock-bit interface (Dareing, 1984, 1985).

Using an energy balance, we will show that power input at the bit is controlled by AE $u_x u_t(0,t)$ where A is the cross-sectional area, E is Young's modulus, and u(x,t) is the axial displacement. Calculated examples using postulated rock-bit mechanical impedance conditions are discussed.

Improved MWD tools and mud motors. MWD tools and mud motors are now designed to operate at higher temperatures and pressures in more corrosive environments. Because they run in longer and deeper wells, and necessarily undergo reduced maintenance, they need to be more reliable and sturdy.

Operating in deviated and horizontal wells, these smaller tools undergo higher build rates and experience higher torques and forces; also, the advent of steerable drilling systems means increased bending stresses and bearing loads. Advances in materials must accompany our improving knowledge of the vibration environment.

While building angle, drill bits will slip, bounce, jolt, and clog. These uncontrollable events, plus shock sub noise, will cause pressure surges that create signal *synchronization* problems with received MWD signals at the surface. Understanding bit and mud column dynamics will assist in obtaining uninterrupted real-time MWD information.

We indicated that bit dynamics affects bearing life and related operational issues (Fear, Thorogood, Whelehan, and Williamson (1992) discuss drilling economics, cost and risk analysis models). Recent developments in real-time measurement can reduce these risks; Falconer, Burgess, and Wolfenberger (1986) show that tooth wear and bearing failure can be diagnosed by MWD downhole-weight-on-bit and torque sensors.

Understanding downhole vibrations helps us design better sensors and interpretation models. For example, what proportions of a measured strain are due to axial versus lateral flexure? Where should different sensor types be ideally placed and mounted? It goes without saying that improved MWD designs and improved understanding of the environment go hand-in-hand.

Formation imaging. Inferring medium properties from incident and scattered waves is standard in ultrasonic, medical, and X-ray imaging. Compton's classical experiments in scattering, for example, contributed greatly to our knowledge of atomic structure. Our 1D drillstring is geometrically simple; it does, though, support several classes of waves, but their properties are different enough that undesired waves may be filtered. Imaging is one area where technology transfer can reap significant gains.

Psychological discomfort. Many will acknowledge the irritation that loose change in clothes dryers causes. Imagine what visions that rattling drillpipes must evoke in terms of potential blowouts, lost time and money, not to mention psychological discomfort!

This factor appears on few lists of worrisome factors, but history has shown that as technology advances, environmental demands increase; this was so in jet engine "noise abatement" design, intended to lessen "noise pollution" near airport communities!

ELASTIC LINE MODEL OF THE DRILLSTRING

Attempts at developing simulators modeling *both* drillstring dynamics and formation response have been made. For example, Eronini, Somerton, and Auslander (1982) describe an early dynamic model for rotary drilling, using simple *lumped parameter ordinary differential equation* analysis, as opposed to *distributed parameter partial differential equation* analysis.

Early efforts. The GEODYN program in Baird, Caskey, Tinianow, and Stone (1984) is another effort. This finite element model simulated the 3D transient response of a PDC bit interacting with a nonuniform formation. Modeling parameters included hole size, bottom surface shapes, and rock nonuniformities.

Ultimately, the detailed response of the BHA and the entire drillstring would be considered; areas for development included modeling rock penetration, bit torque, side forces, directional tendencies. Baird, Caskey, Wormley, and Stone (1985) subsequently considered BHA behavior due to PDC bits; inputs included stabilizer, bit cutter, and formation models.

Although the effort undertaken in such analyses is credible, there are disadvantages to this line of attack. Computer times required for stand-alone runs were approximately one *hour* on a Cray supercomputer. Significant labor and facilities resources are needed, rendering any product beyond the reach of many drillers. *In this book, reliable methods requiring seconds and minutes on 486 class PCs are developed.*

Elastic line simplifications. Despite the apparent generality of the cited work, all of the 3D details in drillstring simulation need not be treated equally. In Chapter 3, we showed that fine scale phenomena in 3D waveguides (*here*, the drillstring and BHA) are likely to decay without ever seeing propagation as a plane wave.

While drillstrings undergo extremely complicated, dynamically coupled vibrations, their geometries are fortunately simple because they are much longer than typical diameters. As most wavelengths are likewise large, we are additionally justified in modeling the drillstring as a lineal system.

Thus, modeling the inserts on a tricone bit is unnecessary if the important large-scale effects along the drillstring are alone sought. The vertical response of a bit on a given formation, determined from vibration-isolated lab tests, such as those published by Amoco, suffice for a macroscopic description of the rock-bit interaction. In fact, we follow this approach.

We will model axial, torsional, and lateral components as vibrations of an idealized *elastic line*, where the only spatial independent variable is the length coordinate x. This is *not* a single degree-of-freedom system: we allow static and dynamic superpositions of axial, torsional, and two modes of lateral vibrations, everywhere along our elastic line.

Cross-sectional details appear by way of properties such as area, rectangular and polar moments of inertia, while lengthwise variations are easily incorporated by allowing explicit dependences on x. Of course, the correct conservation form of the differential equations must be used for *gradual* changes; *sudden* variations are modeled using global matching conditions that correctly fix reflection and transmission coefficients.

This elastic line approach will be generalized to include nonlinear coupling between axial and transverse vibrations, while retaining the simplification that planar cross-sections remain so upon bending. We also introduce nonlinear boundary conditions for wall contact and rock-bit interaction simulation.

Historical precedents. This approach is not new; for example, this philosophy is used in the vibration modeling of machine shaft oscillations. The origin behind elastic line models can be found in the classic work of Love (1944). Applications appear the petroleum industry, in offshore engineering problems involving pipelines, risers, and mooring lines.

For example, Nordgren (1974) formulated a nonlinear, three-dimensional, large-amplitude, transient problem, and solved it by finite differences. Garrett (1982) gave a finite element model that modeled large amplitude deflections, finite rotations, and tension variation along the length.

Walker and Friedman (1977) proposed, for directional drilling applications, a *static* mathematical model of the bottomhole assembly that handles arbitrary hole inclination, geometrical and material properties. We will develop the dynamic extension to their work.

We emphasize that modern vibrations, to include its sophisticated mathematics, is well developed; however, the applications are acoustical and aerospace oriented. Classic rod and beam papers, for example, include Clark and Reissner (1951), Abramson, Plass, and Ripperger (1958), Eisner (1964), and Clough and Penzien (1975).

Our focus. The above works cannot be reformulated to deal with oilfield boundary conditions, e.g., rock-bit interactions, drill bit modeling, and penetration rate analysis. In Part II, ideas from acoustic and mechanical impedance modeling, and earthquake seismology, are introduced, and combined with rod and beam theory, to model real drillstring motions.

In our discussions of elastic line modeling, we focus on integral elements of the drilling process. For example, in axial vibrations, we separately study the surface system, the drillbit kinematics, the rock-bit interaction, and reflections at area discontinuities, in this one-dimensional limit.

The same approach applies to torsional oscillations. While torsional and axial *wave equations* remain independent of each other, dynamic coupling is achieved at the bit by means of boundary conditions. When the bit bounces, a torsional free end exists; when it drills ahead, the dynamic torque-on-bit is proportional to the dynamic weight-on-bit as calculated above.

Axial and torsional calculations are taken beyond conventional analysis by allowing energy transfer between mean and dynamic states. Thus, increases in torsional strain energy due to initial wind-up, its subsequent release to drive dynamic rotations, plus bit friction effects, will manifest themselves through calculated results that display axial bit bounce and torsional stick-slip.

Once distributed axial and torsional loads are available at any instant, they are used two couple lateral vibration modes. We give a stable numerical algorithm which solves the complete problem; in an illustrative calculation, we demonstrate how the simultaneous effects of drillstring precession, stick-slip, bit bounce, and nonzero rate-of-penetration can be realistically modeled.

Numerical models, being computational, do not shed insight into physical processes. Whenever possible, analytical solutions to simpler problems are given, which highlight the character of the equations. The extent to which these solutions are consistent with the proposed physical model determines the utility of the mathematical model.

OBJECTIVES AND DISCUSSION PLAN

Researchers traditionally *force-fit* axial, torsional, and lateral vibration problems into simple categories treated in elasticity books. Early papers applied rod and beam models to downhole vibrations, examining classical resonances and mode shapes, but recent MWD data, suggests that downhole phenomena are more complicated dynamically than simple models suggest.

As we have indicated, researchers have attempted to drill at the calculated classical resonances with little success; they have tried to relate violent vibrations to the same resonances, but little evidence supports the idea that natural frequencies seen in field bear any relation to existing theory.

Work in wave propagation and the stability of dynamical systems has proceeded rapidly in other disciplines, especially in hydrodynamic stability (e.g., see Chapter 3), and workers now concede that instabilities, being many-faceted, must not be restricted to limiting classical resonances.

In Chapter 6, we reevaluate the meaning of stability, develop a physical model that explains the mysteries of violent lateral vibrations, and integrate inter-disciplinary ideas into drillstring modeling. We combine axial, torsional, and lateral vibrations with insight acquired through practical experience.

In addition to the work cited earlier, we will survey the literature, consolidating only those relevant papers that address important unifying themes. We will cite, though, only a portion of the literature if only to give a *flavor* of drilling problems: our review is *not* a detailed literature survey.

We mainly stress physical ideas and fundamental modeling concepts, some introduced long ago, having stood the test of time and the benefits of hindsight. Thus, Part II is not about directional drilling, nor is it a "how-to" manual on damage avoidance; we will not catalogue the voluminous experimental data published in recent years.

We will, in our presentations, emphasize *time* rather than *frequency domain* analysis, since drillstring vibrations appear to be highly transient. The final *algorithms* cover, of course, dynamically steady oscillations; if such exist under sinusoidal excitation, our computational procedures *will* model them.

But we will not attempt to generalize, understanding well that exact details depend on the bottomhole assembly, bit wear and design, and formation properties. To keep our discussions fundamental and introductory, we will not translate our findings into stabilizers, reamers, drilling operations procedures, build and walk rates, or inclinations, directions, and depths. Nor will we design shock subs and improved drill bits.

On the modeling side, our mathematics is rigorous and readable, covering only those points needed to explain the application. A background in calculus, plus exposure to differential equations is helpful. Numerical methods are similarly developed so that their limitations, strengths, and conceptual foundation are understood. We will not dwell on integration rules, or improved spline-fit algorithms. Rather, we demonstrate how the appropriate equations are physically correct, and how they are differenced, programmed, and solved.

Once the formalism for vertical wells is successfully tested against empirical observation, qualitatively anyway, the required modifications extending the framework to deviated and horizontal wells are presented. No claim is made to have "solved" the general problem.

In fact, our limited objective aims at providing only the analytical and computational tools needed to model new phenomena and evolving concepts. This approach is necessarily so with a subject that, despite its long-standing interest, is nonetheless seeing rapid development.

Unlike the author's books, *Borehole Flow Modeling in Horizontal, Deviated, and Vertical Wells* dealing with a *narrow* specialty on eccentric annuli and hole-cleaning, and *Modern Reservoir Flow and Well Transient Analysis*, describing a *mature* but broad subject, the final word is not yet in (see Chin, 1992, 1993).

Drillstring vibration, as presently envisioned, draws from numerous disciplines: mechanical vibrations, partial differential equations, earthquake seismology, numerical analysis, rock mechanics, acoustics, and evolving empirical knowledge available from MWD measurements taken near the bit.

Drillstring vibration modeling, in conclusion, is an *integrated* science, and hence, from it derives the personal challenge that I have undertaken: a readable, usable, and down-to-earth volume that brings state-of-the-art physical, mathematical, and numerical ideas to practical engineers, without emphasizing prerequisites, formality, or academic credentials.

REFERENCES

Abramson, H.N., Plass, H.J., and Ripperger, E.A., "Stress Wave Propagation in Rods and Beams," in *Advances in Applied Mechanics, Vol. V*, edited by Hugh Dryden and Theordore von Karman, Academic Press, New York, 1958.

Baird, J.A., Caskey, B.C., Tinianow, M.A., and Stone, C.M., "GEODYN: A Geological Formation/Drillstring Dynamics Computer Program," SPE Paper No. 13023, presented at the 59th Annual Technical Conference and Exhibition, Houston, Texas, Sept. 16-19, 1984.

Baird, J.A., Caskey, B.C., Wormley, D.N., and Stone, C.M., "GEODYN2: A Bottomhole Assembly/Geological Formation Dynamic Interaction Computer Program," SPE Paper No. 14328, presented at the 60th Annual Technical Conference and Exhibition of the Society of Petroleum Engineers, Las Vegas, Nevada, Sept. 22-25, 1985.

Chin, W.C., *Borehole Flow Modeling in Horizontal, Deviated, and Vertical Wells*, Gulf Publishing, Houston, 1992.

Chin, W.C., *Modern Reservoir Flow and Well Transient Analysis*, Gulf Publishing, Houston, 1993.

Clark, R.A., and Reissner, E., "Bending of Curved Tubes," in *Advances in Applied Mechanics, Vol. II*, edited by Richard von Mises and Theordore von Karman, Academic Press, New York, 1951.

Clough, R.W., and Penzien, J., *Dynamics of Structures*, McGraw-Hill, New York, 1975.

Dareing, D.W., "Vibrational Forces and Motions in Rotary Drill Strings," presented at the 1972 Rotary Drilling Conference, International Association of Drilling Contractors, Dallas, Texas, Feb. 23-25, 1972.

Dareing, D.W., "Vibrations Increase Available Power at the Bit," Oil & Gas Journal, March 5, 1984, pp. 91-96.

Dareing, D.W., "Vibrations Increase Available Power at the Bit," Journal of Energy Resources Technology, Transactions of the ASME, March 1985, pp. 138-141.

Dareing, D.W., and Radzimovsky, E.I., "Effect of Dynamic Bit Forces on Bit Bearing Life," Society of Petroleum Engineers Journal, Dec. 1965, pp. 272-276.

Eisner, E., "The Design of Resonant Vibrators," in Physical Acoustics, Vol. 1, Part B, edited by W.P. Mason, Academic Press, New York, 1964.

Eronini, I.E., Somerton, W.H., and Auslander, D.M., "A Dynamic Model for Rotary Rock Drilling," Journal of Energy Resources Technology, Transactions of the ASME, June 1982, pp. 108-119.

Falconer, I.G., Burgess, T.M., and Wolfenberger, E., "MWD Interpretation Tracks Bit Wear," Oil & Gas Journal, Feb. 10, 1986, pp. 55-59.

Fear, M.J., Thorogood, J.L., Whelehan, O.P., and Williamson, H.S., "Optimization of Rock-Bit Life Based on Bearing Failure Criteria," SPE Drilling Engineering Journal, Sept. 1992, pp. 163-166.

Garrett, D.L., "Dynamic Analysis of Slender Rods," Journal of Energy Resources Technology, Transactions of the ASME, Dec. 1982, pp. 302-306.

Jansen, J.D., "Whirl and Chaotic Motion of Stabilized Drill Collars," SPE Drilling Engineering Journal, June 1992, pp. 107-114.

Love, A.E.H., *A Treatise on the Mathematical Theory of Elasticity*, Dover Publications, New York, 1944.

Nordgren, R.P., "On Computation of the Motion of Elastic Rods," Journal of Applied Mechanics, Transactions of the ASME, Sept. 1974, pp. 777-780

Vandiver, J.K., Nicholson, J.W., and Shyu, R.J., "Case Studies of the Bending Vibration and Whirling Motion of Drill Collars," SPE/IADC Paper No. 18652, presented at the 1989 SPE/IADC Drilling Conference, New Orleans, Louisiana, February 28-March 3, 1989.

Walker, B.H., and Friedman, M.B., "Three-Dimensional Force and Deflection Analysis of a Variable Cross-Section Drill String," Journal of Pressure Vessel Technology, May 1977, pp. 367-373.

5
Axial Vibrations

Axial or *longitudinal* vibrations, terms used interchangeably, result in displacements parallel to the drillstring axis. In this chapter, we formulate the general boundary value problem for longitudinal vibrations, to include drillbit kinematics, time-varying rate-of-penetration, and rock-bit interaction. But first, we review existing models, and explain their strengths and limitations.

Axial vibrations and stresses are considered first for several reasons. For one, their solutions draw upon the classical wave equation analyses given in Chapter 1, but some inadequacies of these techniques motivate the development of new mathematical methods in a natural setting.

Second, an understanding of axial vibrations is essential to properly modeling lateral vibrations, and in particular, *why* the latter can be devastating downhole, yet undetectable at the surface. Often, nonlinearities are blamed for these violent instabilities, while their mysterious disappearance is attributed to borehole wall contact.

In Chapter 6, the mystery behind lateral vibrations is solved, and a single linear model resolving both of the foregoing field observations is given, which yields to analytical solution via *kinematic wave theory* (see Chapter 2). Then, a nonlinear model which couples both axial and lateral vibrations is formulated. Numerical finite difference algorithms for axial and lateral vibrations are given in both chapters which simulate fully transient motions.

PIONEERING AXIAL VIBRATION STUDIES

With three decades of hindsight, it is correct to say that many early findings on vibration control are over-simplified. Nonetheless, several studies remain of lasting value for their key physical insight. One such work is the paper by Bailey and Finnie (1960).

They modeled axial vibrations, and their natural frequencies and resonances, using the linear partial differential equation governing the undamped longitudinal oscillations of a bar. At the surface, a mass-spring (but damper*less*) boundary condition was postulated, where the mass of the swivel, traveling block, and kelly was viewed as a *concentrated mass* M

The cables and derrick were treated as a spring, and simple experiments were suggested to determine its stiffness. The spring constant was inferred by observing the free oscillations of the system, once it had been set into motion by various means. Displacements were assumed to be zero at the drillbit.

The longitudinal bar was formed by the drillpipe and collar. At their interface, continuity of displacement and force were assumed in order to connect the sinusoidal wave solutions obtained for each side. These, taking an unwieldy "$(A \sin \omega x/c + B \cos \omega x/c)(D \sin \omega t + E \cos \omega t)$" form, were further related by a transcendental equation which required graphical solution.

A closely related torsional vibration analysis, which satisfied similar governing equations, was also undertaken by the authors. They suggested that the surface boundary condition might be taken as a fixed end, treating the bottom as more or less free. At the drill pipe and collar interface, the authors invoked continuity of angular displacement and torque.

Finnie and Bailey (1960) described experimental techniques developed for measuring time-dependent force, torque, axial and rotational motions at the top of a drillstring. Some observations agreed with theoretical predictions, but many results did not.

Paslay and Bogy (1963) studied the axial vibration nature of intermittent bit teeth and bottomhole contact, showing that appreciable bit load variations may be detected by *surface* instruments; they also demonstrated that downhole excitations may induce oscillating bit forces with amplitudes of the same order as those of the static bit load.

Bailey and Finnie (1960) assumed an undamped drillstring to simplify the mathematics. Later, Angona (1965) studied wave attenuation in drillstrings by measuring the decay of normal stress pulses propagating along them. These were created by falling weights which struck the top end of the string. Strain gauges measured pulse amplitudes as they executed multiple reflections.

Angona obtained the relationship between attenuation and frequency by Fourier analyzing the pulse signal. Typically, three round trips were possible before complete attenuation, indicating that the system was lengthy enough to be regarded as wave-like, as opposed to being a lumped mass. The local decay, taking an "$e^{-\alpha x}$" character, was weakly exponential with space; this demonstrates, following Chapter 1, that linear modeling sufficed.

This early paper importantly showed that the wave form of the pulse changes very little with distance, that is, that the wave propagation was nondispersive; again, the linear theory afforded by the classical wave equation appeared to be valid. Finally, the author suggested drilling methods driven by surface vibrators, which took advantage of controlled resonant interactions in order to increase rate-of-penetration. The design of resonant vibrators, incidentally, is given in the summary paper of Eisner (1964).

An important suite of vibration data, collected by research workers at (then) Esso Production Research Company over a two year span, was published three years later in the late 1960s (e.g., see Koch (1967), and Deily, Dareing, Paff, Ortloff, and Lynn (1967)).

"Downhole Measurements of Drill String Forces and Motions," by Deily, Dareing, Paff, Ortloff, and Lynn (1968), provided the first detailed description of the real environment downhole, pointing to the significance of dynamic behaviour in drilling operations.

The paper described a self-contained, specially-designed downhole recording instrument that was used to measure and capture drillstring forces, moments, and motions at the source. Eight signals were recorded by "pulse-width modulation" on magnetic tape, namely, axial, torsional, and bending loads, axial, angular, and lateral accelerations; and internal pipe and external annular pressure.

The hardened mechanical device was used over several years to collect downhole data, under a wide range of field drilling conditions, for numerous wells. A complete dataset would collect approximately ten minutes of cumulative recordings, at which point the tool would be retrieved and returned to the surface. There, the data was converted from magnetic tape format to one suitable for oscilloscope display (selected portions of the dataset were digitized for analysis purposes, and studied in the cited paper).

The authors noted that variations in measured downhole bit load ranged from 25 to 50 percent of the mean value, indicating that weight-on-bit was by no means unchanging. Maximum bit loads reached 3.5 times their mean values at times, accompanied by large annular pressure variations. Frequencies of weight, torque and bending traces showed evidences of rock bit tooth action, of cone action, of rotation, and pump pulsations.

The paper also documented cases of *bit bounce*, provided exact measured magnitudes for a suite of forces and accelerations, and demonstrated the coupling between bit weight and torque. This dynamic information, plus details on the mechanical construction of the downhole recorder, would no doubt be of value to MWD designers still a decade away.

Cunningham (1968), studying the Esso data further, noted irregular bit rotations, torque oscillations, periodic beating, bit bouncing, bit weight fluctuations, mud pump pressure fluctuations, and large losses of rotary horsepower along the drill stem.

The model of Bailey and Finnie (1960), which assumed an undamped mass-spring surface system and an undamped longitudinally vibrating rod, and Angona (1965), who experimentally characterized the dissipative nature of the drillstring, would converge with the work of Dareing and Livesay (1968).

This paper solved a more realistic vibrations model satisfying the damped wave equation $\rho\, \partial^2u/\partial t^2 + \zeta\, \partial u/\partial t - E\, \partial^2u/\partial x^2 = 0$, which replaced the simpler $\rho\, \partial^2u/\partial t^2 - E\, \partial^2u/\partial x^2 = 0$ model. Mass-spring surface boundary conditions were again assumed to be undamped; also, matching conditions again assumed continuity of displacement and force at the pipe-to-collar interface.

The reader is encouraged to study the above references, since they emphasize the *fundamental physics* of downhole vibrations. This background is important to understanding modern papers, which stress real-time data acquisition methods, and more complicated phenomena such as *stress-reversals*, *stick-slip* torsional oscillations, and *bit whirl*.

GOVERNING DIFFERENTIAL EQUATION

Axial vibrations are significant because drillstrings *make hole* parallel to the drillstring axis. Many drilling related results appear in the literature on resonant vibrators, pneumatic jackhammers, and area-contoured wave amplification devices; however, none are relevant to modeling rotary drilling.

Here, the fundamental governing equations are given; recent literature is reviewed in the context of specific issues. Key ideas introduced in other technical disciplines will be adapted for use in our work; these include ideas in earthquake seismology and physical acoustics. Then, applying methods from finite difference analysis, the general formulation is solved numerically.

Damped wave equation. To introduce the basic concepts, let us consider the classical equation for longitudinal or axial vibrations

$$\rho\, \partial^2u/\partial t^2 + \zeta\, \partial u/\partial t - E\, \partial^2u/\partial x^2 + \rho g = F_e(x,t,u,\partial u/\partial x,\partial u/\partial t) \qquad (5.1)$$

where ρ is the *mass density per unit volume*, ζ is the damping factor, E is Young's modulus, and g is the acceleration due to gravity. In this equation, u(x,t) is the *longitudinal displacement* of a material element from its equilibrium position at x for a particular time t.

Note that Dareing and Livesay (1968) employ the form $\rho\ \partial^2 u/\partial t^2 + \gamma\ \partial u/\partial t - AE\ \partial^2 u/\partial x^2 + \rho g = 0$ where ρ is mass per *unit length*; thus, the two damping factors are related by $A\zeta = \gamma$, with ours being $\zeta = \gamma/A$, where A is the cross-sectional area. The *vertical* drillstring is schematically shown in Figure 5-1, where x = 0 describes the formation contact point, with x increasing upwards.

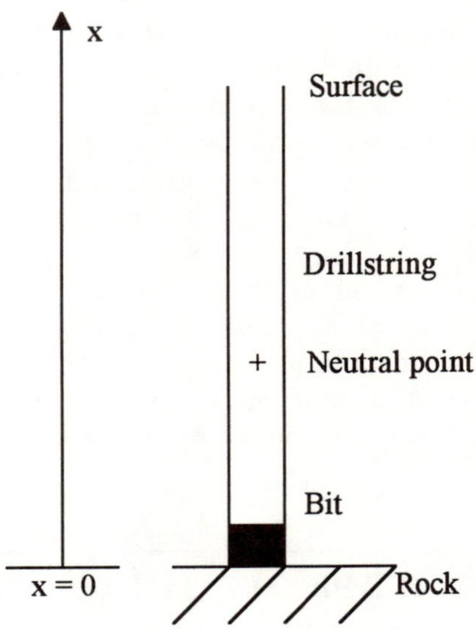

Figure 5-1. Drillstring coordinate system.

Equation 5-1 follows from Newton's "**F=ma**", which relates the resultant acceleration to the applied external force. The "ma" is just "$\rho\ \partial^2 u/\partial t^2$ " on a per unit volume basis, while the "$E\ \partial^2 u/\partial x^2$" arises from the difference between the normal stress $E\ \partial u/\partial x$ acting on one face of the control mass, and $E\ \partial u/\partial x + \partial(E\ \partial u/\partial x)/\partial x$ acting on the opposite. Their difference leads to $\partial(E\ \partial u/\partial x)/\partial x$ or $E\ \partial^2 u/\partial x^2$.

The "$\zeta\ \partial u/\partial t$" refers to viscous dissipation (we'll demonstrate this later), while "ρg" refers to static weight. Our use of "x" does not require a straight hole, but drillstrings in deviated wells should be modeled using only that component of weight along the borehole axis. We will elaborate on this in Chapter 8.

For now, the drillstring is assumed to be uniform, without any changes in properties in the x direction, although these restrictions will be relaxed later. Thus, the properties ρ, ζ, E, and A are *constants*. Later, we will discuss *impedance matching conditions* that must be enforced at *discontinuites*, and develop numerical models that simulate *partial reflections and transmissions*.

Note that A represents the cross-sectional area of the drillstring, that is, the area occupied by actual metal if the pipe is hollow, or the entire area if the drillstring is an idealized solid rod. To keep the discussion simple, we will not consider the effects of the mud inside or outside of the drillpipe, nor will we consider the effects of buoyancy.

To obtain "typical" strain estimates, consider a drillstring with cross-sectional area $A = \pi/4\ (4^2 - 3.34^2) \approx 4\ \text{in}^2$ and $E = 30 \times 10^6$ psi, so that $AE = 120 \times 10^6$ lbf. If the F in "$F = AE\ \partial u/\partial x$" were 30,000 lbf, then $\partial u/\partial x = 0.0003$ approximately. We might speculate that $\partial u/\partial x \approx O(10^{-4})$, give or take an order of magnitude, depending on static and dynamic conditions.

External forces and displacement sources. We have also introduced an axial force excitation $F_e(x,t,u,\partial u/\partial x,\partial u/\partial t)$ which may depend linearly or nonlinearly on the displacement function and its derivatives. This force may be concentrated, e.g., at mild *dogleg*; or, it may be distributed, e.g., when *continuous wall friction* exists. For brevity, our "F_e" designation will represent both of these effects, plus others to be discussed, collectively.

It turns out that F_e will play more than its conventional role in describing *force* excitation. We introduce its use as a *couple* internal to the drillstring, following the ideas developed in Chapter 1, to create *displacement source* effects that model the up-and-down kinematics of rotating tricone bits.

This displacement source completely describes the *3 x rpm* motions imposed by the rotary table, freeing up the *boundary condition* (applied at x = 0 by early authors) so that *rock-bit interactions* (a.k.a., *mechanical impedance boundary conditions*) can be prescribed there instead.

Dynamic and static solutions. Equation 5-1, we emphasize, models the *complete dynamic and static longitudinal fields* for the *total displacement* u(x,t). We will also refer, from time to time, to *dynamic* and *static* as *AC* and *DC* when analogies to electric circuits are suggested.

In mechanical engineering analyses, these AC and DC component fields are (correctly) modeled separately, as we will demonstrate, and subsequently superimposed. But in drilling vibrations where rock-bit interactions are allowed, the two dynamically interact through non-standard boundary conditions, and the advantages of any AC/DC separation are lost.

For example, AC motions will produce changes in DC level; thus, *the need to combine both displacements into a single unambiguous variable is seen*. This is the case when we model transient shocks, bit-bounce, and rate-of-penetration, and especially so, when nonlinear back-interaction effects due to high amplitude lateral vibrations are considered.

Free-fall as a special solution. Our sign convention for the gravity term is chosen consistently with the *free-fall* limit: any particular "dx slice" satisfies "$\rho\ \partial^2u/\partial t^2 = -\ \rho g$", which leads to "$u = -1/2\ gt^2$" if the remaining formulation terms were absent. This situation represents a drillstring dropping unimpeded, in the limit of vanishing formation support.

The free-fall limit is used as a check on the numerical scheme, to ensure that the drillstring drills ahead (or, falls) under obvious conditions. In modeling rate-of-penetration, the sign convention $\partial u/\partial t < 0$ means *making hole*, in which case empirical rock-bit boundary conditions are used, while $\partial u/\partial t > 0$ means *bit bounce*, in which case a stress-free end is assumed.

The presence of the *full* "ρg" body force term in Equation 5-1, usually neglected in dynamic analyses, is again consistent with the need to consider the complete dynamic and static field, that is, with the requirement to have the drillstring fall freely in the limit when the formation disappears.

More on AC/DC interactions. Static *DC* axial displacements (which are, again, affected by AC solutions) are important in our analysis for two reasons. For one, the end result of successful drilling is decreased drillbit elevation, a DC effect if the bit is to advance on the average.

This philosophy contrasts with the specification "$u(0,t) = u_0 \sin \omega t$" typically used, in which the drillbit executes purely periodic motions, but does not drill. Second, the results of Chapter 6 demonstrate, consistently with limited observations, that the static component of the axial displacement field is responsible for violent bending oscillations near the neutral point. Treating the *total* AC/DC displacement allows us to model this effect naturally.

In Chapter 7, similar arguments are given for studying the complete dynamic and static *torsional* displacement rather than each separately. But the reason is quite different. In order to model *stick-slip* oscillations, the cyclic transformation of mean potential strain energy to disturbance kinetic energy, and vice-versa, suggests that it is simpler and more natural to deal with a single entity.

Thus, in this book, unless specifically noted, all axial, torsional, and lateral disturbances are modeled *in their totality* without book-keeping them into distinct AC and DC components. But in the next section, we *will* separate AC and DC solutions, in order to develop conventional ideas on mechanical vibrations *and* to demonstrate possible pitfalls incurred by this separation.

CONVENTIONAL SEPARATION OF AC/DC SOLUTIONS

To see how the usual ideas about static and dynamic formulations follow from Equation 5-1, which describes the complete problem, we write the *total* longitudinal displacement u(x,t) in the form

$$u(x,t) = u_s(x) + u_d(x,t) \tag{5.2}$$

where *s* and *d* denote *static* and *dynamic*. Substitution in Equation 5-1 yields

$$\rho\, \partial^2\{u_s(x) + u_d(x,t)\}/\partial t^2 + \zeta\, \partial\{u_s(x) + u_d(x,t)\}/\partial t - E\, \partial^2\{u_s(x) + u_d(x,t)\}/\partial x^2 + \rho g = F_e \tag{5.3}$$

or

$$\rho\, \partial^2 u_d(x,t)/\partial t^2 + \zeta\, \partial u_d(x,t)/\partial t - E\, \partial^2\{u_s(x) + u_d(x,t)\}/\partial x^2 + \rho g = F_e \tag{5.4}$$

Now, we *choose* the function $u_s(x)$ so that it satisfies the ordinary differential equation

$$d^2u_s(x)/dx^2 = \rho g/E \tag{5.5}$$

The accounting implicit in Equation 5-5 assumes that static wall friction, if any, is *not* book-kept in $u_s(x)$; the appearance of frictional effects is conspicuously absent. This is not to say that friction does not exist, only that any friction will need to be modeled when considering $u_d(x,t)$. This book-keeping is natural for vertical wells, where borehole wall contact is minimal. Thus, combination of Equation 5-5 and Equation 5-4 yields

$$\rho\, \partial^2 u_d(x,t)/\partial t^2 + \zeta\, \partial u_d(x,t)/\partial t - E\, \partial^2 u_d(x,t)/\partial x^2 = F_e \tag{5.6}$$

The static equation for our assumed $u_s(x)$ integrates to

$$du_s(x)/dx = (\rho g/E)\, x + C_1 \tag{5.7}$$

$$u_s(x) = (\rho g/2E)\, x^2 + C_1 x + C_2 \tag{5.8}$$

where C_1 and C_2 are constants of integration.

Sign conventions. Suppose that a concentrated mass M rests on top of a vertical drillstring of length L, and that the system in Figure 5-2, in static equilibrium, is entirely supported by the formation. We require that the axial force AE $du_s(x)/dx$ equal -Mg at x = L. Thus, Equation 5-7 can be written in the form

$$AE\, du_s(L)/dx = A\rho g L + AE\, C_1 = -Mg \tag{5.9}$$

leading to $C_1 = -Mg/AE - \rho gL/E$ or

$$AE\, du_s(x)/dx = A\rho gx - AE\{Mg/AE + \rho gL/E\} \tag{5.10}$$

The left hand side of Equation 5-10 is a force; in particular, $AE\, du_s(x)/dx = -AE\{Mg/AE + \rho gL/E\}$ or $-Mg - A\rho gL < 0$ at the bit $x = 0$, while $AE\, du_s(x)/dx = -Mg < 0$ at the surface $x = L$ (the $-A\rho gL$ arises from drillstring weight). Our convention assumes that $AEu_x < 0$ for compression, with $AEu_x > 0$ for tension; here, the negative signs indicate that the drillstring is entirely in compression.

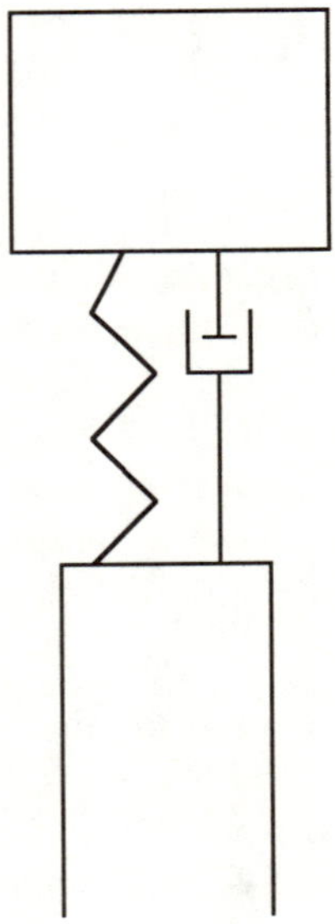

Figure 5-2. Drilling surface system.

Static weight on bit. In reality, the weight "$-Mg-A\rho gL$" is not completely applied at the bit, since the overwhelming force would precipitate drillbit bearing damage and excessive torques required for rotation. Instead, a tensile force $P > 0$ is applied at the surface that reduces compressive loads everywhere. That is, "$-Mg - A\rho gL$" is reduced, or rendered less negative, by the addition of a positive P, so that Equation 5-10 is replaced by

$$AE\, du_s(x)/dx = A\rho gx + \{P - (M + A\rho L)g\} \tag{5.11}$$

The force AE $du_s(x)/dx$ at the surface x = L is now P-Mg, while the corresponding force AE $du_s(0)/dx$ at the drillbit x = 0 is now the "weight-on-bit" P - (M +AρL)g.

For our purposes, the transition point where the sign of the axial stress reverses is the *neutral point* defined by setting the right side of Equation 5-11 to zero, that is,

$$x_{neutral\ point} = -\{P-(M+A\rho L)g\}/A\rho g \tag{5.12}$$

Other definitions exist in the drilling literature, but Equation 5-12 complements the objectives of Chapter 6. Also, it will be convenient to define

$$W_0 = P-(M+A\rho L)g \tag{5.13a}$$

so that

$$AE\ du_s(x)/dx = A\rho gx + W_0 \tag{5.13b}$$

where W_0 is the *negative* of the static *weight-on-bit WOB*.

Again, the weight-on-bit in Equation 5-13a is not necessarily the value observed because Equation 5-5 was selected to simplify Equation 5-4. In doing so, we ignored the possibility that external forces exist which contain static components of friction (such forces will need to be included in Equation 5-6). These *would* alter the force balance as implied by Equation 5-13a, which does not contain distributed borehole wall friction or concentrated forces.

Equation 5-5 was chosen to simplify the mathematics. This static WOB is *not* the dynamic value measured at the bit, which may vary from highly compressive to *zero* when the drillbit bounces off the formation. Finally, Equation 5-13b, under these restrictive conditions, can be integrated to give

$$\begin{aligned} u_s(x) &= 1/2\ \rho gx^2/E + W_0x/AE + C_2 \\ &= 1/2\ \rho gx^2/E + W_0x/AE \end{aligned} \tag{5.14}$$

where C_2 vanishes because the drillbit is rigidly fixed to the bottom x = 0 at t = 0 (that is, we assume that $u_s(0) = 0$). Since $u(0,t) = u_s(0) + u_d(0,t) = u_d(0,t)$, the $u_d(0,t)$ computed later is the exact static plus dynamic position of the bit.

Recapitulation. Let us summarize our discussion thus far. If the effects of borehole wall friction and contact are disregarded in the static model, then the static axial force distribution is given by

$$AE\ du_s(x)/dx = A\rho gx + \{P-(M+A\rho L)g\} = A\rho gx - WOB \tag{5.15}$$

Dynamic oscillations superimposed on this static field satisfy

$$\rho\ \partial^2u_d(x,t)/\partial t^2 + \zeta\ \partial u_d(x,t)/\partial t - E\ \partial^2u_d(x,t)/\partial x^2 = F_e \tag{5.16}$$

Having explained the conventional separation of static and dynamic forces, we now consider the solutions to some earlier dynamical models.

Equation 5-16 governs $u_d(x,t)$. It is the classical damped wave equation considered in Chapter 1, and it must be solved subject to initial and boundary conditions. Different types of auxiliary conditions are possible. For example, the *initial* conditions $u_d(x,0) = 0$ and $\partial u_d/(x,0)/\partial t = 0$ describe a drillstring at rest at $t = 0$. The *boundary* condition $u(0,t) = A \sin \omega t$ at the bit $x = 0$, *not* used in this book, describes *prescribed* sinusoidal displacements with amplitude A and frequency ω, while $\partial u_d(0,t)/\partial x = 0$ models a free-end during bit bounce.

We later show *why* the $\zeta\ \partial u_d(x,t)/\partial t$ in Equation 5-16, and a similar term in the surface condition, represent damping (dissipation in the draw works and power swivel is small). This crude *frequency-independent model* (Chapters 1 and 2) accounts for energy losses due to drillstring movement, formation and mud attenuation, internal viscous losses in the metal, and sound radiation, if significant. Damping and power transfer at the bit are accounted for directly through rock-bit interaction, which we model through an impedance condition.

BOUNDARY CONDITIONS: OLD AND NEW IDEAS

We will discuss three topics: surface boundary conditions, rock-bit interactions, and drillbit kinematic modeling. The first is straightforward; the latter two, however, require some development. In each case, we describe existing methods, critique them, and extend them to remove any limitations.

Surface boundary conditions. At the surface, the derrick and draw works are usually modeled by a mass-spring-damper system, this assumption resulting in the boundary condition

$$M\, \partial^2 u/\partial t^2 + \beta\, \partial u/\partial t + AE\, \partial u/\partial x + ku + Mg = 0 \qquad (5.17)$$

where the damping factor β is generally small. As in Equation 5-1 for $u(x,t)$, our sign convention for gravity is chosen so that in the free-fall limit, the reduced equation $M\partial^2 u/\partial t^2 = - Mg$ leads to the familiar

$$u(t) = -1/2\ gt^2 \qquad (5.18)$$

from particle mechanics.

Recently, Clayer, Vandiver, and Lee (1990) reevaluated this surface model in light of field data and numerical simulations. The authors conclude that the mass-spring-damper model is sufficiently accurate for surface modeling, and probably suffices for most engineering analyses. Equation 5-17 appears to be satisfactory, and its use is retained in this book.

Conventional bit boundary conditions. In their early paper, Dareing and Livesay (1968) introduced the "obvious" drill bit boundary condition

$$u_d(0,t) = u_0 \sin \omega t \tag{5.19}$$

at $x = 0$, where ω is related to the "3 x rpm" frequency typical of tricone bits, and u_0 is the peak-to-peak cone vertical displacement (alternatively, we might have taken $u_d(0,t) = u_0 \cos \omega t$).

The frequency ω may be nonzero or zero. If nonzero, the *displacement* $u_0 \sin \omega t$ (or, $u_0 \cos \omega t$) is marked by a zero time average, that is,

$$\int_{period} u_d(0,t)\, dt = u_0 \int_{period} \sin \omega t\, dt = 0 \tag{5.20}$$

Thus, the drillstring *never* penetrates the formation on the average! If zero, we have at best a fixed nonzero bit displacement, yielding uninteresting solutions.

The use of $u_d(0,t) = u_0 \sin \omega t$, of course, only partially addresses problems satisfying transient boundary conditions of the form $u_d(0,t) = f(t)$. Also, noticeably absent from it is any rock-bit interaction: the most important part of the *physical* problem is absent. Thus, the boundary condition in Equation 5-19 implicitly assumes *hard rock* with zero mean rate-of-penetration.

But the very fact that *something* has been assumed at $x = 0$ raises the most concern. If we wish to specify rock-bit boundary conditions instead of Equation 5-19, then how do we *additionally* introduce "sin ωt" to characterize bit motion? It seems that only one, but not both models, is possible; fortunately, it will be possible to fix this shortcoming using ideas developed in Chapter 1.

Other inconsistencies are found with Equation 5-19, particularly with respect to predicted downhole resonances. Researchers have attempted to drill at the resonances predicted by the simple theory, in order to increase rate-of-penetration, but without success.

Dareing (1972) explained why *drillstrings do not experience resonant vibrations* frequently, and gave three possible reasons. For one, he noted that the right excitation source does not appear all too often. And second, when it does, the frequency bandwidth for drill string resonance is narrow - thus, delicately maintaining a constant rotary speed that is *just right* is essential (the width of the frequency band reduces with drillstring length). Finally, the irregularities which create the excitation source are often drilled away before the resonant vibration mode has time to develop.

The possibility that other transient vibration modes exist, though, was raised in his paper. We might note that if these exist, the actual drilling process itself would have to be modeled in order to simulate reality, while somehow retaining the periodic, kinematic character of the rotating drillbit.

Dareing (1972) importantly observed that harmonic motions may not be the most general downhole, and that they may not be the predominant form of vibration while drilling. He noted that large dynamic forces can occur in the drillstring and at the bit, at non-resonant frequencies - in fact, the bit forces measured by the downhole recorder *were* at non-resonant frequencies. Dareing noted that vibration modes do not develop instantaneously, because time is required to establish normal modes; but while acknowledging that *general* transient motions may well predominate, he did not propose any other *physical mechanism* that may be responsible for *dangerous* transient motions.

In summary, fixed displacement boundary conditions only describe drilling in hard rock, or possibly, drilling over short time scales, or both. Any resonances predicted by the model are likely to have disappeared by the time required for the harmonically steady solutions to establish themselves, since the conditions needed to create them will have been drilled away. In formulating a realistic model, the boundary conditions must be generalized, and transient, rather than harmonic calculations should be performed.

We have agreed that the differential equation drillstring model in Equation 5-1, and the surface boundary condition model in Equation 5-17, are likely to apply to real situations. But Equation 5-19 is not acceptable, since predicted resonances do not exist physically, not to mention its complete failure in modeling forward drilling. We are left with a dilemma posed by *two* simple wishes: how can we model *both* drilling ahead *and* drillbit kinematics?

Since Equation 5-19 does *not* model forward drilling, penetration rate optimization *cannot* be studied with it; nor can we model bit-bounce, which clearly satisfies stress-free boundary conditions, when it does occur. Before developing a unified model, further discussion of existing models is useful.

Modeling rock-bit interactions. We address the former question first, but we need to return to fundamentals, examining available options. In Chapter 1, we studied models of the form $u(0,t) = 0$ and $u_x(0,t) = 0$; the first describes rigid ends, whereas the second simulates stress free ends.

These models are extreme limits representing hard and soft formations. In general, one might postulate a relationship connecting the penetration depth $u(0,t)$ to the instantaneous rate-of-penetration $u_t(0,t)$ and the local strain $u_x(0,t)$,

$$G(u,u_x,u_t) = 0 \qquad (5.21)$$

which may be linear or nonlinear.

The particular form that $G(u,u_x,u_t) = 0$ takes will depend on many factors, e.g., the type of drillbit used, the kind of rock being drilled, and intermediate conditions such as bit and bearing wear. *Time-dependence* might be included if mud invasion beneath the bit affects formation hardness.

Azimuthal θ-dependence may be useful (once axial and torsional analyses are coupled) to model formation dip, bedding plane or anisotropic rock effects. Models represented by Equation 5-21 are impedance boundary conditions; the impedance relationship $G(u,u_x,u_t) = 0$ is just a rock-bit interaction model.

We will discuss possibilities for Equation 5-21 later, but for numerical solutions, few restrictions on functional form need to be imposed (Fortran *if-then* statements can be used). For analytical solutions, we can take

$$\alpha u_x + \beta u_t + \lambda u = 0 \tag{5.22}$$

which yields to linear analysis, modeling the relative effects of instantaneous elastic rock displacement, impact velocity, and dynamic strain.

From Chapter 1, we know that Equation 5-22 can promote interactions between AC and DC solutions; thus, there are optimistic reasons to believe that *penetrating values* of *bit elevation* u(0,t) can be found for certain values of α, β, and γ: solutions for u(0,t) are *not* likely to be purely sinusoidal.

But Equation 5-22 is not complete, since it does not apply when the drillstring bounces upward. Thus, we will refine our model, invoking Equation 5-22 *only* when the velocity $\partial u(0,t)/\partial t < 0$ is downward; however, we insist that

$$\partial u(0,t)/\partial x = 0, \textit{ if } \partial u(0,t)/\partial t > 0 \tag{5.23}$$

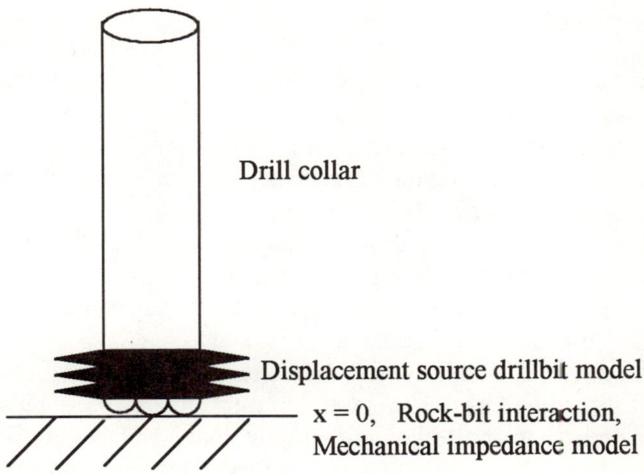

Figure 5-3. Downhole model near the bit.

This represents a stress-free end when bouncing off-bottom; the logic tests, for example, can be carried out using previous time step values of u(0,t) in a numerical scheme. The boundary conditions in Equations 5-22 and 5-23 are applied at x = 0 as shown in Figure 5-3. Henceforth, the statement in Equation 5-21 should be interpreted as including Equation 5-23.

Empirical notes on rock-bit interaction. Penetration rate into a formation depends upon the entire drillstring dynamics: the same bit drilling the same rock will penetrate differently with different drillstring lengths. The dynamics is controlled by local interactions at the bottom of the hole. This is clear to any home carpenter who has drilled with new versus dull bits.

In vibrations modeling, we view the simulation problem as the result of two sub-problems. The first is *macroscopic;* this includes the governing partial differential equation and the surface boundary condition. Its description is determined by rig, drillpipe, and bottomhole assembly details.

The second is a *microscopic* view related to events at the bottom x = 0. These details, known as *rock-bit interactions*, are empirical: they depend on drillbit geometry, bit and bearing wear, and formation hardness. What we seek is an "absolute" statement about local interactions which can be used to show how different drillstrings drill differently with bit and formation held constant.

If this statement is inferred from laboratory rig tests, the final result must be carefully expressed in a manner relating bit and rock only. The effects of the drilling assembly must be removed (or, *deconvolved*) so that the boundary condition can be used with other drilling assemblies.

Laboratory drillbit data. It is difficult to obtain true rock-bit boundary conditions if the dynamics of the drilling assembly (i.e., the system transfer function) cannot be characterized accurately. Controlled drilling where vibration isolates the bit from reverberant oscillations is ideally desired.

At least in principle, information along these lines does appear in the drilling literature. For example, Rowley, Howe, and Deily (1960) performed laboratory drilling tests with hard formation bits under rock and borehole fluid pressures simulating a deep hole in a impermeable dolomite.

Bit weight, torque, and rotary speed were carefully controlled on a hydraulically driven drilling machine that was shock and vibration isolated. Individual parameter effects on penetration rate were obtained and curve-fitted to polynomial relations that relate our u, u_t, and u_x.

Warren (1981) similarly gave laboratory derived drilling models for full-scale soft formation bits, relating weight-on-bit, rotary speed, bit size, bit type, and rock strength to penetration rate. In his study, the laboratory rig WOB and rotary speed were servo-controlled so that they were constant during the test.

Warren (1983) explored efficiency factors affecting tricone bit performance, introducing torque as another variable into his model, while Warren (1984), dealing with roller cone bits, studied penetration rate performance as influenced by hole cleaning.

Gray, Armstrong, and Gatlin (1962) studied 2D rock breakage in drag-bit drilling at atmospheric pressure. They varied cutting tip rake angles, depths of cuts, and cutting speeds in a comprehensive test matrix, and obtained motion pictures of the cutting process at camera speeds of 5,000 to 8,000 frames per second, shedding insight into the brittle failure mechanism in rocks. Again, detailed functional relationships connecting u, u_t, and u_x were obtained.

Single-tooth impact results. Podio and Gray (1965), and Yang and Gray (1967), gave analogous relationships for idealized *single-blow bit-tooth impact tests* on saturated rocks under confining pressure. These papers offered excellent reproductions of oscilloscope traces showing simultaneous force, velocity, and displacement at different initial instances of time. Figure 5-4 provides a crude sketch, typical of the authors' experimental results.

```
                 +
             +        +
          +              +
        +                  +
++++++++                     ++++++++++
                 *         Displacement
              *   *
            *     *
          *        *
         *          *
        *           *
*********            *****************
                           Force

         +  +
       +      +
+++++++         +         +++++++++++++
                  +      + Velocity
                     + +

--------------------------------------->
                           Time
```

Figure 5-4. Simple rock-bit interaction time traces.

In these papers, brittle, transitional, and plastic collisions were considered through a range of pore pressures and fluid saturations. Results were obtained for impacts parallel and perpendicular to bedding planes.

Observed relationships connecting u, u_t, and u_x were typically nonlinear, and not exactly simple; however, *similarities in shape demonstrated repeatedly over a range of conditions indicate that well-defined rock-bit interaction models do exist*, possibly taking forms not unlike Equation 5-22 under limited displacement ranges.

Others dealing with similar subjects include Outmans (1960), who provided a theoretical framework for instantaneous rate of penetration; and Maurer (1962), who gave rate versus weight-on-bit formulas derived from rock-cratering mechanisms under *perfect cleaning* conditions. Also, Maurer (1965) described experimental findings on crater formation, and Eronini (1982) characterized the dynamic interaction between an impacting tool and rock with models derived from fracture mechanics.

These papers deal with single-tooth impacts, demonstrating that detailed rock-bit measurements *can* be made, but it is the net drillbit performance *in the large* that is important. Again, many laboratory rig results are available.

Warren's early lab tests, and the work of Grimes, Felderhoff and Brown (1992), for example, describe how faster penetration rates can be achieved by using heavy weight rock bits, with special cone, cutter and bearing structures, designed specifically for drilling with high WOB in hard rock formations.

Finally, Langeveld (1992) provides insight into PDC bit dynamics, explaining their poor performance in hard formations, and suggesting means for improvement. He emphasizes the importance of the boundary condition at $x = 0$ in controlling the dynamics of the entire drillstring, particularly with respect to its ability to make hole efficiently.

Bit design factors, e.g., cutter number and orientation, play important roles in determining drillstring vibration and directional characteristics such as *bit walk* and *vibration mode coupling*. Controlled tests should be performed for axial, torsional, and bending moment characteristics in order to determine the precise effects of the bits themselves.

We have emphasized that the *physical relationship* between u, u_t, u_x obtained at impact is the missing link needed for rock-bit interaction and correct boundary condition definition. The exact forms for Equation 5-21 and 5-22 will require experimental study; several models, together with their analytical and numerical consequences, are considered later.

Modeling drillbit kinematics. We still have not discussed the "$u_0 \sin \omega t$" needed to describe rotating drillbit kinematics, stating only that Equation 5-19 should *not* be used at $x = 0$. In general, a periodic vertical excitation characterizes the geometry of the bit, its cones, offsets, and inserts, and the way it is rotated by the rotary table from the surface or downhole by a mud motor.

The rotation rate and bit geometry affect the displacement excitation imparted to the formation and the power inputted into the system. Ma and

Azar (1985, 1986) provide excellent examples of studies focusing upon the kinematics and dynamics of the overall *bit* motion. These and similar results are can provide useful data for displacement source modeling.

Let us take a conceptual view, following ideas developed in Chapter 1, and consider the "accordion" in Figure 5-3. This accordion can expand and contract vertically with prescribed length variations, simulating the up-and-down motion of a rotating tricone bit. This displacement creation is exactly the *displacement source* model, used whenever the bit is drilling ahead.

The desired physical effect is a periodic expansion and contraction cycle which *creates* a length $[u] = u_0 \sin \omega t$ at the location x_{bit} of the bit centroid (this notation is discussed in Chapter 1). Displacement source modeling allows us to prescribe sinusoidal excitations *without* prescribing actual bit positions, which *should* be found as part of the solution.

Now, Equation 5-1 (that is, $\rho\ \partial^2u/\partial t^2 + \zeta\ \partial u/\partial t - E\ \partial^2u/\partial x^2 + \rho g = 0$) describes longitudinal *displacements*. From Chapter 1, the application of a *single force* associated with a single delta function $\delta(x-x_{bit})$ will create a sudden change in the value of *the spatial derivative* $\partial u/\partial x$. This implies that an external force exists, but it doesn't. However, the use of a *couple*, with equal and opposite forces situated infinitesimally close will cause a change in *the function u*, representing length. Thus, the kinematics of the drillbit can be easily modeled by an external couple with a *jump in displacement* or *strength*

$$[u(x_{bit},t)] \propto \sin \omega t \tag{5.24}$$

provided the drillbit is in contact with the formation. This model is flexible since the exact level of $u(0,t)$ is not itself prescribed at $x = 0$; the final determination of bit position is made *jointly* with Equations 5-21 to 5-23.

This formulation for drillbit kinematics, together with earlier rock-bit interaction ideas, meets our modeling objectives. Importantly, its solutions *will* include problems satisfying the "$u(0,t) = u_0 \sin \omega t$" *if* they evolve as part of a transient solution (that is, for hard enough rock, if the bit does not bounce). But in general, computed values of $u(0,t)$ are unlikely to be purely sinusoidal; they will contain fluctuating AC/DC components, a consequence of the complicated dynamic displacements at the bit.

This usage was first reported in Chin (1988a,b) where computed solutions showing rate-of-penetration and bit bounce were given; detailed results for a stabler scheme are given later in Chapter 8. However, the rudimentary ideas underlying displacement sources are not new; as noted in Chapter 1, they are well known in earthquake seismology, where different types of source singularities model symmetric and antisymmetric triggering events.

The first use of the displacement source concept in drilling actually appeared in Lutz, Raynaud, Gstalder, Quichaud, Raynal, and Muckleroy

(1972). Drawing upon extensive field and test bench measurements, and detailed kinematical interpretation of drillbit motions, the authors concluded that the tricone bit was more than a passive creator of longitudinal and torsional vibrations.

Unlike earlier studies which modeled bit excitation with prescribed vertical displacement histories, e.g., Equation 5-19, Lutz *et al* reiterated the role of the the tricone as a *source of axial displacement* or as a *displacement generator* characterized by the up-and-down lift involved in passing from one tooth to the other.

This concept is important because *sources*, following Chapter 1, adhere to different rules than do *boundary conditions*, and lend greater modeling flexibility. The power of the idea was not immediately realized, and its only application since then were this author's 1988 papers.

The notion of displacement sources was recently revived in Clayer, Vandiver, and Lee (1990), which clearly described the excitation process at the bit. These authors specifically described the action of a tricone bit as a relative *displacement source* situated between the rock and the end of the drillstring.

As the cones and teeth roll over irregular formations, relative movement between the body of the bit and a fixed, unaffected, faraway reference point in the rock is observed. Close to the drill bit, the rock may deform elastically or inelastically, failing in a small region close to the teeth. The action of a tricone bit parallels that of a displacement source, e.g., the accordion shown in Figure 5-3, inserted between the bit and a flexible formation, which is modeled with an equivalent spring and damper.

This type of source can model mathematically both the periodic lifting action of the teeth as well as the displacement caused by the tri-lobed pattern which develops in the rock when one uses tri-cone bits, while at the same time allowing some rock-bit interaction.

Analogies from earthquake seismology. Displacement sources were first used to model seismic wave generation due to explosion-induced volume changes in the earth; underground rumblings and wave interactions admit multiple reflections, not unlike those in drillstrings, that ultimately pass transparently through the source on re-reflection.

The seismology text by Aki and Richards (1980), for example, introduces the subject of displacement discontinuities and volume sources in a readable fashion, especially with respect to symmetries and antisymmetries in the displacement and stress variables about the source point.

For example, over large times, the normal modes of the earth's oscillations depend on whether they are excited by the slippage of a fault plane or by the volume creation due to a hidden nuclear blast. The type of source (or, generating mechanism) is inferred from initial measured data; once known, the strength of the source and the subsequent vibration history can be determined.

Of course, seismologists deal with 3D problems, utilizing *vector* forces, couples and dipoles to model different kinds of excitation. Hudson (1980) provides a discussion, showing how various kinds of point loadings combine to model physical phenomena having different symmetries or antisymmetries.

Similarly, the axial vibrations of a rod excited *internally* will depend upon the nature of the source; e.g., if it is a diaphragm that moves to-and-fro, or if it is an "accordion" that creates volume. In the former, stress is antisymmetric about the diaphragm; a net force acts on the system. In the latter, no net force acts: equal and opposite forces are present, which create nonzero displacement.

Consider, as an example, air-driven jackhammers where pneumatic effects create internal displacement sources without creating net applied force: these sources are not associated with nonzero external applied forces (it not necessary to have a nonzero force in order to drill). Of course, one can assist the drilling process by yanking on the drill with one's foot, in which case the net excitation is the superposition of a displacement source and an external applied force!

The excitation process. Again, the displacement source nature of rotating tricone bits is modeled, applying two equal and opposite periodic axial forces at the bit centroid, assuming *displacement* as the dependent variable (on the other hand, a single delta function can be used if the ϕ - model in Chapter 1 is used).

Summary. We have invoked a rock-bit interaction model $\alpha u_x + \beta u_t + \lambda u = 0$ as a boundary condition at $x = 0$. The displacement source, unlike Equation 5-19, is not a boundary condition, but an *excitation* creating axial length without specifying bit position, while allowing the transparent passage of waves. Also, we have emphasized the need to describe AD/DC solutions by a single transient dependent variable. These discussions were detailed because the approaches are new. This *unified* approach will generally simulate drillbit kinematics, rock-bit interaction, and nonzero rate-of-penetration.

GLOBAL ENERGY BALANCE

We now summarize our *momentum-based* differential equations, but before considering solutions, it is instructive to consider *energy arguments*, since they provide insight into the physical mechanisms at work.

Formulation summary. The *complete* AC/DC axial displacement u(x,t) satisfies Equation 5-25, together with Equation 5-26 at the surface, and the rock-bit interaction in Equation 5-27.

$$\rho\ \partial^2u/\partial t^2 + \zeta\ \partial u/\partial t - E\ \partial^2u/\partial x^2 + \rho g = F_e \qquad (5.25)$$

$$M\ \partial^2u/\partial t^2 + \beta\ \partial u/\partial t + AE\ \partial u/\partial x + ku + Mg = 0 \text{ at } x = L \qquad (5.26)$$

$$G(u, \partial u/\partial x, \partial u/\partial t) = 0 \text{ at } x = 0 \qquad (5.27)$$

Equations 5-26 and 5-27 represent *boundary* conditions. Additional *initial* conditions are needed to "start" the time integrations. If the axial vibrations are initiated from rest, Equation 5-14 might be used at $t = 0$; that is,

$$u(x,0) = u_s(x) = 1/2\ \rho g x^2/E + W_0 x/AE \qquad (5.28)$$

$$\partial u(x,0)/\partial t = 0 \qquad (5.29)$$

The boundary condition in Equation 5-27 describes the mechanical impedance condition at the rock-bit interface; it does not model the kinematics of the drillbit. As discussed, the "accordion-like" displacement variations due to the bit are naturally modeled by displacement sources, i.e.,

$$[u]_{x=xbit} = f(t) \ldots \{\text{e.g., } u_0 \sin \omega t\} \qquad (5.30)$$

where x_{bit} represents the centroid of the bit, and f(t) describes the vertical displacement history of the bit.

Energy considerations. Differential equations describe dynamics from a *momentum* viewpoint, a by-product of their "**F** = ma" origins. We now reinterpret our drillstring formulation from an *energy* perspective; this sheds insight into the type of constraints needed to reproduce observed phenomena.

It is clear that numerical solutions to formulations that appear reasonably correct can lead to disaster if lurking sign errors go undetected. Energy checks provide additional safety; if both momentum and energy descriptions are consistent, greater confidence can be placed in the results.

The drillstring. By learning how energy enters and exits the system, we understand the ingredients required for modeling rock-bit interaction and other quantities of physical interest. For this purpose, consider the integral

$$\Gamma(t) = \int \{1/2\ \rho\ (\partial u/\partial t)^2 + 1/2\ E\ (\partial u/\partial x)^2 + \rho g u\}dx \qquad (5.31)$$

representing the total *distributed* kinetic, elastic strain and gravitational potential energy of the drillstring per unit cross-sectional area.

For brevity, "$\int \ldots dx$" denotes integration over $0 < x < L$, where the drillstring length L is assumed to be constant. During drilling, L will vary with time, as sections of pipe are added or removed. However, this occurs when drilling terminates and vibrations are absent.

Thus, while dL/dt vanishes in our modeling, we need to remember that L does increase during the drilling; it is *piecewise constant* with time. Next, time differentiation yields yields the rate of change

$$\Gamma'(t) = d/dt\ \{\int \{1/2\ \rho\ (\partial u/\partial t)^2 + 1/2\ E\ (\partial u/\partial x)^2 + \rho g u\}dx\ \} \quad (5.32)$$

$$= \rho \int u_t u_{tt}\, dx\ + E \int (\partial u_t/\partial x)\ u_x\, dx\ + \rho g \int u_t\, dx \quad (5.33)$$

$$= \int u_t (\rho u_{tt} - E u_{xx} + \rho g) dx\ + E\ u_x u_t(L,t) - E\ u_x u_t(0,t) \quad (5.34)$$

where we have integrated by parts, collected distributed properties within the integrand, and separated end effects from the integral.

At this point, we invoke the partial differential equation governing the drillstring, that is, Equation 5-25, so that

$$\Gamma'(t) = \int u_t (-\zeta u_t + F_e)\, dx\ + E\ u_x u_t(L,t) - E\ u_x u_t(0,t) \quad (5.35)$$

$$= -\zeta \int u_t^2\, dx\ + \int u_t F_e\, dx + E\ u_x u_t(L,t) - E\ u_x u_t(0,t) \quad (5.36)$$

The right side of Equation 5-36 contains all the mechanisms by which energy can exit or enter the system. We will describe these physically, but first, we derive an important surface result.

The surface. We digress and reconsider the surface mass-spring-damper boundary condition

$$M\ \partial^2 u/\partial t^2\ + \beta\ \partial u/\partial t\ +\ AE\ \partial u/\partial x\ + ku\ + Mg = 0 \quad (5.26)$$

which is valid at $x = L$ for all times t. We can rewrite this local momentum balance in a different but equally enlightening form, by multiplying Equation 5-26 throughout using $\partial u/\partial t$,

$$M\ u_{tt} u_t + \beta\ u_t^2 + AE\ u_x u_t + kuu_t + Mgu_t = 0 \quad (5.37)$$

so that

$$d/dt\ \{1/2\ Mu_t^2 + 1/2\ ku^2 + Mgu\} + \beta\ u_t^2\ + AEu_x u_t = 0 \quad (5.38)$$

This energy description for the traveling block at $x = L$ is analogous to Equations 5-35 and 5-36 describing the drillstring.

Combined drillstring/surface system. Now, we combine Equations 5-32, 5-35, and 5-36, and multiply each term by A to give

$$d/dt\ \int \{1/2\ \rho\ (\partial u/\partial t)^2 + 1/2\ E\ (\partial u/\partial x)^2 + \rho g u\}\ A\, dx - AE\ u_x u_t(L,t) \quad (5.39)$$

$$= -\zeta \int u_t^2\ A\, dx\ - AE\ u_x u_t(0,t) + \int u_t F_e\ A\, dx \quad (5.40)$$

Simplifying with Equation 5-38 leads to

$$\begin{aligned} d/dt\ \int \{1/2\ \rho\ (\partial u/\partial t)^2 + 1/2\ E\ (\partial u/\partial x)^2 + \rho g u\}\ A\, dx \\ + d/dt\ \{1/2\ Mu_t^2 + 1/2\ ku^2 + Mgu\}_{x=L} \\ = \quad -\zeta \int u_t^2\ A\, dx\ - \beta\ u_t^2\, |_{x=L} - AE\ u_x u_t(0,t) + \int u_t F_e\ A\, dx \end{aligned} \quad (5.41)$$

The *left side* represents the timewise rate of change of total energy for the combined surface system and drillstring. For the drillstring, the contributions in the integrand are the distributed kinetic, elastic strain and gravitational potential energies. For the surface system, we have the kinetic energy of the traveling block, the spring potential energy, and the block gravitational potential energy, in that order.

The *right side*, shows how work interaction changes the total energy balance as the complete system evolves in time. Since ζ and β are both positive, as are the *positive-definite* integrals multiplying them, the first two (negative) terms represent dissipation due to internal viscous effects.

The $AEu_xu_t(0,t)$ term represents the work introduced at the formation as a result of rock-bit interaction. Here, AEu_x is the *dynamic weight-on-bit* (equal to the reactive force on the formation), and $u_t(0,t)$ is the instantaneous bit velocity; their product represents *power input* due to rate of penetration.

This power is not strictly positive or negative. Its sign depends on relative cone-to-rock displacement pattern phasing, whose details are affected by cone geometry, formation hardness, drill rate, and surface rpm. This local argument is only part of the explanation; the effect of drillstring dynamics, whose qualitative features are less obvious, are also important (of course, when the bit bounces off bottom, vanishing stress implies zero power transfer).

Finally, $\int u_tF_eA\ dx$ represents the work done by external forces other than those applied at the bit. If distributed wall friction or point contacts (such as doglegs) are present, these will contribute to the integral. But the displacement source describing drillbit motions does not directly contribute to net energy, since the sum of equal and opposite forces completely vanishes; of course, it does dictate the form of the wave patterns created along the drillstring.

Detailed bit motions. Of interest is the form for the displacement function f(t) in Equation 5-30. In field operations, rotation rates are specified in *rpm*, where 1 rpm represents one revolution per 60 seconds, or 1/60 *th* Hz.

If N_{rpm} is the drillstring rpm, and a factor of 3 is introduced to simulate tricone effects, it follows that the frequency satisfies $f = 3\ N_{rpm}/60$ Hz. But since $\omega = 2\pi f$, the accordion-like length variation at the bit may be written as

$$[u]_{x=xbit} = u_0 \sin \omega t,\ \omega = 2\pi\ (3N_{rpm})/60 \tag{5.42}$$

where u_0 is the maximum axial displacement due to the cones. Higher harmonic motions may be introduced as necessary.

PDC bits rely on the shearing action of the cutters to make hole, and differ significantly from roller cone bits in form and in cutting mode. Equation 5-42 still applies, with the adjustment factor (instead of 3) varying anywhere from 9 to 20 depending on bit design.

The amplitude u_0 is also likely to change; it may be necessary to include higher order harmonics, taking the displacement source function in the form $f(t) = u_0 \sin \omega t + u_1 \sin 2\omega t + ...$ Note that PDC axial vibrations, excited by flatter bit surfaces, are characterized by stronger DC modes than are "3 x rpm" roller-cone-driven drillstring vibrations.

SIMPLE SOLUTION FOR RATE-OF-PENETRATION

Clayer, Vandiver, and Lee (1990) recently studied the effects of surface and downhole boundary conditions on drillstring vibration. The authors' experimental and numerical work shows that *simple displacement boundary conditions do not describe observed events* well, but *suitably chosen linear models will capture the essential physics.*

These conclusions were based on multi-channel surface data for a drillstring (with a 17.5 inch tricone drilling in hard limestone), acquired with an instrumented sub inserted just below the power swivel. This sub collected high-bandwidth force, torque, acceleration and rotary speed data, and direct impedance measurements of the drilling rig in axial and torsional motion.

Field motivation. It turns out that the real-time bottom boundary condition is a transient function of several parameters. In brief, *the effective stiffness and damping of the rock depends strongly on the dynamic WOB* and related parameters such as rpm, mud flow rate, and bit type.

Comparisons between experiment and numerical simulations suggest that damping at the drillbit due to formation interaction, closely related to rock fracture mechanisms, significantly affects drillstring dynamical response.

The authors suggest that a simple *massless* spring-dashpot model might suffice. The spring emulates the elasticity of the rock as the bit pushes against it, while the damper represents energy loss mechanisms related to fracture, rock and cuttings displacement. All of these observations lend support to Equation 5-22, first proposed in Chin (1988a,b) as a simple rock-bit interaction model.

Simple analytical solution. Proper boundary conditions are important in modeling correct energy partitioning at the bit: part of the wave energy destroys the rock, another part reflects into the drillstring toward the surface, while the remainder dissipates in the form of heat. How much of each is the problem.

We first attempt to understand the consequences of end boundary conditions used in mechanical engineering, and demonstrate why they are unsuitable to drillstring vibrations. For simplicity, the kinematics of the bit are ignored: we simply examine the reflections of an incident pulse and study its implications insofar as penetration is concerned.

Our results are qualitative, since surface interactions and finite drillstring length are neglected for now. The results are intended to demonstrate only the range of possibilities available with different mathematical models.

In Chapter 1, we showed how f(ct+x) and g(ct-x) represent axial displacement waves propagating in opposite directions. Here, we will suppose that f(ct+x) represents a known wave heading towards $x = 0$. We will construct the complete solution comprising of both incoming and outgoing waves, and from it, determine and explain different elements of the bottom response.

This exploratory work ascertains the form of boundary conditions needed to model desired physical effects. In particular, we seek classes of boundary conditions for which sinusoidal inputs will create DC levels of displacement change. For brevity, we refer to u_x as the stress (that is, the normal stress, or Eu_x). To motivate the final result, we consider some inappropriate models first in order to explain their limitations.

Classic fixed end. The rigid end boundary condition $u(0,t) = 0$ is implemented by the construction $u(x,t) = f(ct+x) - f(ct-x)$, which solves $u(0,t) = f(ct) - f(ct) = 0$ at $x = 0$. Consequently, the speed $u_t(0,t)$ vanishes for all time. The stress $u_x(x,t) = f'(ct+x) + f'(ct-x)$ takes on the value $2f'(ct)$ at $x = 0$, which is twice the $f'(ct)$ value due to the incoming wave alone. In this *zero* rate-of-penetration *rigid wall* model, the power input term $u_x u_t$ vanishes for all time.

Classic free end. Here, we satisfy $u_x(0,t) = 0$. The construction $u(x,t) = f(ct+x) + f(ct-x)$ leads to $u_x(x,t) = f'(ct+x) - f'(ct-x)$, which satisfies $u_x = 0$ at $x = 0$. The speed $u_t = cf'(ct+x) + cf'(ct-x) = 2cf'(ct)$ at $x = 0$, which is twice the $cf'(ct)$ value due to the incident wave. The displacement at the bit is $u(0,t) = 2f(ct)$, which is generally nonzero; however, for a given sinusoidal Fourier wave component, its time-averaged value over a period identically vanishes.

Other possibilities. Other models are often used in pipe acoustics. For example, $\sigma u_{tt} + \alpha u_x = 0$ describes the force balance due to an end stress with a mass. On the other hand, $\alpha u_x + \lambda u = 0$ describes the interaction between an end stress and a spring, while $\alpha u_x + u_t = 0$ describes that between a damper and a stress. Later, we will study the model $\alpha u_x + \beta u_t + \lambda u = 0$ (see Equation 5-22); first, we consider the third model discussed in this paragraph.

Simple derivative model, $\alpha u_x = u_t$ at $x = 0$. With f(ct+x) given, the solution takes the form

$$u(x,t) = f(ct+x) + \{(\alpha-c)/(\alpha+c)\}\, f(ct-x) \quad (5.43)$$

(see Morse and Feshbach (1953)). Since, $\alpha u_x = u_t$, it is clear that the product $u_x u_t = \alpha\, u_x^2$ is nonzero, so that nontrivial energy transfer does occur at the bit and formation interface. However, the solution shows that

$$u(0,t) = \{1 + (\alpha-c)/(\alpha+c)\}\, f(ct) \quad (5.44)$$

which vanishes on the average for a sinusoidal drillbit excitation. Thus, on the average, the rate-of-penetration is unfortunately zero.

The general impedance model, $\alpha u_x + \beta u_t + \lambda u = 0$. This boundary condition, stated in Equation 5-22, *will* model nonzero rate-of-penetration and bit-bounce. It contains the λu *spring* and βu_t *damper* elements suggested by Clayer, Vandiver, and Lee (1990), first postulated in Chin (1988a,b) as part of a more general model. The effects of static and transient WOB appear in the term αu_x .

The presence of odd *and* even order derivatives produces nonzero DC displacement levels when the incident excitation is sinusoidal. This is crucial to modeling nonzero rate-of-penetration. Consider now the linear impedance boundary condition

$$\alpha u_x + \beta u_t + \lambda u = 0 \quad (5.45)$$

at x = 0, a subset of the nonlinear statement in Equation 5-21.

Again with f(ct+x) given, we seek the function g(ct-x) such that u(x,t) = f(ct+x) + g(ct-x) satisfies Equation 5-45 at x = 0. To this end, we substitute u(x,t) = f(ct+x) + g(ct-x), $u_x(x,t) = f'(ct+x) - g'(ct-x)$, and $u_t(x,t) = cf'(ct+x) + cg'(ct-x)$ into Equation 5-45, and evaluate the resulting expression at x = 0.

This leads to the ordinary differential equation $(\beta c+\alpha)f' + \lambda f + (\beta c-\alpha)g' + \lambda g = 0$ for g(ct). If F(s) and G(s) denote Laplace transforms of f and g with respect to $\tau = ct$, we obtain

$$G(s) = - \{(\beta c+\alpha)s + \lambda\} F(s) / \{(\beta c-\alpha)s+\lambda\} \quad (5.46)$$

which can be conveniently reexpressed as the sum of two contributions

$$\begin{aligned} G(s) = &- \{(\beta c+\alpha)/(\beta c-\alpha)\} F(s) \\ &- \lambda \{\{1 - (\beta c+\alpha)/(\beta c-\alpha)\}/(\beta c-\alpha)\} \{1/\{s - (-\lambda/(\beta c-\alpha))\}\} F(s) \end{aligned} \quad (5.47)$$

The first line represents a reflection identical in form to the original incoming wave, except for possible phase and amplitude changes. The second line embodies more subtle effects, which are best explored by inverting G(s) and returning to the time domain.

Now, "$\{1/\{s - (-\lambda/(\beta c-\alpha))\}\}$ x F(s)" can be viewed as the product of two transforms. The curly-bracketed term is a valid transform because it vanishes as s tends to infinity; its inverse is the exponential $\exp(-\lambda\tau/(\beta c-\alpha))$. The inverse of F(s) is just $f(\tau)$, so that the inverse of the product in the time domain follows from the convolution integral

$$\begin{aligned} g(\tau) = &- \{(\beta c+\alpha)/(\beta c-\alpha)\} f(\tau) \\ &- \lambda \{\{1 - (\beta c+\alpha)/(\beta c-\alpha)\}/(\beta c-\alpha)\} \int_0^{\tau} f(\xi)\, e^{-\lambda(\tau-\xi)/(\beta c-\alpha)}\, d\xi \end{aligned} \quad (5.48)$$

In this general form, the subtleties are still unclear. To understand axial displacement effects induced by drillbit rotation, let us consider a sinusoidal wave incident upon x = 0, having the Fourier wave component

$$f(\tau) = A \sin \omega\tau/c \tag{5.49}$$

The result in Equation 5-48 specializes to

$$\begin{aligned} g(\tau) = &- A \{(\beta c+\alpha)/(\beta c-\alpha)\} \sin \omega\tau/c \\ &- A\lambda \{\{1 - (\beta c+\alpha)/(\beta c-\alpha)\}/\{(\beta c-\alpha)\{(\lambda/(\beta c-\alpha))^2 + (\omega/c)^2)\}\} \\ &\quad X \{(\lambda/(\beta c-\alpha)) \sin \omega\tau/c - (\omega/c) \cos \omega\tau/c + (\omega/c)\, e^{-\lambda\tau/(\beta c-\alpha)}\} \end{aligned} \tag{5.50}$$

The solution containing both incoming and outgoing waves is obtained by construction, replacing the dummy variable τ in $f(\tau)$ and $g(\tau)$ by ct+x and ct-x as required in $u(x,t) = f(ct+x) + g(ct-x)$. Thus, we have the entire solution

$$\begin{aligned} u(x,t) = &\ A \sin \omega(ct+x)/c - A \{(\beta c+\alpha)/(\beta c-\alpha)\} \sin \omega(ct-x)/c \\ &- A\lambda \{\{1 - (\beta c+\alpha)/(\beta c-\alpha)\}/\{(\beta c-\alpha)\{(\lambda/(\beta c-\alpha))^2 + (\omega/c)^2)\}\} \\ &\ x \{(\lambda/(\beta c-\alpha)) \sin \omega(ct-x)/c -(\omega/c) \cos \omega(ct-x)/c +(\omega/c)\, e^{-\lambda(ct-x)/(\beta c-\alpha)}\} \end{aligned} \tag{5.51}$$

The outgoing wave contains expected phase changes to the incoming sinusoidal input due to elastic-resistive boundary interactions, but it also includes a nonharmonic DC contribution whose average over integer periods does *not* vanish. At the x = 0, this DC level can be written as

$$\begin{aligned} DC\{u(0,t)\} = &-A\lambda\{\{1 -(\beta c+\alpha)/(\beta c-\alpha)\}/\{(\beta c-\alpha)\{(\lambda/(\beta c-\alpha))^2 +(\omega/c)^2)\}\} \\ &\ x \{(\omega/c)\, e^{-\lambda ct/(\beta c-\alpha)}\} \end{aligned} \tag{5.52}$$

where $DC\{u(0,t)\}$ *is* the *non-periodic penetration depth* that the drill bit descends with.

Modeling the constants α, β, and λ. The constants in $\alpha u_x + \beta u_t + \lambda u = 0$ are not arbitrary. Consider $\alpha = 0$, for which

$$u_t + \lambda/\beta\, u = 0 \tag{5.53}$$

This has the exponential solution

$$u(0,t) \propto e^{-(\lambda/\beta)t} \tag{5.54}$$

which admits unrealistic "lifting" elevations when $\lambda/\beta > 0$. In general, the coefficients are constrained by the mechanics of rock fracture; an example in Chapter 8 shows that their relative values dictate smooth versus rough drilling.

FINITE DIFFERENCE MODELING

Exact analytical solutions to practical engineering problems are rare, and recourse to numerical solutions is often necessary. *Finite element*, *finite difference*, and *boundary integral* (a.k.a., *panel*) methods have been successfully used to solve complicated problems.

In this section, we will discuss finite difference techniques; these extremely powerful methods can be mastered with a minimum amount of expertise in higher math. We will introduce and develop the fundamental ideas, and rapidly progress to the formulation of state-of-the-art algorithms for wave equations.

Elementary considerations. Let us consider a differentiable function f(x) at three consecutive equidistant locations x_{i-1}, x_i , and x_{i+1}, where i-1, i, and i+1 are indexing parameters. In this book, we will assume that all grids are uniformly separated by the constant gridblock distance Δx. Now, it is clear from Figure 5-5 that the first derivative at an intermediate point A between x_{i-1} and x_i is

$$df(x_A)/dx = (x_i - x_{i-1})/\Delta x \tag{5.55}$$

while the first derivative at an intermediate point B between x_i and x_{i+1} is

$$df(x_B)/dx = (x_{i+1} - x_i)/\Delta x \tag{5.56}$$

Hence, the second derivative of f(x) at x_i satisfies

$$d^2f(x_i)/dx^2 = \{df(x_B)/dx - df(x_A)/dx\}/\Delta x \tag{5.57}$$

or

$$d^2f(x_i)/dx^2 = \{f_{i-1} - 2f_i + f_{i+1}\}/(\Delta x)^2 + O(\Delta x)^2 \tag{5.58}$$

Taylor series analysis shows that Equation 5-58 for the second derivative is *second order accurate* in Δx. The $O(\Delta x)^2$ describes the order of the truncation error; if Δx is *small*, then $O(\Delta x)^2$ may be regarded as being *very small*. Likewise, it can be shown that

$$df(x_i)/dx = \{f_{i+1} - f_{i-1}\}/(2\Delta x) + O(\Delta x)^2 \tag{5.59}$$

for the first derivative, is second order accurate. Equations 5-58 and 5-59 are *central difference* representations for the respective quantities at x_i because they involve left and right quantities at x_{i-1} and x_{i+1}.

Note that the *backward difference* formula

$$d^2f(x_i)/dx^2 = \{f_i - 2f_{i-1} + f_{i-2}\}/(\Delta x)^2 + O(\Delta x) \tag{5.60}$$

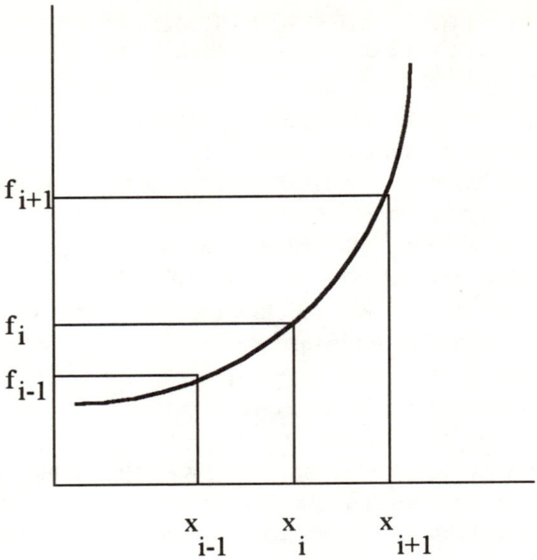

Figure 5-5. Finite difference discretizations.

for the second derivative is *not* incorrect. It is not as accurate as the central difference formula, since it turns out to be *first order accurate,* the error being only *somewhat small*. Similar comments apply to the *forward differencing*

$$d^2f(x_i)/dx^2 = \{f_i - 2f_{i+1} + f_{i+2}\}/(\Delta x)^2 + O(\Delta x) \tag{5.61}$$

Alternative representations for Equation 5-59 are the backward and forward difference formulas

$$df(x_i)/dx = \{f_i - f_{i-1}\}/\Delta x + O(\Delta x) \tag{5.62}$$

$$df(x_i)/dx = \{f_{i+1} - f_i\}/\Delta x + O(\Delta x) \tag{5.63}$$

Appropriate backward or forward difference formulas must be used at boundaries of computational domains; at such boundaries, central difference formulas (e.g., Equations 5-58 and 5-59) require values of i that are outside the domain, and hence, undefined.

Transient finite difference modeling. The governing equation for idealized wave disturbances in the absence of external forces and internal dissipation is the classical wave equation

$$\partial^2u/\partial t^2 - c^2\ \partial^2u/\partial x^2 = 0 \tag{5.64}$$

where c is the sound speed.

This model provides a useful vehicle for introducing basic ideas, and in testing difference schemes for use in *forward simulation*, that is, in modeling events as they evolve in time, for given parameters and auxiliary conditions. Fully transient models, again, are necessary for modeling vibrations with shocks, and in simulating drilling with rate-of-penetration, since harmonic analysis is less valid.

We will solve Equation 5-64 by approximating it with algebraic equations at the nodes formed by a net of coordinate lines in the x-t plane. If we replace our space-time continuum with independent variables formed by a discrete set of spatial points $x_i = i\ \Delta x$, where $i = 1, 2, 3, \ldots, i_{max}$, and a discrete set of timewise points $t_n = n\ \Delta t$, we can represent the function as $u(x,t) = U_{i,n}$.

We expect that, at any time t_n, the function u at any point x_i will be influenced by its left and right neighbors, so that the central difference formula

$$u_{xx}(x_i,t_n) = (U_{i+1,n} - 2U_{i,n} + U_{i-1,n})/(\Delta x)^2 \qquad (5.65)$$

holds. The same cannot be said of the second time derivative. Since *causality* require that events must depend on past and not future history, backward differences necessarily apply. Thus, following Equation 5-60, we must write

$$u_{tt}(x_i,t_n) = (U_{i,n} - 2U_{i,n-1} + U_{i,n-2})/(\Delta t)^2 \qquad (5.66)$$

The solution methodology. Thus, substitution of Equations 5-65 and 5-66 in Equation 5-64 shows that a difference scheme proposed for the second-order wave equation can be represented by the formula

$$(U_{i,n} - 2U_{i,n-1} + U_{i,n-2})/(\Delta t)^2 - c^2 (U_{i+1,n} - 2U_{i,n} + U_{i-1,n})/(\Delta x)^2 = 0 \qquad (5.67)$$

which is $O\{(\Delta x)^2\}$, but only $O(\Delta t)$ correct.

More accurate discretization schemes are available, but the one suggested, implemented later for axial and torsional vibrations, is extremely stable. This stability is essential since "rough drilling" will be encountered computationally, that may degrade less robust schemes. Stability is particularly required when coupling axial with violent lateral vibrations.

It will be useful to define the dimensionless parameter

$$m = c\ \Delta t/\Delta x \qquad (5.68)$$

so that the difference model can be more conveniently written as

$$(U_{i,n} - 2U_{i,n-1} + U_{i,n-2}) = m^2 (U_{i+1,n} - 2U_{i,n} + U_{i-1,n}) \qquad (5.69)$$

or, as we will show, in the preferred form

$$U_{i-1,n} - (2 + 1/m^2)\ U_{i,n} + U_{i+1,n} = (-2U_{i,n-1} + U_{i,n-2})/m^2 \qquad (5.70)$$

The "recipe" inferred from Equation 5-70 is straightforward. For convenience, let us assume that $t = 0$ begins with $n = 3$, and that *identical* static solutions are available and assumed to hold at time steps $n = 1$ and 2. These two solutions "start" our transient calculations from a state of rest.

Now, the transient solution will be computed one time level at a time, with n fixed, beginning with $n = 3$. When a particular value of n is so "frozen", the specialized finite difference equation is written for each of the internal nodal points $i = 2, 3, 4, \ldots$ and so on, until $i = i_{max} - 1$.

The indexes $i = 1$ and i_{max} are specifically excluded from this process since their finite difference equations require U values to the *left* of $i = 1$ and to the *right* of $i = i_{max}$ which do not exist. At this point we have i_{max} unknowns, but only $i_{max} - 2$ linear equations; two more equations, to be obtained from boundary conditions, are needed to complete the formulation.

Consider a six-node system for illustrative purposes, where the index i varies from 1 to 6. Corresponding to $i = 2, 3, 4$ and 5, we have the respective equations

$$U_{1,3} - (2 + 1/m^2)U_{2,3} + U_{3,3} = (-2U_{2,2}+U_{2,1})/m^2 \quad (5.71)$$

$$U_{2,3} - (2 + 1/m^2)U_{3,3} + U_{4,3} = (-2U_{3,2}+U_{3,1})/m^2 \quad (5.72)$$

$$U_{3,3} - (2 + 1/m^2)U_{4,3} + U_{5,3} = (-2U_{4,2}+U_{4,1})/m^2 \quad (5.73)$$

$$U_{4,3} - (2 + 1/m^2)U_{5,3} + U_{6,3} = (-2U_{5,2}+U_{5,1})/m^2 \quad (5.74)$$

which provides four equations for the six unknowns to the left of the equal signs (those on the right are known quantities from previous time steps).

As discussed, two additional equations are obtained from boundary conditions. For example, the equation that would appear just before Equation 5-71 might take the form $U_{1,3} = 0$ when modeling a fixed end, or $U_{1,3} - U_{2,3} = 0$ when assuming a free end, since $\partial u/\partial x = 0$. The analogous equations that would apply just after Equation 5-74 are $U_{6,3} = 0$ and $U_{5,3} - U_{6,3} = 0$. Thus, the final system of equations might take the form,

$$U_{1,3} - U_{2,3} = 0 \quad (5.75a)$$

$$U_{1,3} - (2 + 1/m^2)U_{2,3} + U_{3,3} = (-2U_{2,2}+U_{2,1})/m^2 \quad (5.75b)$$

$$U_{2,3} - (2 + 1/m^2)U_{3,3} + U_{4,3} = (-2U_{3,2}+U_{3,1})/m^2 \quad (5.75c)$$

$$U_{3,3} - (2 + 1/m^2)U_{4,3} + U_{5,3} = (-2U_{4,2}+U_{4,1})/m^2 \quad (5.75d)$$

$$U_{4,3} - (2 + 1/m^2)U_{5,3} + U_{6,3} = (-2U_{5,2}+U_{5,1})/m^2 \quad (5.75e)$$

$$U_{5,3} - U_{6,3} = 0 \quad (5.75f)$$

This short example illustrates the *set-up* process only. Equations 5-75a to 5-75f, as shown, will yield zero solutions for zero initial conditions, because an excitation source has not *yet* been introduced. In general, it is clear that when i_{max} spatial nodes define the drillstring, corresponding to (i_{max} -1) grid blocks, the resulting equations take the *tridiagonal* matrix form

$$\begin{vmatrix} B_1 & C_1 & & & \\ A_2 & B_2 & C_2 & & \\ & A_3 & B_3 & C_3 & \\ & & \cdots\cdots & & \\ & & & & \\ & & A_{imax-1} & B_{imax-1} & C_{imax-1} \\ & & & A_{imax} & B_{imax} \end{vmatrix} \begin{vmatrix} V_1 \\ V_2 \\ V_3 \\ .. \\ \\ V_{imax-1} \\ V_{imax} \end{vmatrix} = \begin{vmatrix} W_1 \\ W_2 \\ W_3 \\ \ldots \\ \\ W_{imax-1} \\ W_{imax} \end{vmatrix} \quad (5.76)$$

Thus, this difference scheme is said to involve *tridiagonal equations.* Equation 5-76 can be solved computationally by *calling* the Fortran subroutine in Figure 5-6 below. Tridiagonal forms are important to numerical analysis because an order N system (that is, N equations in N unknowns) requires only 3N *multiplies* or *divides* for inversion. By contrast, inefficient *Gaussian elimination* (for full matrixes) requires N^3 such operations.

```
      SUBROUTINE TRIDI(A,B,C,V,W,N)
      DIMENSION A(100), B(100), C(100), V(100), W(100)
      A(N) = A(N)/B(N)
      W(N) = W(N)/B(N)
      DO 100  I = 2,N
      II = -I+N+2
      BN = 1./(B(II-1)-A(II)*C(II-1))
      A(II-1) = A(II-1)*BN
      W(II-1) = (W(II-1)-C(II-1)*W(II))*BN
100   CONTINUE
      V(1) = W(1)
      DO 200  I = 2,N
      V(I) = W(I)-A(I)*V(I-1)
200   CONTINUE
      RETURN
      END
```

Figure 5-6. Tridiagonal matrix solver.

It is also important that the tridiagonal structure is *diagonally dominant*, that is, the absolute value of the diagonal coefficient $(2+1/m^2)$ exceeds the sum of the (unity) coefficients of the side diagonals. This property lends itself to *numerical stability*, meaning that iterative solutions are not likely to "blow up" as a result of *truncation* and *round-off errors*. This does not guarantee that the computed solutions are correct, but it does buttress the accepted (but questionable) philosophy that "any solution is better than no solution."

Only two previous time levels of solution need to be stored at any given step of the solution process, so that three levels of information are required in total. Thus, the Fortran associated with our scheme can be written using three *dimensioned scalar arrays* UN(1000), UNM1(1000), and UNM2(1000) only, representing $u_n(i)$, $u_{n-1}(i)$ and $u_{n-2}(i)$, where the *dimension* 1000 might signify 1000 closely spaced nodes. At the end of each time step, we copy UNM1 into UNM2, and UN into UNM1, and repeat the procedure until termination.

It is not necessary (or advisable) to have computer memory allocated for a complete *field* U(1000,500), say, representing 1000 nodes, and 500 time steps. Intermediate results, e.g., bit location, velocity, and stress, can be written to output files for subsequent post-processing and display.

Also note that the matrix coefficients **A**, **B**, and **C** need not be recomputed for subsequent time steps, since they are constants defined once and for all (for example, refer to the Fortran listing in *Figure 5-8* for matrix coefficient definition). While the matrix solver given in Figure 5-6 *will* destroy **A**, **B**, and **C** at the end of each inversion, thus requiring redefinition of these coefficients prior to each integration, other solvers are available which retain their input values, at the expense of increased memory requirements.

Stability of the scheme. We can address the *stability* issue quantitatively, through *von Neumann analysis*. Let us refer to Equation 5-70. Since the difference equation is linear, with constant coefficients, exact solutions are sought in the *separated* form

$$U_{i,n} = \beta^n \exp[(-1)^{1/2} \alpha\, i] \tag{5.77}$$

This assumption (or, *Ansatz*) applies only if α and β are related by the equation

$$(1 + 4m^2 \sin^2 \alpha/2)\, \beta^2 - 2\beta + 1 = 0 \tag{5.78}$$

a requirement found on direct substitution in Equation 5-70. If we now set

$$\cos\theta = (1 + 4m^2 \sin^2 \alpha/2)^{-1/2} \tag{5.79}$$

that is,

$$\theta = \tan^{-1}(2m \sin \alpha/2) \tag{5.80}$$

it follows that

$$\beta = \exp[\pm(-1)^{1/2}\,\theta]\cos\theta \tag{5.81}$$

For all real values of α and m, the *amplification factor* β clearly satisfies

$$|\beta| \leq 1 \tag{5.82}$$

so that the finite difference scheme is *absolutely* or *unconditionally stable*.

Numerical stability does not guarantee that a computed solution tends to a solution of the differential equation. This issue is a subtle and important one, and forms the subject of ongoing researches. Let us consider a practical illustrative example for this nonconvergence. In the present case, an exact solution u(x,t) of the *differential* equation is found to be

$$u_{i,n} = \exp[(-1)^{1/2}\,\lambda\,(x_i \pm t_n)] \tag{5.83}$$

for real λ, where $x_i = i\,\Delta x$ and $t_n = n\,\Delta t$. This is easily compared with Equation 5-77 if we rewrite it with $\alpha = \Delta x\,\lambda$ in the form

$$U_{i,n} = \exp[(-1)^{1/2}\,\lambda\,(x_i \pm \psi\,t_n)]\cos^n\theta \tag{5.84}$$

where

$$\psi = \theta/(\lambda\Delta t) = \{1/(\Delta x\,\lambda m)\}\tan^{-1}(2m\sin\Delta x\,\lambda/2) \tag{5.85}$$

If Δx is small and λ is independent of Δx, we have by Taylor series expansion

$$\psi = 1 - \{(1+8m^2)/24\}\lambda^2\Delta x^2 + O(\Delta x^4) \tag{5.86}$$

and

$$\begin{aligned}\cos^n\theta &= \exp[n\log\cos\theta]\\ &= \exp[(-n/2)\log(1+4m^2\sin^2\alpha/2)]\\ &= \exp[-\Delta x\,m\lambda^2 t_n/2 + O(\Delta x^3 t_n)]\end{aligned} \tag{5.87}$$

Now, the real and imaginary parts of the exact solution oscillate periodically in time without damping. On the other hand, the corresponding terms of $U_{i,n}$ possess damped oscillations, the amplitude being attenuated like $\exp[-\Delta x\,m\lambda^2 t_n/2 + O(\Delta x^3 t_n)]$. Also, the period is increased in the ratio $1/\psi = 1 + \{(1+8m^2)\lambda^2\Delta x^2/24$.

Grid sizes, time steps, and convergence. All of this means that care must be taken in judiciously selecting space and time steps. It does not imply that numerical solution are undesirable; often, they are the only means of obtaining solutions. From a different perspective, we must also understand that $\rho\,\partial^2u/\partial t^2 + \zeta\,\partial u/\partial t - E\,\partial^2u/\partial x^2 + \rho g = F_e$ is limited as a model for the physical system; little real benefit is obtained by solving with very fine discretization. For example, $\zeta\,\partial u/\partial t$ damping, chosen for mathematical simplicity, does not model the frequency-dependent dissipation that characterizes real drillstrings.

Practical errors in characterizing F_e may exceed those associated with truncation errors. Thus, subjective value judgments are often warranted.

Again, absolute stability is a desirable trait for numerical schemes. However, it provides no guarantee that computations will converge to solutions of the *differential equation* being modeled, even in the limit of vanishing Δx and Δt, if the ratio $\Delta x/\Delta t$ happens to be inappropriate. *Convergence* has long intrigued mathematicians, and remains an area of active research.

One requirement for convergence and physical resolution is a solid understanding of the (possibly, multiple) length and time scales characterizing the problem, e.g., the wavelength and period of a typical propagating disturbance (longer space-time scales will arise from dissipative effects).

For example, consider a drillstring rotating at 60 rpm (that is, 60 cycles/60 sec or 1 Hz); assuming a tricone bit, the axial frequency may be a nominal "3 x rpm" or 3 Hz. If the sound speed is 15,000 ft/sec, the wavelength is λ = (15,000 ft/sec)/(3 Hz) = 5000 ft. On this basis, the choice Δx = 5% λ or 250 ft *might* suffice; if the numerical scheme used is *conditionally stable*, Δx must be chosen so that it satisfies both *stability* and *resolution requirements*.

Once Δx is selected, suitable time steps Δt are needed. For any particular transient scheme, the temptation to choose unrealistically large Δt's exists. On an exact basis, $\Delta t = (\Delta x)/c$. Thus, any attempt to choose larger steps will violate the physics, and usually, will lead to instabilities. The restriction to $\Delta t < (\Delta x)/c$ is known as the *CFL* or *Courant-Friedrichs-Lewy criterion*.

We emphasize that high excitation frequencies will lead to large truncation errors unless time steps small compared to a period are used (a *5% rule* is recommended). Such errors appear as losses in amplitudes; the phase errors noted earlier affect computed propagation speeds, in such a way that they will differ from the "c" in $\partial^2 u/\partial t^2 - c^2\, \partial^2 u/\partial x^2 = 0$. Series solutions may provide some improvement, but in this author's experience with drillstring vibration and MWD telemetry problems, *hundreds* of terms may be necessary.

COMPLETE FORMULATION AND NUMERICAL SOLUTION

In this section, we summarize the complete boundary value problem, and give the finite difference formulation and numerical algorithm needed to model axial vibrations with rock-bit interaction, bit-bounce, rate-of-penetration, and drillbit kinematics. Displacements, velocities, and stresses everywhere along the drillstring can be computed, with the main quantities of interest being those at the formation contact point x = 0.

The boundary value problem. We recapitulate the boundary value problem for the partial differential equation governing the complete *static plus dynamic* displacement u(x,t), to include simple damping, body forces, external dipole drillbit excitation, surface and rock-bit interaction boundary conditions, and initially quiescent starting conditions. A uniform drillstring with constant properties is assumed, but modifications for nonuniformity are given later. Again, we have

Partial differential equation for uniform drillstring:

$$\rho\, \partial^2 u/\partial t^2 + \zeta\, \partial u/\partial t - E\, \partial^2 u/\partial x^2 + \rho g = F_e \qquad (5.25)$$

Displacement source model for tricone bit:

$$[u]_{x=xbit} = f(t) \ldots \{e.g.,\ u_0 \sin \omega t\} \qquad (5.30)$$

Surface boundary condition:

$$M\, \partial^2 u/\partial t^2 + \beta\, \partial u/\partial t + AE\, \partial u/\partial x + ku + Mg = 0 \text{ at } x = L \qquad (5.26)$$

Rock-bit interaction model at formation contact:

$$G(u, \partial u/\partial x, \partial u/\partial t) = 0 \text{ at } x = 0 \qquad (5.27)$$

$$\partial u(0,t)/\partial x = 0, \textit{ if } \partial u(0,t)/\partial t > 0 \qquad (5.23)$$

Initial conditions for a static drillstring at rest:

$$u(x,0) = u_s(x) = 1/2\ \rho g x^2/E + W_0 x/AE \qquad (5.28)$$

$$\partial u(x,0)/\partial t = 0 \qquad (5.29)$$

Computational objective. We wish to calculate the *displacement* u(x,t) for $0 < x < L$ and $t > 0$, and from it, determine the *velocity* $\partial u(x,t)/\partial t$ and the *strain* $\partial u(x,t)/\partial x$ everywhere; the *acceleration* is just $\partial^2 u(x,t)/\partial t^2$, while the *force* equals AE $\partial u(x,t)/\partial x$. These quantities are obtained by post-processing the array $U_{i,n}$ using the forward, backward, and central difference formulas given earlier.

Of particular engineering interest are speed, force, and displacement *at the bit*; and, for MWD designers, maximum accelerations and forces along the drillstring. The formulation above completely specifies the boundary value problem; we now apply the general finite difference ideas developed earlier to the set of equations given here. We will also discuss special details related to modeling displacement sources and rock-bit interaction.

Difference approximations. Equation 5-64, used as a simple model for discussion, does not contain damping or body forces, but similar ideas apply to Equation 5-25 (with $F_e = 0$). Again, i = 1 represents the rock-bit contact point downhole, while $i = i_{max}$ represents the surface. We have, corresponding to Equation 5-67, the discretization

$$\rho\,(U_{i,n} - 2U_{i,n-1} + U_{i,n-2})/(\Delta t)^2 + \zeta\,(U_{i,n} - U_{i,n-2})/(2\Delta t) - E\,(U_{i+1,n} - 2U_{i,n} + U_{i-1,n})/(\Delta x)^2 + \rho g = 0 \tag{5.88}$$

From this equation follows the tridiagonal form

$$U_{i-1,n} - \{2 + (\rho/E)(\Delta x/\Delta t)^2 + (\zeta/2E)((\Delta x)^2/\Delta t)\}\,U_{i,n} + U_{i+1,n} = -(\rho/E)(\Delta x/\Delta t)^2\,(2U_{i,n-1} - U_{i,n-2}) - (\zeta/2E)((\Delta x)^2/\Delta t)\,U_{i,n-2} + \rho g(\Delta x)^2/E \tag{5.89}$$

analogous to Equation 5-70.

The surface boundary condition in Equation 5-26 can be differenced as

$$M(u_{imax,n} - 2u_{imax,n-1} + u_{imax,n-2})/(\Delta t)^2 + \beta\,(u_{imax,n} - u_{imax,n-2})/(2\Delta t) + AE\,(u_{imax,n} - u_{imax-1,n})/\Delta x + k\,u_{imax,n} + Mg = 0 \tag{5.90}$$

This result can be rearranged so that it occupies the bottom row of the coefficient matrix shown in Equation 5-76, taking the form

$$(-AE/\Delta x)\,u_{imax-1,n} + \{M/(\Delta t)^2 + \beta/(2\Delta t) + AE/\Delta x + k\}\,u_{imax,n} = \{2M/(\Delta t)^2\}u_{imax,n-1} + \{\beta/(2\Delta t) - M/(\Delta t)^2\}u_{imax,n-2} - Mg \tag{5.91}$$

The boundary condition imposed at $x = 0$ occupies the top row of the matrix in Equation 5-76. It is somewhat more complicated than the surface condition above. Because the drillstring may *forge ahead* or *retreat upward*, the boundary model describes *either* rock-bit interaction *or* a stress-free end. Since it is impossible to know *a priori* which boundary condition applies at any instant, the choice itself must be determined as part of the numerical solution. If, based on results from the preceding time step, the *drillbit* velocity $\partial u/\partial t$ is *upward*, then the *stress-free* end condition $\partial u/\partial x = 0$ will be imposed, that is,

$$U_{1,n} - U_{2,n} = 0 \tag{5.92}$$

following Equation 5-23. On the other hand, if the *drillbit* velocity $\partial u/\partial t$ is *downward* into the formation, then the *rock-bit boundary condition* in Equations 5-22 or 5-45 (or, $\alpha u_x + \beta u_t + \lambda u = 0$) will be discretized in the form

$$\{-\alpha/\Delta x + \beta/\Delta t + \lambda\}\,U_{1,n} + \{\alpha/\Delta x\}\,U_{2,n} = \{\beta/\Delta t\}\,U_{1,n-1} \tag{5.93}$$

In implementing this switching criterion, we obviously take advantage of the capabilities offered by Fortran *if-then* logic. Any other programming language may be used, but in this book, subroutines are conveniently given in Fortran.

We have referred to α, β and λ as constants, for the purposes of analytical solution, but these quantities can in fact represent *functionals* of $u(x,t)$ and its space and time derivatives, if empirically justified. This change is trivial numerically. The extension from a linear model to *any* nonlinear one, e.g., $G(u, \partial u/\partial x, \partial u/\partial t) = 0$ at $x = 0$, is simple from a programming point of view.

So far, we have imposed *boundary conditions* at the surface $x = L$ and downhole at the rock-bit interface $x = 0$. We have yet to specify the *displacement source (ds) excitation* that models the axial reciprocation due to the rotating drillbit. Following Chapter 1, with reference to Equation 1-120, we require that the *jump in displacement* or the *length generation* satisfy

$$[u]_{x=xbit} = - F_{ds}\, dx / E \tag{5.94}$$

Again, the excitation in Equation 5-94 is formed by the action of two forces of magnitude F_{ds}, acting in opposite directions, and separated by a distance Δx. We will model these forces directly, with *each force* numerically modeled as follows.

Since *a single force is responsible for a discontinuity in the first spatial derivative* (refer to Chapter 1 for a detailed discussions), we have in the notation of the present chapter

$$F_{ds}/E = u_x^+ - u_x^- = (U_{ibit+1,n} - U_{ibit+,n})/\Delta x - (U_{ibit-,n} - U_{ibit-1,n})/\Delta x \tag{5.95}$$

where $i = i_{bit}$ represents the spatial index of the bit centroid $x = x_{bit}$.

Now, the displacement values $U_{ibit+,n}$ and $U_{ibit-,n}$ to the *immediate* right and left of $i = i_{bit}$ are identical, since the displacement function is continuous for isolated forces. Thus, we denote $U_{ibit+,n}$ and $U_{ibit-,n}$ by the single-valued quantity $U_{ibit,n}$. Hence, the action of a single *force* leads to

$$U_{ibit-1,n} - 2U_{ibit,n} + U_{ibit+1,n} = F_{ds}\Delta x/E = - [u]_{x=xbit} \tag{5.96}$$

where we have used Equation 5-94.

The second *force* required to complete the construction of the desired *couple* can be taken a single grid block away, with the required opposite sign, in the form

$$U_{ibit,n} - 2U_{ibit+1,n} + U_{ibit+2,n} = -F_{ds}\Delta x/E = + [u]_{x=xbit} \tag{5.97}$$

The *strength* of the displacement source $[u]_{x=xbit}$ can, for simplicity, be taken in the form

$$[u]_{x=xbit} = u_0 \sin \omega t,\ \omega = 2\pi\, (3N_{rpm})/60 \tag{5.98}$$

where u_0 represents (the absolute value of) the maximum axial displacement generation due to the drillbit. Here, N_{rpm} is the drillstring *rpm*, and the factor 3 models the tricone nature of the bit. PDC bits, on the other hand, may require factors anywhere from ranging 9 to 20, with smaller values of u_0.

Finally, initial conditions are required to start the transient calculations, as discussed earlier. Because the right side of Equation 5-89 indicates that information from two earlier time steps are required before the time-iterations may begin, we initialize *both* of these initial solutions using the static displacement function given by Equation 5-28 (this automatically satisfies the static requirement in Equation 5-29).

MODELING PIPE-TO-COLLAR AREA CHANGES

So far, we have assumed a uniform drillstring to illustrate the basic ideas. If area discontinuities and changes in material properties exist, e.g., at the pipe-to-collar interface, the partial differential equations (PDEs) to the left and right break down *at the interface* because spatial derivatives taken through it do not exist.

In reality, the drillstring is composed of numerous sections, such as drillpipe, shock sub, jarring devices, and drill collar, each being characterized by different cross-sectional properties. For such problems, separate PDEs apply to each section, say, with different subscripted terms ρ_1, ζ_1, E_1, A_1, and u_1, and ρ_2, ζ_2, E_2, A_2, and u_2.

Matching conditions. The solutions at either side are connected by special matching conditions $u_1 = u_2$ and $A_1E_1\partial u_1/\partial x = A_2E_2\partial u_2/\partial x$ (in Part III, we show how these do *not* apply to fluid columns). The first, continuity of displacement, applies because the system does not "tear" at the interface. The second, continuity of force, holds because there is no additional external loading applied at the point; stress, incidentally, is double-valued or discontinuous. Double-valued stresses are well known in *strength of materials*; normal stresses abruptly change at changes in cross-sectional area.

Let us consider, for example, a two-section drillstring consisting of the pipe (p) and collar (c). Each section of drillpipe, as discussed, is governed by its own unique partial differential equation, that is,

$$\rho^{(p)}\, \partial^2 u^{(p)}/\partial t^2 + \zeta^{(p)}\, \partial u^{(p)}/\partial t - E^{(p)}\, \partial^2 u^{(p)}/\partial x^2 + \rho^{(p)} g = 0 \qquad (5.99)$$

$$\rho^{(c)}\, \partial^2 u^{(c)}/\partial t^2 + \zeta^{(c)}\, \partial u^{(c)}/\partial t - E^{(c)}\, \partial^2 u^{(c)}/\partial x^2 + \rho^{(c)} g = 0 \qquad (5.100)$$

Again, at the pipe-to-collar interface, the PDEs break down and do not apply, since sudden changes in the medium lead to rapid wave variations for which the spatial derivatives may not exist. There they are replaced by matching conditions, that allow each wave solution to *analytically continue* into the other, while conserving physically correct quantities. Now let us derive the difference model.

Finite difference model. The condition describing *force continuity* can be expressed as

$$A^{(p)}E^{(p)}\, \partial u^{(p)}/\partial x = A^{(c)}E^{(c)}\, \partial u^{(c)}/\partial x \tag{5.101}$$

We have used *Hooke's law* $\sigma = E\varepsilon$, which relates the stress σ to the strain ε (in our notation, $\sigma = E\, \partial u/\partial x$, and force varies like σA). In addition, since the drillpipe and drill collar are rigidly held together, the *displacement must remain continuous*, so that

$$u^{(p)} = u^{(c)} \tag{5.102}$$

Now, we importantly *require* that Equations 5-101 and 5-102 adapt to the finite difference formulation derived earlier for uniform drillstrings in a *stable* manner, so that divergent numerical results do not appear during the iterations. Ideally, the matching condition will be similar in appearance to the difference equation for the PDE and retain *diagonal dominance*.

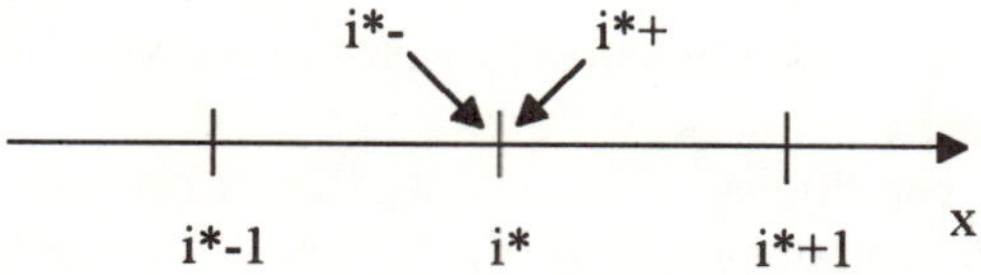

Figure 5-7 . Pipe-to-collar indexing convention.

Let us designate by i* the x-index of the interface, as shown in Figure 5-7, and assume uniform grids Δx. Equation 5-101 can be differenced as follows,

$$A^{(p)}E^{(p)}\,(u_{i^*+1,n} - u_{i^*+,n})/\Delta x = A^{(c)}E^{(c)}\,(u_{i^*-,n} - u_{i^*-1,n})/\Delta x \tag{5.103}$$

where i*+ and i*- represent infinitesimally close positions on either side of i* (we again assume that the index increases from i = 1 at the bottom x = 0, to i = i_{max} at the top of the drillstring x = L). Equation 5-102 indicates that the displacement function is continuous through the interface, so that both $u_{i^*+,n}$ and $u_{i^*-,n}$ can be written as as $u_{i^*,n}$. If we multiply by Δx throughout, we can rewrite Equation 5-103 in the form

$$A^{(c)}E^{(c)}\, u_{i^*-1,n} - (A^{(p)}E^{(p)} + A^{(c)}E^{(c)})\, u_{i^*,n} + A^{(p)}E^{(p)}\, u_{i^*+1,n} = 0 \tag{5.104}$$

Generalized formulation. We modify our earlier finite difference recipe as follows. *First*, we replace Equation 5-88 for uniform drillstrings with the differenced forms of Equations 5-99 and 5-100 as required for different values of the index i, noting that the coefficients in these equations are different. *Second,* at the index i = i* corresponding to the pipe-to-collar interface, we replace the PDE-based difference equation we (may) have written, with Equation 5-104 above.

Third, above the pipe-collar interface at the surface, Equation 5-90 applies to $u^{(p)}$. *Fourth,* below the pipe-collar interface, our rock-bit boundary conditions and displacement source model given by Equations 5-92, 5-93, 5-96, and 5-97 apply to $u^{(c)}$.

Because Equation 5-104, by construction, is diagonally dominant, we are assured that the resulting time integrations are stable. This does not guarantee convergence, but again, stability is at least essential to having a robust algorithm. Near the interface, the quantities $u_{i*-1,n}$ and $u_{i*+1,n}$ refer to collar and pipe values, respecticely, while $u_{i*,n}$ applies equally to both sections.

Alternative boundary conditions. The downhole boundary condition formulations we have given are quite general, but for completeness, we will list without discussion other models that have been considered in the literature.

Fixed end:

$$u(0,t) = 0 \tag{5.105}$$

Free end:

$$\partial u(0,t)/\partial x = 0 \tag{5.106}$$

Mass loaded end:

$$m\, \partial^2 u(0,t)/\partial t^2 = AE\, \partial u(0,t)/\partial x \tag{5.107}$$

Spring loaded end:

$$\alpha u_x + \lambda u = 0 \tag{5.108}$$

Dashpot end:

$$\alpha u_x + \beta u_t = 0 \tag{5.109}$$

EXAMPLE FORTRAN IMPLEMENTATION

Part II of this book deals with *drillstring vibrations*; in particular, Chapters 5-8 deal with *axial, lateral, torsional* and *fully coupled vibrations,* in that order. The numerical algorithms discussed have been extensively tested, but we defer sample numerical calculations until Chapter 8.

There, we will describe a fairly general Fortran program developed for coupled vibrations, expanding on that of Chin (1988a,b). The research algorithm allows users to decouple or couple any component modules as desired and to introduce general rock-bit boundary condition models as needed.

The *code fragment* shown in Figure 5-8 was extracted from this program to illustrate how our difference equations are coded and solved by the tridiagonal matrix solver in Figure 5-6. In this program, the following nomenclature applies (for now, disregard θ, V and W).

`AE` ... Product of area and Young's modulus
`BETA` ... Damping factor in surface cables
`DUDX` ... $\partial u(0,t)/\partial x$ at bit
`ELAST` ... Young's modulus
`G` ... Acceleration due to gravity
`GAMA` ... Damping factor in drillstring
`POWER` ... Power at bit
`RBALPH` ... " α " rock-bit interaction parameter
`RBBETA` ... " β " rock-bit interaction parameter
`RBLAMB` ... " λ " rock-bit interaction parameter
`RHO` ... Mass density
`SPRING` ... Surface spring constant
`TBMASS` ... Mass of traveling block system
`UZERO` ... u_0
`VEL` ... Velocity at bit
`WMEGA` ... ω
`WZERO` ... W_0
`X` ... X position
`XDERIV` ... $\partial u(0,t)/\partial x$ at bit
`XS` ... Drillstring length L
.
`I` ... Spatial x index
`IBIT` ... Spatial index at drillbit centroid
`IMAX` ... Spatial index at surface end of drillstring
`DX` ... Δx
.
`T` ... Time
`N` ... Time step index
`DT` ... Δt

`A(I)` ... " A " matrix coefficient in Equation 5-76
`B(I)` ... " B " matrix coefficient in Equation 5-76

`C(I)` ... " C "matrix coefficient in Equation 5-76
`W(I)` ... " W" matrix coefficient in Equation 5-76
`VECTOR(I)` ... " V "matrix coefficient in Equation 5-76

.

`UN(I)` ... $U_{i,n}$ axial displacement
`UNM1(I)` ... $U_{i-1,n}$ axial displacement
`UNM2(I)` ... $U_{i-2,n}$ axial displacement

.

`TN(I)` ... $\theta_{i,n}$ angular displacement
`TNM1(I)` ... $\theta_{i-1,n}$ angular displacement
`TNM2(I)` ... $\theta_{i-2,n}$ angular displacement

.

`VN(I)` ... $V_{i,n}$ lateral displacement
`VNM1(I)` ... $V_{i-1,n}$ lateral displacement
`VNM2(I)` ... $V_{i-2,n}$ lateral displacement

.

`WN(I)` ... $W_{i,n}$ lateral displacement
`WNM1(I)` ... $W_{i-1,n}$ lateral displacement
`WNM2(I)` ... $W_{i-2,n}$ lateral displacement

Not shown in Figure 5-8 are the Fortran *dimension statements* for the arrays used and the interactive input queries for specific simulation parameters. The first block initializes the drillstring to simple compression, without torsional twist or lateral deflection.

Then, time iterations commence. At each step in the time loop, the matrix coefficients corresponding to Equation 5-89 are given in the 200 sub-loop. This is followed, respectively, by program logic for the *surface boundary condition* in Equation 5-91, the *rock-bit interaction model* in Equations 5-92 and 5-93, and the *displacement source model* in Equations 5-96 and 5-97.

Once the tridiagonal matrix solver is *called* and executed, the solution for the axial displacement $U_{i,n}$ is available for post-processing as discussed. The solution *at the bit* will be used to determine appropriate torsional bit boundary conditions, to be discussed in Chapter 7; the *complete axial solution* is used to evaluate variable coefficients in the bending equations of Chapters 6 and 8.

The reader should experiment with our Fortran. For example, by removing the rock-bit and displacement source logic, and replacing it with B(1) = 1.0, C(1) = 0.0, and W(1) = UZERO*SIN(WMEGA*T), deleting the gravity term in the difference equation, and initializing with zero displacements, a transient implementation of the Dareing and Livesay (1968) model is obtained.

```
.
.
C
C
C     INITIALIZE AXIAL, ANGULAR AND BENDING DISPLACEMENTS
      DO 130  I=1,IMAX
      X = XS*(I-1)/(IMAX-1)
      UNM1(I) = RHO*G*(X**2)/(2.*ELAST) +WZERO*X/AE
      UNM2(I) = UNM1(I)
      TNM1(I) = 0.
      TNM2(I) = 0.
      VNM1(I) = 0.
      VNM2(I) = 0.
      WNM1(I) = 0.
      WNM2(I) = 0.
  130 CONTINUE
C     BEGIN TIMEWISE INTEGRATION
      T = 0.
      DO 900  N=1,NMAX
      T = N*DT
C     AXIAL VIBRATIONS
      DO 200  I=2,IMAXM1
      A(I) = 1.
      C(I) = 1.
      B(I) = -2.-RHO*DX*DX/(ELAST*DT*DT)-GAMA*DX*DX/(2.*ELAST*DT)
      W(I) = -RHO*DX*DX*(2.*UNM1(I)-UNM2(I))/(ELAST*DT*DT)
     1       -GAMA*DX*DX*UNM2(I)/(2.*ELAST*DT)+RHO*G*(DX**2)/ELAST
  200 CONTINUE
      A(IMAX) = -AE/DX
      B(IMAX) = TBMASS/(DT**2) +BETA/(2.*DT) +AE/DX +SPRING
      C(IMAX) = 99.
      W(IMAX) = 2.*TBMASS*UNM1(IMAX)/(DT**2)
     1       +(BETA/(2.*DT) -TBMASS/(DT**2))*UNM2(IMAX)-TBMASS*G
      A(1) = 99.
      VEL = (UNM1(1)-UNM2(1))/DT
      IF(VEL.LE.0.) GO TO 205
      B(1) = -1.
      C(1) = +1.
      W(1) =  0.
      GO TO 215
  205 B(1) = -RBALPH/DX +RBBETA/DT +RBLAMB
      C(1) = +RBALPH/DX
      W(1) = +RBBETA*UNM1(1)/DT
  215 XDERIV = (UNM1(2)-UNM1(1))/DX
  235 IF(VEL.GT.0.) GO TO 240
      A(IBIT) = 1.
      B(IBIT) = -2.
      C(IBIT) = 1.
      W(IBIT) = -UZERO*SIN(WMEGA*T)
      A(IBIT+1) = 1.
      B(IBIT+1) = -2.
      C(IBIT+1) = 1.
      W(IBIT+1) = +UZERO*SIN(WMEGA*T)
```

Figure 5-8. Axial vibration code listing.

```
 240   CALL TRIDI(A,B,C,VECTOR,W,IMAX)
       DO 250  I=1,IMAX
       UN(I) = VECTOR(I)
 250   CONTINUE
       POWER = -AE*VEL*XDERIV
       DO 270  I=1,IMAX
       UNM2(I) = UNM1(I)
       UNM1(I) = UN(I)
 270   CONTINUE
       DUDX(1) = (UN(2)-UN(1))/DX
       DUDX(IMAX) = (UN(IMAX)-UN(IMAXM1))/DX
       DO 275  I=2,IMAXM1
       DUDX(I) = (UN(I+1)-UN(I-1))/(2.*DX)
 275   CONTINUE
C
.
```

Figure 5-8. Axial vibration code listing (cont'd).

Modeling dynamically steady problems. Drillstrings undergo fully transient vibrations, e.g., bit bounce, rate-of-penetration, high shock loadings, intermittent drillbit chatter, and so on. But it may well be that periodic external excitations do lead to periodic motions for limited ranges in time.

When a drillstring is excited harmonically at constant frequency ω and periodicity is observed, the drillstring motion is *dynamically steady* because its standing wave pattern remains unchanged with time. If such a situation exists, it *will* be computed by the transient algorithm; *if it does not, it won't.*

Thus, a transient formulation is more general and less restrictive; for the general algorithm in Chapter 8, *Intel*™ 386/486 class personal computers more than suffice for the most demanding computations. This, again, is so for two reasons. First, *elastic line modeling* reduces the number of unknowns; second, the *absolute stability* and *second order spatial accuracy* of our axial, torsional, and lateral vibration algorithms allows us to select large time steps.

Therefore, we will not formulate special dynamically steady models in this book. We will, however, review the basic analytical ideas, if only to demonstrate the mathematics (and complications) involved. Let us consider *uniform* drillstrings satisfying the governing partial differential equation

$$\rho\, \partial^2 u_d(x,t)/\partial t^2 + \zeta\, \partial u_d(x,t)/\partial t - E\, \partial^2 u_d(x,t)/\partial x^2 = 0 \qquad (5.110)$$

for the transient part $u_d(x,t)$ of the complete displacement field $u(x,t)$

If the surface can be represented by a stress-free end with $\partial u_d(L,t)/\partial x = 0$, while the drillbit is forced to oscillate with the usual imposed displacement $u_d(0,t) = u_0\, e^{i\omega t}$, where the amplitude u_0 is real, simplifications are possible.

The motivation for complex exponentials was given in in Chapter 1. Again, the notation $u_0\, e^{i\omega t}$ means that only the real part $u_0 \cos \omega t$ (that is, *Real* $\{u_0\, e^{i\omega t}\}$) of the complete expression is significant. This device, owing to linearity, allows us to express constraints in a simple manner.

We next consistently formulate the displacement problem in terms of an amplitude function U(x) satisfying $u_d(x,t) = e^{i\omega t}\, U(x)$. Equation 5-110 reduces to an ordinary differential equation, namely,

$$U''(x) + \{(\rho\omega^2 - i\zeta\omega)/E\}U(x) = 0 \tag{5.111}$$

whose general solution takes the form

$$U(x) = A \sin \{\sqrt{(\rho\omega^2 - i\,\zeta\omega)/E}\}\, x + B \cos \{\sqrt{(\rho\omega^2 - i\,\zeta\omega)/E}\}\, x \tag{5.112}$$

where A and B are integration constants.

At the bit, we therefore find that $u_d(0,t) = e^{i\omega t}\, U(0) = e^{i\omega t}\, B = u_0\, e^{i\omega t}$, so that $B = u_0$. The fact that $U'(L) = A\{\sqrt{(\rho\omega^2 - i\,\zeta\omega)/E}\}\cos\{\sqrt{(\rho\omega^2 - i\,\zeta\omega)/E}\}L - B\{\sqrt{(\rho\omega^2 - i\,\zeta\omega)/E}\}\sin\{\sqrt{(\rho\omega^2 - i\,\zeta\omega)/E}\}L = 0$ implies that the remaining constant satisfies $A = B \tan\{\sqrt{(\rho\omega^2 - i\,\zeta\omega)/E}\}L = u_0 \tan\{\sqrt{(\rho\omega^2 - i\,\zeta\omega)/E}\}L$. The solution for U(x) is completely determined.

From Equation 5-112, it is clear that U(x) must be complex, that is, $U(x) = U_r(x) + i\, U_i(x)$ where $U_r(x)$ and $U_i(x)$ are real. We had started with the convention that *Real* $\{u_d(0,t)\}$ = *Real* $\{u_0\, e^{i\omega t}\} = u_0 \cos \omega t$. Correspondingly, we have the solution *Real* $\{e^{i\omega t}U(x)\}$ = *Real* $\{U_r\cos \omega t - U_i\sin \omega t + i\, U_i\cos \omega t + iU_r\sin \omega t\} = U_r\cos \omega t - U_i\sin \omega t$ where, again, the (complex) function U(x) is obtained by simplifying Equation 5-112 as required.

It is apparent that the formulation simplifications afforded by timewise periodicity do *not* justify the obvious analytical complexities; more difficult boundary conditions, for example, $\alpha u_x + \beta u_t + \lambda u = 0$, are not easily handled.

A fully transient numerical formulation, as suggested here, will yield dynamically steady solutions if and when they exist, but it will predict bit bounce and shock loadings as they arise. The difference model deals with a real dependent variable whether or not damping exists, and is simpler in this regard. Robust numerical approaches, therefore, offer simplicity and complete generality for both linear and nonlinear vibrations.

Next, consider *nonuniform* drillstrings. We cite the work of Dareing and Livesay (1968), who consider vertical drillstrings consisting of two sections. The first section, marked by "1" subscripts, represents the drill collar occupying $0 < x < L_1$, where L_1 is the collar length. The second section, denoted by "2" subscripts, represents the drillpipe in the interval $L_1 < x < L_2$. Thus, our Equation 5-110 is replaced by the pair of partial differential equations

$$\rho_1\, \partial^2 u_{d,1}(x,t)/\partial t^2 + \zeta_1\, \partial u_{d,1}(x,t)/\partial t - E_1\, \partial^2 u_{d,1}(x,t)/\partial x^2 = 0 \tag{5.113a}$$

$$\rho_2\, \partial^2 u_{d,2}(x,t)/\partial t^2 + \zeta_2\, \partial u_{d,2}(x,t)/\partial t - E_2\, \partial^2 u_{d,2}(x,t)/\partial x^2 = 0 \tag{5.113b}$$

The authors then assume that

$$u_{d,1}(x,t) = U_1(x)\, e^{i\omega t} \tag{5.114a}$$

$$u_{d,2}(x,t) = U_2(x)\, e^{i\omega t} \tag{5.114b}$$

and obtain

$$d^2U_1(x)/dx^2 + \{(\rho_1\omega^2 - i\,\zeta_1\omega)/E_1\}\, U_1 = 0 \tag{5.115a}$$

$$d^2U_2(x)/dx^2 + \{(\rho_2\omega^2 - i\,\zeta_2\omega)/E_2\}\, U_2 = 0 \tag{5.115b}$$

Their respective solutions are

$$U_1(x) = C_1 \sin\{\sqrt{(\rho_1\omega^2 - i\,\zeta_1\omega)/E_1}\}\, x + D_1 \cos\{\sqrt{(\rho_1\omega^2 - i\,\zeta_1\omega)/E_1}\}\, x \tag{5.116a}$$

$$U_2(x) = C_2 \sin\{\sqrt{(\rho_2\omega^2 - i\,\zeta_2\omega)/E_2}\}\, x + D_2 \cos\{\sqrt{(\rho_2\omega^2 - i\,\zeta_2\omega)/E_2}\}\, x \tag{5.116b}$$

The integration constants C_1, C_2, D_1 and D_2 are then obtained assuming that the displacement is continuous at the pipe-to-collar position $x = x_{p-c}$, that is,

$$u_{d,1}(x_{p-c},t) = u_{d,2}(x_{p-c},t) \tag{5.117a}$$

$$U_1(x_{p-c}) = U_2(x_{p-c}) \tag{5.117b}$$

and, that force is continuous, with

$$A_1E_1\partial u_{d,1}(x_{p-c},t)/\partial x = A_2E_2\partial u_{d,2}(x_{p-c},t)/\partial x \tag{5.118a}$$

$$A_1E_1\, dU_1(x_{p-c})/dx = A_2E_2\, dU_2(x_{p-c})/dx \tag{5.118b}$$

The drill bit displacement boundary condition, noting our objections, is

$$u_1(0,t) = u_0\, e^{i\omega t} \tag{5.119a}$$

$$U_1(0) = u_0 \tag{5.119b}$$

at $x = 0$. Details are available in Dareing and Livesay (1968), but the limitations of the analytical approach are severe, as discussed in Chapter 1, when more general boundary condition models must be considered.

Jarring issues and stuck pipe problems. *Stuck pipe*, due to *dog-legs* and/or poor *hole cleaning* in deviated and horizontal wells, is an industry-wide problem. Drilling or *fishing jars*, typically mounted close to the bit, are employed in the bottomhole assembly of a drillstring in order to free the string should it become immobilized during drilling operations.

Traditionally, jarring calculations are rough and approximate; however, since expensive sensors are now incorporated in downhole MWD tools, jarring actions must be sensitively designed to minimize their potential damage. Drilling personnel are routinely involved in jar selection and optimization. Typical operational parameters include free contraction speed, hammer speed, impact force, and stuck point placement.

Since present day BHAs are becoming increasingly complex for analysis by closed form solution methods, numerical methods have been developed by recent authors to model what are, essentially, longitudinal stress waves of the type studied in this chapter.

Wang, Kalsi, Chapelle, and Beasley (1987) and Kalsi, Wang, and Chandra (1988), for example, underscore the necessity for detailed numerical simulation. Because the tools must be designed to survive the harsh vibration environment during drilling, as well as the maximum shock transmitted during drilling, high resolution simulations were needed. Thus, they used a general purpose commercial finite element program to perform nonlinear transient dynamic analyses, which tracked stress and displacement wave propagation throughout the drillstring.

The effects of jarring on stuck pipe depend on the level of borehole friction acting on the drillpipe. *Spotting fluids* are often added to the mud in the annulus in order to lubricate the string, that is, decrease the net averaged apparent viscosity, to facilitate its movement. The resulting coefficient of friction is an important input parameter for such jarring calculations.

The calculation of friction coefficient is complicated by the *eccentricity* of the hole and the *nonlinear mud rheology*. Also, the *preventive* problem concerned with avoiding stuck pipe due to poor hole cleaning, indicates that borehole friction should addressed in the drilling plan early on.

These issues are discussed in the book *Borehole Flow Modeling in Horizontal, Deviated, and Vertical Wells* (Chin, 1992a). The book gives solutions for eccentric holes and concentric ones with rotating pipe, for different mud types. It discusses practical applications ranging from cuttings transport to spotting fluid analysis, and problems ranging from drilling to cementing. The book and companion software for IBM-compatible PCs, *Petrocalc 14: Horizontal and Vertical Annular Flow Modeling* (Chin, 1992b), are offered by Gulf Publishing Company, Houston, Texas.

DRILLSTRING AND FORMATION IMAGING

Why study waves traveling from downhole? The basic idea is central to all applications of *imaging*: incident waves, altered upon impact, reflect with different amplitudes and phases. The latter contain information that, when properly interpreted (or, *deconvolved*) using boundary condition models, describes the target.

For example, turn-of-the-century *Compton scattering* experiments helped determine the structure of the atom. *Ultrasonic imaging* detects *structural flaws* in *nondestructive testing* or fetus shapes in medical applications. *X-ray* and *Catscan* methods play similar roles in medicine and engineering.

Drillstring imaging. Why look ahead of the bit? One application for vibration modeling is the detection of mechanical drillstring dysfunctions while drilling. In Chapter 6, the detection of large downhole lateral vibrations using axial surface measurements is discussed. Here we describe some diagnostic tools that have been field-tested with optimistic results.

Henneuse (1992) cites case histories using Elf Aquitaine's "Drilling Dynamics Control Unit", a vibration monitoring sub located at the top of the drillstring, dedicated to data acquisition and signal analysis. Dynamic surface measurements were useful in detecting downhole problems such as bit wear, bit bouncing, blocked cones, shock sub performance, stick-slip, stabilizer hangup, backwhirl, and bit and stabilizer balling.

Rappold (1993a,b) noted that additional Elf objectives included optimization of the rock destruction process, elimination of premature pullout of the bit, reduction in the number of drillstring fatigue failures, tracking of the rock formations drilled without time lag, and optimization of mud circulation, cleaning, and lubrication.

When *stick-slip* torsional oscillations are progressing, the bit stalls and accelerates cyclically as the surface rotary speed remains constant (see Chapter 7 for a full discussion). During stick-slip drilling, penetration rates are low and the bit wears faster.

According to Rappold, stick-slip oscillations can cause downhole rotational speeds up to ten times the specified surface speed; such fluctuations last from 1 sec to 40 sec, depending on the BHA. Momentarily high torques can develop, exceeding the elastic limit of drillpipe. This can cause premature aging of the drillstring, and damage MWD equipment, downhole motors, and shock subs.

Dufeyte and Henneuse (1991), delving deeper into the problem, analyse simultaneous downhole and surface dynamic measurements to characterize stick-slip accurately.

Surface measurements of static and dynamic axial and torsional forces provide useful indicators of existing mechanical conditions far downhole. Combined with real-time MWD data, e.g., root-mean-square *rms* information lateral shock load and frequency, a more complete and accurate interpretation of real-time environment is possible.

Seeing ahead of the bit. The ability to detect potential downhole dysfunctions along the drillstring before they become dangerous is an important objective of vibrations modeling. Another objective that has intrigued industry professionals is *formation imaging*, or put simply, *seeing ahead of the bit*. Different approaches are possible, and we discuss some of these.

MWD-VSP. A mature technology known as *vertical seismic profiling* (or, *VSP*) has been used over the past two decades to provide seismic earth images near a wellbore. Recently, researchers have speculated about the application of *signal sources* other than those traditional to the seismic industry. For example, Rector and Marion (1989a,b) propose an *MWD-VSP*, where *checkshot surveys* use the drillbit as a downhole source of seismic energy.

In this approach, the vibrations induced by a drillbit *while drilling* are used to produce VSP data. But unlike MWD methods based on *mud pulse telemetry*, which are relatively expensive, low in data rate, and prone to mechanical failure, downhole instrumentation is not required to acquire data. Also, the data acquisition process does not interfere with drilling operations.

The continuous signals generated by the drillbit are monitored with a reference sensor attached to the top of the drillstring. These are processed together with signals from receivers placed at or near the earth's surface to produce VSP data; in a marine environment, the same sensors would be deployed on the ocean floor.

This idea deserves further development, as problematic issues remain. A quick scan through *Geophysics*, or *Geophysical Prospecting*, and other seismological journals, reveals many papers addressing subtleties due to acoustic source characterization, modeling, and interpretation.

Commonly used sources range from "near impulsive" dynamic blasts, to "2D" seismic lines towed by geophysical ships, to *Vibroseis*™ land signatures generated by controlled thump trucks. The acoustic source signature generated by the drillbit, here complicated by coupled mode drillstring vibrations as well as changing drillstring length, is not likely to be any cleaner.

Vibration logging of the formation. The application of drillstring vibrations to formation evaluation is as old as rotary drilling itself. Drillers will reminisce with stories about "rough running" and "twist-offs", which indicate hard rock formations. On the other hand, high penetration rates with minimal bit wobbling are suggestive of soft, unconsolidated sands.

Drilling engineers today monitor the bottomhole environment by studying downhole and surface measured changes to weight-on-bit and torque-on-bit. Changes in such measurements yield, in addition, information about dogleg severity, drillbit wear, lost cones, and borehole friction.

Recently, research workers dealing with detailed downhole vibrations data, for instance, have noted strong, positive, correlation between vibration logs and resistivity logs as the bit penetrates alternating sand and shale sequences.

The use of vibration data for detailed logging is not new. A successful early effort dealing with dynamical effects is found in the very interesting paper of Lutz, Raynaud, Gstalder, Quichaud, Raynal, and Muckleroy (1972) entitled "Instantaneous Logging Based on a Dynamic Theory of Drilling".

With a simplified analysis involving only longitudinal acceleration measurements taken at the top of the string, the authors were able to produce usable logs which gave information on the hardness of the rock being drilled.

As noted in Lutz *et al*, workers at the ***Societe Nationale des Petroles*** *d'Aquitaine* observed that static weight-on-bit only partially accounted for penetration rate. The complete WOB signature, therefore, *had* to contain lithology information; hence, the *SNAP Log*.

This log recorded raw axial vibration data at the top of the string. It is expected to be important in detecting the approach to, and the entrance into, abnormally pressured formations since it reflects the greater hardness of the caprock above them and the lesser hardness and density of the abnormally pressured shales themselves.

One log taken in an upper soft shale formation of the Aquitanian Basin showed that vibration measurements are able to detect shale compaction in a manner very similar to sonic logs. The unprocessed data correlated well with neutron, sonic and gamma ray results in holes made through compacted shales and limestones. Lutz *et al* provides a large number of field examples of instantaneous logging for comparative study.

Booer and Meehan (1992, 1993), for example, regard the drillstring as a system with layered impedances arising from differences in collar and pipe sizes. Their objective was to separate or *deconvolve* drillstring resonances from measured surface vibrations, in order to obtain the signal created at the bit.

While their block diagrams showed basic elements such as "drillstring transfer function", "bit signature analysis", and "lithology bit boundary condition", the papers offered few of the details needed in deconvolution modeling and signal processing.

But the authors importantly observe that in the frequency band typically generated by a rock bit, up to 100 Hz, approximately, changes in lithology *are* detectable, and that drillstring axial or torsional imaging can detect major structural changes.

Notes on rock-bit interaction. We can *see ahead of bit* while drilling, using the same ideas introduced earlier in forward simulation. Again, this requires us to integrate existing technologies in laboratory rock/bit interaction modeling, drillstring vibration analysis, real-time data acquisition, and downhole signal processing.

A large amount of necessarily empirical work on *rock/bit interaction* is available in the literature, as we indicated in our discussion on its role in rate-of-penetration. The cited works were driven by the need to understand cuttings formation, jet hydraulics and bottomhole cleaning.

Outmans (1959) developed one of the earliest theoretical models for rock/bit interaction. His work showed that the relationship between drillbit

WOB and instantaneous penetration rate is linear over a range of WOBs. This applied to multiple rock types, at different rotation rates and depths. Results for a typical bit in different formations are shown in Figure 5-9. Strong rocks like gray granite and pink quartzite, for example, are seen to drill more slowly.

```
Penetration
Rate, ft/hr
or, ut(0,t)

  14 +        *
     |       *          * Indiana limestone
  12 +       *          $ Carthage marble
     |      *           % Black dolomite
  10 +     *            # Gray granite
     |     *            + Pink quartzite
   8 +    *
     |   *
   6 +   *
     |  *
   4 +  *          $     %
     | *      $    %       #+      +
   2 +     $  %  # +
     |*$% # +                          WOB
   0 +--------------------------------------

     0      10,000   20,000    30,000  lbs
```

Figure 5-9. Linear WOB versus $u_t(0,t)$.

Again, similar results due to Maurer (1962), Warren (1981), Podio and Gray (1965), and Yang and Gray (1967) show definitive correlation between rate-of-penetration and weight-on-bit. The latter two studies gave detailed time oscillograph traces relating displacement, force and impact velocity

These results indicate that predictive models characterizing drillstring response to elastic or brittle rock contact exist and can be defined; for this reason, it is not surprising that the *SNAP Log* proved successful in its lithology correlations.

Taken together, these studies suggest a drillstring response at the bit taking the general linear form $\alpha u_x(0,t) + \beta u_t(0,t) + \lambda u(0,t) = 0$, at least over a limited range of displacements, as suggested in Equation 5-22. Such relationships are often used in jack-hammer and resonant drilling rod design; they are mechanical impedance models which relate force, displacement, and velocity through constants characteristic of the rock and drilling rod.

However, the use of impedance models is new to *inverse* formation evaluation problems, and Equation 5-22 is significant in two respects: the model *does* describe reality and it *is* linear. From a practical viewpoint, it is also important that the relation can be obtained in the laboratory under closely controlled conditions absent of drillstring reverberation.

Basic mathematical approach. The interaction between the rock bit and the formation was assumed in Equations 5-22 and 5-45 to take the general form $\alpha u_x + \beta u_t + \lambda u = 0$, where u denotes the value u(0,t) at the bit. In reality, the actual condition may be nonlinear, but this would preclude analytical solution; of course, complicated boundary conditions can be used numerically.

In our forward simulation, the constants α, β and λ were assumed to be known from detailed laboratory studies where various types of bits are drilled into different rocks using vibration isolated apparatus; thus, $\alpha u_x + \beta u_t + \lambda u = 0$ provides a characteristic signature for the particular rock-bit pair.

In *formation imaging*, the *inverse problem* determining α, β and λ given measured values of u, u_x and u_t is considered. The problem is difficult if it is tackled using Equations 5-25 to 5-30 in their entirety. Even if a closed form solution were available, uncertainties will arise because input parameters (e.g., our surface mass-spring-damper constants) cannot be characterized accurately.

From this viewpoint, an improved understanding of the vibration process only leads to increased uncertainty in the results. But without a useful host solution, seeing ahead of the bit is destined to failure. The crucial idea is the use of a single limited and stable aspect of the total problem that relies on a minimum of uncertain information.

To motivate the ideas, we consider a semi-infinite drillstring, in order to develop a simple inverse model. Recall that undamped axial vibrations satisfy the classical wave equation for u(x,t), whose general solution takes the form u = f(ct-x) + g(ct+x); here, g represents a downgoing incident wave, and f represents the reflected upgoing wave.

Suppose that processed information from all necessary sensors is available, and that in this first example, the data indicates a reflected displacement wave equal in magnitude to the incident, but out-of-phase by 180°. Thus, we would choose f(ct-x) = -g(ct-x); from the wave solution u(x,t) = -g(ct-x) + g(ct+x) that follows, we find that u(0,t) = 0. Thus, we infer that the drillstring is in contact with rigid, unyielding, hard rock.

If in our second example, we detect a reflected displacement wave equal in magnitude but perfectly in phase, we would select f(ct-x) = g(ct-x). In this case, the solution $u(x,t) = g(ct-x) + g(ct+x)$ similarly constructed would lead us, upon differentiation with respect to x, to the conclusion that the strain satisfies $u_x(x,t) = -g'(ct-x) + g'(ct+x)$ or $u_x(0,t) = 0$. Thus, one would infer a free end, or soft, unconsolidated formation.

We discussed our above input information in terms of *displacements*. We could have posed the problem via *stress*. The corresponding interpretations would be made from the rule derived in Chapter 1: stresses reflect with like signs at rigid (hard) boundaries, and opposite signs at free (soft) boundaries.

These simple but exact thought experiments drew upon several idealizations. First, we used known "rigid" and "fixed end" results from classical theory to assist in our interpretation. Second, we assumed semi-infinite drillstrings in order to circumvent issues related to multiple reflections.

One way of extending this approach to more real conditions requires a two-pronged attack. First, we need to supplement our limited rigid and fixed end models with more general classes of reflection solutions. Second, we need to provide for (simulated) semi-infinite conditions so that simple comparisons between incident and reflected waves can be made. We discuss these two issues separately.

More rock-bit interaction models. So far, we have noted simple phase changes, but changes other than 0° and 180° are possible, corresponding to rocks of intermediate hardness. Our amplitudes have remained unchanged, but in general, amplitude changes and shape distortions are both possible.

We will give formulas that allow us to determine α, β and λ from measured displacement, impact speed, and tip stress data. Two separate impact models are considered. From these values, rock type is inferred from (to be) catalogued properties for the particular drillbit used.

An inelastic impact model. To develop the necessary formulas, we follow a differential equation framework similar to one in Morse and Feshbach (1953). We first substitute the general solution $u(x,t) = f(ct-x) + g(ct+x)$ into the proposed rock/bit interaction model

$$\alpha u_x(0,t) + \beta u_t(0,t) + \lambda u(0,t) = 0 \tag{5.120}$$

to obtain

$$(c\beta - \alpha)df(z)/dz + (c\beta + \alpha)dg(z)/dz + \lambda f(z) + \lambda g(z) = 0 \tag{5.121}$$

If the incident wave $g(ct+x)$ is assumed to be known, the above can be viewed as an equation for the function f, which has the solution

$$f(z) = \exp(-\kappa z) \int_a^z \{\gamma dg(\sigma)/d\sigma - \kappa g(\sigma)\} \exp(+\kappa\sigma)\, d\sigma \tag{5.122}$$

where

$$\kappa = \lambda/(c\beta - \alpha) \tag{5.123}$$

$$\gamma = (\alpha + c\beta)/(\alpha - c\beta) \tag{5.124}$$

and "a" is an arbitrary constant. This exact solution allows us to extract the values of the rock-bit interaction constants α, β, and λ when discrete sampled values of F and G are available.

For clarity, we first consider the *inelastic* limit where $\lambda = 0$. This case describes problems where forces resulting from elastic displacements are insignificant. Here, the downhole boundary condition reduces to

$$u_t(0,t) = -(\alpha/\beta)\, u_x(0,t) \tag{5.125}$$

an interaction model depending on the single ratio α/β only. The resulting formula $f(z) = \gamma\, g(z)$, using Equation 5-122, leads to the complete solution

$$u(x,t) = g(ct+x) + \gamma\, g(ct-x) \tag{5.126}$$

showing that, as a consequence of the boundary condition in Equation 5-125, the reflected waveform propagates with its shape unchanged, although its amplitude and sign (that is, phase) may be different.

Equation 5-126 contains a reflected wave that is identical with the incident one (that is, both terms contain the same *g*), except modified by a (positive or negative constant) factor γ: its amplitude is generally *not* unity.

How are inverse solutions obtained? Assuming that the net contributions $g(ct+x)$ and $\gamma g(ct-x)$ propagating in *opposite* directions are individually available (we will demonstrate how in the next section), γ is completely determined as follows.

Since the incident wave $g(ct+x)$ is known, the function $g(ct-x)$ can be constructed. Since $\gamma g(ct-x)$ is known, and $g(ct-x)$ is available, γ isobtained by division. Once γ is known, the definition $\gamma = (\alpha + c\beta)/(\alpha - c\beta)$ obtained from Equation 5-124 yields

$$\alpha/\beta = -c(1+\gamma)/(1-\gamma) \tag{5.127}$$

which completely determines rock/bit mechanical impedance condition in Equation 5-125. Reference to the (hypothetical) rock-bit interaction database will then identify the formation corresponding to the known rock bit.

The above paragraph assumes displacement measurements. The same argument applies to stress data. Since $u_x(x,t) = g_x(ct+x) - \gamma\, g_x(ct-x)$, the (known) ratio between $\gamma\, g_x(ct-x)$ and $g_x(ct+x)$ again yields γ. The foregoing derivation contains the limiting results used in our earlier illustrations.

In particular, $\gamma = +1$ leads to a free end having infinite α/β, while $\gamma = -1$ corresponds to a rigid fixed end rock having a vanishing α/β. Intermediate values of α/β and γ model formations of intermediate hardness, whenever elastic effects are unimportant.

Elastic impacts, with stress effects. Downhole or surface signal processing might first determine, in field implementations, if shape distortions between incident and reflected waves exist; if not, one would conclude that the simple rock-bit interaction model of Equation 5-125 holds. If distortions exist, Equation 5-129 below, which includes elastic responses, or its extension allowing for impact velocity, may apply.

In the inelastic rock-bit interaction model of Equation 5-125, space and time derivatives of axial displacement appear, but the exact level of u(x,t) is unimportant. In this limit, the reflected wave shape is undistorted relative to the incident one; general amplitude changes are allowed, but phase changes are restricted to 0° and 180°.

Here, we assume that an elastic response at the bit is balanced purely by the normal stress (impact velocity is ignored). In the foregoing example, we permitted our incident waves to have general shape; now, though, we will need to interrogate our rock-bit interface with a well characterized wave because the primary effect of elastic response is to distort the incident waveform.

For simplicity, we will consider an idealized rectangular displacement pulse, and determine the nature of the reflected signal. The complete solution containing displacement, velocity, and stress, again, is implicit in the results given earlier by Equations 5-45 to 5-52. We will expand upon the details, and obtain specialized formulas for the rectangular pulse, in the process demonstrating the use of Laplace transforms.

Again, let us start with the general solution to the undamped wave equation,

$$u(x,t) = f(ct-x) + g(ct+x) \tag{5.128}$$

but consider the rock-bit interaction model

$$\lambda\, u(0,t) = -\alpha\, \partial u(0,t)/\partial x \tag{5.129}$$

Now, differentiation of Equation 5-128 with respect to x yields, at x = 0,

$$\partial u(x,t)/\partial x \,|_{x=0} = -f\,'(ct-x) + g'(ct+x) \,|_{x=0} = -f\,'(ct) + g'(ct) \tag{5.130}$$

Substitution of Equations 5-128 and 5-130 in Equation 5-129 leads to

$$\lambda\{f(ct) + g(ct)\} = -\alpha\{-f\,'(ct)+g'(ct)\} \tag{5.131}$$

We introduce the dummy variable $\tau = ct$, to obtain

$$f\,'(\tau) - (\lambda/\alpha)f(\tau) = g'(\tau) + (\lambda/\alpha)g(\tau) \tag{5.132}$$

for x = 0.

We now define the Laplace transforms needed to obtain exact closed form solutions. Let us introduce

$$f^*(s) = \int_0^\infty \exp\{-s\tau\}\, f(\tau)\, d\tau \tag{5.133}$$

$$g^*(s) = \int_0^\infty \exp\{-s\tau\}\, g(\tau)\, d\tau \tag{5.134}$$

Assuming zero initial conditions (that is, dynamically steady conditions when the effects of initial conditions have vanished) we have

$$(s - \lambda/\alpha)\, f^* = (s + \lambda/\alpha)\, g^* \tag{5.135}$$

Now we can re-express the right side coefficient in Equation 5-135 as $s + \lambda/\alpha = s - \lambda/\alpha + 2\lambda/\alpha$, so that

$$f^* = g^* + 2(\lambda/\alpha)\,(g^*)/(s - \lambda/\alpha) \tag{5.136}$$

For the inverted solution, we formally write

$$f(\tau) = g(\tau) + 2(\lambda/\alpha)\, L^{-1}\{g^*(s)/(s - \lambda/\alpha)\} \tag{5.137}$$

where L^{-1} denotes the inverse transform operation. At this point, we specialize our results to a single pulsed displacement incident wave having a rectangular shape. In particular, we set

$$g(ct+x) = H\{ct+x-L\} - H\{ct+x-(L+a)\} \tag{5.138}$$

In Equation 5-138, $H(\tau)$ is the *Heaviside step function*, equal to unity for $\tau > 0$ and for $\tau < 0$. Noting that $L(H\{t\}) = 1/s$, we successively obtain

$$g(\tau) = H\{\tau-L\} - H\{\tau-(L+a)\} \tag{5.139}$$

$$g^*(s) = e^{-Ls}/s - e^{-(L+a)s}/s \tag{5.140}$$

$$f(\tau) = g(\tau) + 2(\lambda/\alpha)L^{-1}[e^{-Ls}/\{s(s-\lambda/\alpha)\} - e^{\{-(L+a)s\}}/(s(s-\lambda/\alpha))\}] \tag{5.141}$$

$$f(ct) = g(ct) - 2[1 - e^{\{\lambda(ct-L)/\alpha\}}]\, H(ct-L) + 2[1 - e^{\{\lambda(ct-L-a)/\alpha\}}]\, H(ct-L-a) \tag{5.142}$$

$$f(ct-x) = g(ct-x) - 2[1 - e^{\{\lambda(ct-x-L)/\alpha\}}]\, H(ct-x-L) + 2[1 - e^{\{\lambda(ct-x-L-a)/\alpha\}}]\, H(ct-x-L-a) \tag{5.143}$$

The complete solution obtained from the general wave formula $u(x,t) = f(ct-x) + g(ct+x)$ is therefore

$$\begin{aligned} u(x,t) = {} & H\{ct+x-L\} - H\{ct+x-(L+a)\} + H\{ct-x-L\} - H\{ct-x-(L+a)\} \\ & -2[1 - e^{\{\lambda(ct-x-L)/\alpha\}}]\, H(ct-x-L) \\ & + 2[1 - e^{\{\lambda(ct-x-L-a)/\alpha\}}]\, H(ct-x-L-a) \end{aligned} \tag{5.144}$$

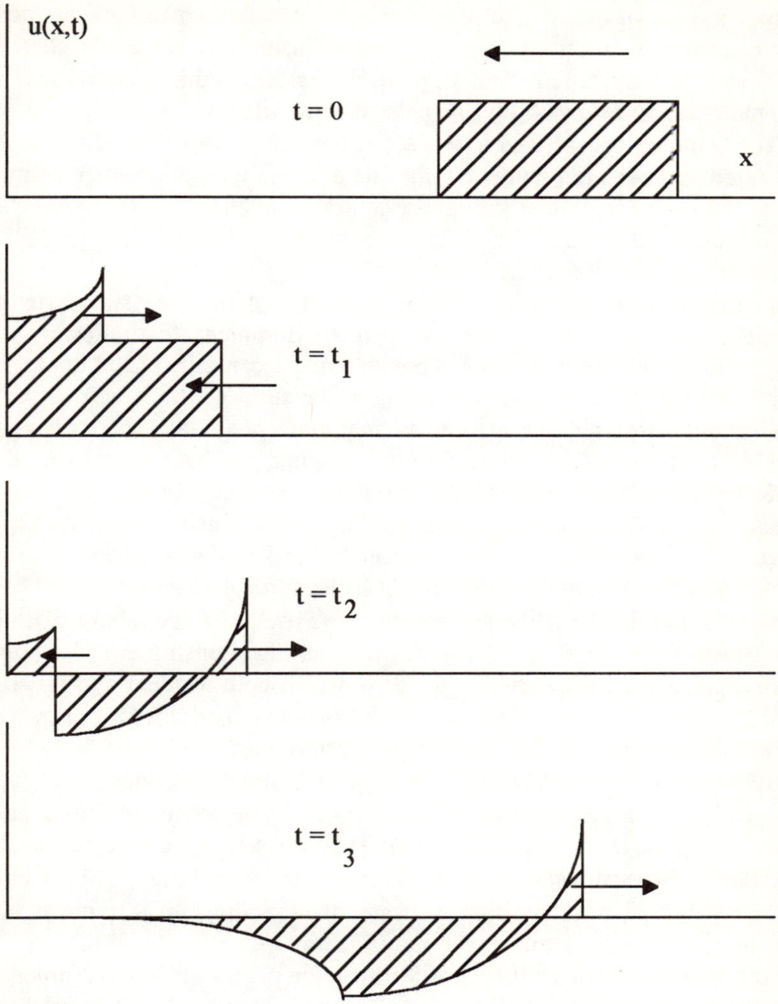

Figure 5-10. Shape distortion due to elastic boundary.

Figure 5-10 qualitatively shows how elastic rock response transforms rectangular pulses into distorted ones with exponentially shaped edges. If the exact exponential and pulse length changes are available from computations, the coefficient λ/α in Equation 5-129 is known. As in the first example, reference to the (hypothetical) rock-bit interaction database will define the formation type, since the drillbit model is known.

Because all incoming pulses will distort when elastic effects are important, as Equation 5-137 clearly shows, it is important to have a simple, well-characterized, incoming pulse so relative distortions are easily observed and conveniently measured. But in general, there is nothing special about a rectangular pulses, as any observable pulse readily suffices.

The two models described in this section do not provide three-dimensional information about the formation, but they do describe the immediate vicinity of the bit. *MWD-VSP* imaging methods, by contrast, are attractive because the use of multiple surface receivers enables the mapping and imaging of geological structures ahead and to the side of the drill bit.

The implicit assumption in using incident and reflected waves, in our derivations, is the availability of a clean environment that allows us to distinguish between the two. This is possible under controlled conditions. For example, we can excite a static drillstring at the surface with a narrow pulse, watch it reflect upwards, and measure all amplitude, phase, and shape changes.

This is more difficult *while drilling*. Earlier, we stressed the "need to provide for (simulated) semi-infinite conditions, so that simple comparisons between incident and reflected waves can be made." Such conditions *can* be achieved, if it is possible to extract incident and reflected waves from transient systems through signal processing. The basic ideas are discussed next.

Separating incident from reflected waves. In the above rock-bit interaction examples, we assumed that we could distinguish between a single incoming pulse and its reflected (altered) wave. If both incident and reflected signals are known, the *transfer function* (or boundary condition, e.g., Equation 5-129) describing rock-bit interaction can be determined.

Let us consider how a hypothetical downhole imaging device might work while drilling. Suppose that the distance-to-bit for a downhole sensor array mounted along the drillstring is known, and that the sound speed of the metal is given; thus, the wave transit time Δt between sensor position and bit can be calculated. We also assume that a means of separating up and downgoing waves (to be derived) is available for real-time use.

At any time t while drilling, the sensor-microprocessor array is turned on momentarily; it determines the up and downgoing waves at that location for that time t, and stores only the downward incident wave. This wave will reflect at the bit, and should return to the sensor in a time $2\Delta t$.

At that instant, the sensor unit self-activates; again, both oppositely traveling waves are extracted, but this time, the upward traveling reflected wave is instead stored. Since the incident wave pulse and its reflection have been isolated, the rock-bit interaction constants are in principle known.

Of course, in any real system, the complications introduced by torsional waves and dispersive lateral waves must be considered. But at least for now, the inverse problem has been reduced to mathematical one which calls for the extraction of up and downgoing waves from a general transient field.

Many techniques are available which separate or *filter* superposed waves that travel in opposite directions; these abound in the seismics, communications, and electrical engineering literature. They are *directional filters*, which must be contrasted with *frequency filters.*

In one-dimensional drillstrings, the filtering problem is actually less demanding than those in 3D earth imaging, despite the strong reverberant fields; Example 1-12 summarizes a well known gas-dynamic model. This is so because, locally anyway, the undamped wave equation $\rho u_{tt} - Eu_{xx} = 0$ applies. We will discuss only the simplest methods, and consider

$$u_{tt} - c^2 u_{xx} = 0 \tag{5.145}$$

Although our discussion will focus on the displacement u(x,t), we emphasize that it applies equally to velocity and stress, since these latter quantities also satisfy the wave equation. This is easily demonstrated. If we differentiate Equation 5-145 with respect to x, we find that $(u_x)_{tt} - c^2 (u_x)_{xx} = 0$. Likewise, if we differentiate respect to t, we have $(u_t)_{tt} - c^2 (u_t)_{xx} = 0$.

Delay line method. In the following schemes, we need only recognize that the general solution to Equation 5-145 takes the form u(x,t) = f(ct-x) + g(ct+x). This shows that waves propagate without distortion along the drillstring or drill collar; if distortions arise, they are due to rock-bit interactions. Since f and g represent waves traveling in opposite directions, our objective is well defined: *extract f and g from measured values of the complete u(x,t).*

It turns out that the wave equation analysis developed to fulfill this objective can be extended to handle wave extractions from more general transient fields of the form

$$u(x,t) = F(t + x/c_1) + G(t + x/c_2) \tag{5.146}$$

Equation 5-146 describes arbitrary waveforms F and G with known speeds c_1 and c_2, respectively, propagating without damping or changes in shape. It suffices to present the derivation for F only, since the one for G is analogous.

We suppose that data for u(x,t) is available at transducer locations $x = x_a$ and x_b for a range of times t. At the former location, Equation 5-146 becomes

$$u(x_a,t) = F(t + x_a/c_1) + G(t + x_a/c_2) \tag{5.147}$$

At x_b, we require those values of u(x,t) shifted backwards in time by an amount τ (to be defined) so that

$$u(x_b,t-\tau) = F(t + x_b/c_1 - \tau) + G(t + x_b/c_2 - \tau) \tag{5.148}$$

We choose $x_a/c_2 = x_b/c_2 - \tau$, so that $\tau = (x_b - x_a)/c_2 > 0$, so that the arguments of G in Equations 5-147 and 5-148 are identical. Subtraction of the two equations leads to a first-order *difference* equation for the function F alone, namely,

$$F(t + x_a/c_1) - F(t + x_b/c_1 - \tau) = u(x_a,t) - u(x_b,t-\tau) \quad (5.149)$$

where x_a, x_b, c_1, c_2 and τ are known constants, and the time functions $u(x_a,t)$ and $u(x_b,t-\tau)$ are available in memory.

Equation 5-149 can be solved analytically using Laplace transforms, electronically by digital methods, or electrically by analogue circuits. In this *delay line* method, there are no restrictions on the values of x_a and x_b. Since F, and similarly G, are known, properties related to the bit-rock interface can be inferred.

In MWD *mud pulse telemetry*, upgoing acoustic *signals* (S) must be extracted from large background pump *noise* (N) propagating downward. Methods such as the one just given are directly applicable, and Equation 5-149, with $c_1 = - c$ and $c_2 = + c$, reduces to

$$F(t - x_a/c) - F(t - x_b/c - \tau) = u(x_a,t) - u(x_b,t-\tau) \quad (5.150)$$

This difference model is the foundation behind many different approaches.

The well known two-transducer wave separation method developed in Foster and Patton (1973), for example, applies to *continuous wave* telemetry. Since the signal, taken as F, is a periodic function with recurring maxima and minima, a robust analogue filter taking advantage of that fact is suggested, and $(x_b - x_a)$ is required to take on certain fractional multiples of a wavelength.

Incidentally, in *communications theory*, the *probability of bit error* depends on the nondimensional *signal-to-noise (S/N) ratio*. We might note that pump disturbances do *not* represent noise in the usual sense; they are *coherent waves* that can be detected and cancelled by signal processing algorithms such as the one given here. Noise, in general, refers to random disturbances.

Differential technique. In this next example, we examine waves with dissipation. In fact, we consider the superposition of a wave $F(t + x/c_1)$ having a dissipation factor $e^{-\eta t}$, and a wave $G(t + x/c_2)$ having the dissipation $e^{-\sigma t}$, where the attenuation constants satisfy $\eta, \sigma > 0$. Thus, we take

$$u(x,t) = e^{-\eta t} F(t + x/c_1) + e^{-\sigma t} G(t + x/c_2) \quad (5.151)$$

We next multiply Equation 5-151 through by $e^{+\sigma t}$, differentiate the resulting expression with respect to t, and then x, to successively obtain

$$ue^{+\sigma t} = e^{(\sigma-\eta)t} F(t + x/c_1) + G(t + x/c_2) \quad (5.152)$$

$$\partial(ue^{+\sigma t})/\partial t = \partial\{e^{(\sigma-\eta)t}F(t + x/c_1)\}/\partial t + G' \quad (5.153)$$

$$\partial(ue^{+\sigma t})/\partial x = \partial\{e^{(\sigma-\eta)t}F(t + x/c_1)\}/\partial x + G'/c_2 \quad (5.154)$$

$$c_2\, \partial(ue^{+\sigma t})/\partial x = c_2 \partial\{e^{(\sigma-\eta)t}F(t + x/c_1)\}/\partial x + G' \qquad (5.155)$$

If we now subtract Equation 5-153 from Equation 5-155, we obtain

$$c_2\partial\{e^{(\sigma-\eta)t}F(t + x/c_1)\}/\partial x - \partial\{e^{(\sigma-\eta)t}F(t + x/c_1)\}/\partial t$$
$$= c_2\, \partial(ue^{+\sigma t})/\partial x - \partial(ue^{+\sigma t})/\partial t \qquad (5.156)$$

and

$$c_2 e^{(\sigma-\eta)t}F'(t + x/c_1)/c_1 - e^{(\sigma-\eta)t}F'(t + x/c_1) - (\sigma-\eta)e^{(\sigma-\eta)t}F(t + x/c_1)$$
$$= c_2\, e^{+\sigma t}\partial u/\partial x - e^{+\sigma t}\, \partial u/\partial t - \sigma e^{+\sigma t}u \qquad (5.157)$$

The $e^{+\sigma t}$ terms cancel throughout, and multiplication by $e^{\eta t}$ gives the required first-order *differential* (not *difference*) equation for F, namely,

$$(c_2/c_1-1)F'(t+x/c_1) - (\sigma-\eta)F(t + x/c_1) = (c_2\, \partial u/\partial x - \partial u/\partial t - \sigma u)\, e^{\eta t} \qquad (5.158)$$

This equation is the sought result, providing a means to determine F from sampled values of u(x,t). Suppose that $\sigma = \eta$ is small, with $c_1 = - c$ and $c_2 = c$. This limit models wave propagation problems where upgoing and downgoing waves have like speeds and dissipation rates. Then, the signal is simply

$$F'(t - x/c) = - \{c\, \partial u/\partial x - \partial u/\partial t - \sigma u\}e^{\eta t}/2 \qquad (5.159)$$

$$= - \{c\, (u_{i+1,n} - u_{i-1,n})/2\Delta x - (u_{i,n} - u_{i,n-1})/\Delta t - \sigma u_{i,n}\}/2 \qquad (5.160)$$

to leading order, where we have used a *finite difference representation* for the right-hand-side of Equation 5-159 (see Chapter 5), with Δx and Δt now being *transducer separation distance* and *sampling time*, respectively.

The left side of Equation 5-160 depends only on present data and single time step old information; thus, the scheme is robust because the effects of history completely disappear.

The scheme is particularly useful because the "c" in Equation 5-160 need only represent the easily measured mud sound determined at the surface. This is crucial because the actual sound speed of the mud varies along the drillstring; mud properties are not constant over the wide range of pressures and temperatures found throughout the flowing system.

We stress that while we have digressed into MWD telemetry, the ideas apply equally to drillstring axial wave formation imaging; only the physical scales, and hence, transducer separations and sampling requirements, change.

The appearance of the indexes i-1 and i+1 indicates that at least two transducers are required; u values at "i" can be taken as arithmetic averages of those at i-1 and i+1. On the other hand, the appearance of n and n-1 indicates that two levels of time storage at least are required.

Whether we deal with MWD telemetry, or drillstring acoustic imaging, what is meant by a derivative is crucial. Suppose, for the latter, that c = 15,000 ft/sec and bit excitations having frequencies up to 15 Hz are considered. Then, the wavelength of 15,000 ft/sec/15 Hz or 1,000 ft suggests a transducer separation of *10% of wavelength* or 100 ft (a *10% of period* rule might apply to the discretization Δt).

More accurate *difference molecules* can be used for increased accuracy; these will require additional transducers in space, more time level memory capacity, and increased computer resources, although the total required work is still modest.

Three-wave formulation. In some physical systems, more than two waves may coexist simultaneously; for example, in borehole geophysics, upgoing and downgoing waves having like speeds may work in the presence of surface waves. The differential method is easily extended, but we need to evaluate derivatives at the next higher order. For example, corresponding to

$$u(x,t) = F(t + x/c_1) + G(t + x/c_2) + H(t + x/c_3) \tag{5.161}$$

are the second derivative formulas

$$F'' + G'' + H'' = u_{tt} \tag{5.162a}$$

$$1/c_1\, F'' + 1/c_2\, G'' + 1/c_3\, H'' = u_{xt} \tag{5.162b}$$

$$1/c_1^2 F'' + 1/c_2^2 G'' + 1/c_3^2 H'' = u_{xx} \tag{5.162c}$$

As before, we obtain an equation for F by eliminating G and H. From elementary algebra, F"can be obtained from the determinant expression

$$F_{tt} = \frac{\begin{vmatrix} u_{tt} & 1 & 1 \\ u_{xt} & 1/c_2 & 1/c_3 \\ u_{xx} & 1/c_2^2 & 1/c_3^2 \end{vmatrix}}{\begin{vmatrix} 1 & 1 & 1 \\ 1/c_1 & 1/c_2 & 1/c_3 \\ 1/c_1^2 & 1/c_2^2 & 1/c_3^2 \end{vmatrix}} \tag{5.163}$$

If $F(\xi)$ in Equations 5-151 or 5-161 is periodic, then so is $F_{tt}(\xi)$. Thus, there is no real need to integrate Equations 5-160 or 5-163, since the derivatives contain the same information. Spatial derivatives in u_{xt} and u_{xx} can be approximated by central difference formulas, while the first and second time derivatives can be modeled using backward differences.

Digital analysis methods. Here we demonstrate still another approach for determining γ in Equation 5-126, that is, in $u(x,t) = g(ct+x) + \gamma\, g(ct-x)$. Once γ is known, the ratio α/β can be obtained from Equation 5-127. This ratio completely fixes the rock-bit interaction parameter in Equation 5-125.

We will need to draw upon a result in Churchill (1941, 1958) for *difference* (not *differential*) equations of the first order constrained by initial conditions. In particular, if an unknown real function Y(t) satisfies

$$Y(t) - \sigma\, Y(t-h) = Z(t) \tag{5.164}$$

$$Y(t) = 0 \text{ when } t < 0 \tag{5.165}$$

where σ and h are constants, h > 0, the real function Z(t) is given, and Z(t) = 0 when t < 0, the solution can be written as

$$Y(t) = \sum_{n=0}^{\infty} \sigma^n Z(t-nh) \tag{5.166}$$

This is *not* an infinite series. For each fixed value of t, this series is *finite* because Z(t-nh) = 0 when t - nh < 0 or t < nh. Thus an alternative form of the solution in Equation 5-166 is

$$Y(t) = Z(t) + \sigma Z(t-h) + \sigma^2 Z(t-2h) + ... + \sigma^m Z(t-mh) \tag{5.167}$$

where m = 0, 1,2,..., when mh < t < (m+1)h. For example, if t = 3h, then $Y(3h) = Z(3h) + \sigma Z(2h) + \sigma^2 Z(h) + \sigma^3 Z(0)$, where Z(t) is prescribed and Z(0) need not be zero.

We return to our inverse problem, and obtain measurements for the displacement function $u(x,t) = g(ct+x) + \gamma\, g(ct-x)$ at some point $x = x^*$. Thus, we can write

$$g(ct+x^*) + \gamma\, g(ct-x^*) = u(x^*,t) \tag{5.168}$$

where the right side can be regarded as a known function of time. If we denote $g(ct+x^*) = g\{c(\mathbf{t} + \mathbf{x}^*/\mathbf{c})\} = Y(t)$, it follows that we can express $g(ct-x^*) = g(ct+x^*-2x^*) = g\{c(\mathbf{t} + \mathbf{x}^*/\mathbf{c} - 2x^*/c)\} = Y(t - 2x^*/c)$. That is, Equation 5-168 becomes

$$Y(t) + \gamma\, Y(t - 2x^*/c) = u(x^*,t) \tag{5.169}$$

If we compare Equation 5-169 with $Y(t) - \sigma\, Y(t-h) = Z(t)$ in Equation 5-164, we have $\sigma = -\gamma$, $h = 2x^*/c > 0$, and $Z(t) = u(x^*,t)$. Thus, the solution to Equation 5-169 becomes

$$g(ct+x^*) = \sum_{n=0}^{\infty} (-\gamma)^n\, u\{x^*, t-2nx^*/c\} \tag{5.170}$$

Since, at any given instant in time t, the incident wave $g(ct+x^*)$ and the total response $u\{x^*,t-2nx^*/c\}$ are both available, Equation 5-170 becomes a polynomial equation in γ that can be solved by algebraic methods. This concludes our discussion on axial vibrations. The author hopes that this chapter will provide a focus for constructive development and discussion.

REFERENCES

Aki, K., and Richards, P.G., *Quantitative Seismology, Volumes I and II: Theory and Methods*, W.H. Freeman and Company, San Francisco, 1980.

Angona, F.A., "Drill String Vibration Attenuation and Its Effect on a Surface Oscillator Drilling System," Journal of Engineering for Industry, Transactions of the ASME, May 1965, pp. 110-114.

Bailey, J.J., and Finnie, I., "An Analytical Study of Drill String Vibration," Journal of Engineering for Industry, Transactions of the ASME, May 1960, pp. 122-128.

Booer, A.K., and Meehan, R.J., "Drillstring Imaging: An Interpretation of Surface Drilling Vibrations," IADC/SPE Paper No. 23889, presented at the 1992 IADC/SPE Drilling Conference, New Orleans, Louisiana, Feb. 18-21, 1992.

Booer, A.K., and Meehan, R.J., "Drillstring Imaging - An Interpretation of Surface Drilling Vibrations, SPE Drilling and Completion Journal, June 1993, pp. 93-98.

Chin, W.C., "Fatal Turbular Bending Motions Difficult to Detect Uphole," Offshore Magazine, April 1988, pp. 32-36.

Chin, W.C., "Why Drill Strings Fail at the Neutral Point," Petroleum Engineer International, May 1988, pp. 62-67.

Chin, W.C., *Borehole Flow Modeling in Horizontal, Deviated, and Vertical Wells*, Gulf Publishing, Houston, 1992.

Chin, W.C., *Petrocalc 14: Horizontal and Vertical Annular Flow Modeling, Petroleum Engineering Software for the IBM PC and Compatibles*, Gulf Publishing, Houston, 1992.

Churchill, R.V., *Fourier Series and Boundary Value Problems*, McGraw-Hill, New York, 1941.

Churchill, R.V., *Operational Mathematics*, McGraw-Hill, New York, 1958.

Clayer, F., Vandiver, J.K., and Lee, H.Y., "The Effect of Surface and Downhole Boundary Conditions on the Vibration of Drillstrings," SPE Paper No. 20447, presented at the 65th Annual Technical Conference and Exhibition of the Society of Petroleum Engineers, New Orleans, La., Sept. 23-26, 1990.

Cunningham, R.A., "Analysis of Downhole Measurements of Drill String Forces and Motions," Journal of Engineering for Industry, Transactions of the ASME, May 1968, pp. 208-216.

Dareing, D.W., and Livesay, B.J., "Longitudinal and Angular Drill String Vibrations with Damping," Journal of Engineering for Industry, Transactions of the ASME, October 1968, pp. 1-9.

Dareing, D.W., "Vibrational Forces and Motions in Rotary Drill Strings," presented at the 1972 Rotary Drilling Conference, International Association of Drilling Contractors, Dallas, Texas, Feb. 23-25, 1972.

Deily, F.H., Dareing, D.W., Paff, G.H., Ortloff, J.E., and Lynn, R.D., "Downhole Measurements of Drill String Forces and Motions," presented at the Petroleum Mechanical Engineering Conference, American Society of Mechanical Engineers, Philadelphia, Pa., Sept. 17-20, 1967.

Deily, F.H., Dareing, D.W., Paff, G.H., Ortloff, J.E., and Lynn, R.D., "Downhole Measurements of Drill String Forces and Motions," Journal of Engineering for Industry, Transactions of the ASME, May 1968, pp. 217-225.

Dufeyte, M.P., and Henneuse, H., "Detection and Monitoring of the Slip-Stick Motion: Field Experiments," SPE/IADC Paper No. 21945, presented at the SPE/IADC Drilling Conference, Amsterdam, Mar. 11-14, 1991.

Eisner, E., "The Design of Resonant Vibrators," in *Physical Acoustics, Vol. 1, Part B*, edited by W.P. Mason, Academic Press, New York, 1964

Eronini, I.E., "On Dynamic Fracture of Rock Under Bit Impact Loads," Journal of Energy Resources Technology, Transactions of the ASME, June 1982, pp. 105-107.

Finnie, I., and Bailey, J.J., "An Experimental Study of Drill String Vibration," Journal of Engineering for Industry, Transactions of the ASME, May 1960, pp. 129-135.

Foster, M.R., and Patton, B.J., "Apparatus for Improving Signal-to-Noise Ratio in Logging-While-Drilling System", United States Patent No. 3,742,443.

Gray, K.E., Armstrong, F., and Gatlin, C., "Two-Dimensional Study of Rock Breakage in Drag-Bit Drilling at Atmospheric Pressure," Journal of Petroleum Technology, Jan. 1962, pp. 93-98.

Grimes, R.E., Felderhoff, F.C., and Brown, H., "Heavy Weight Rock Bits Increase Penetration Rates in Hard Rock," Oil & Gas Journal, May 18, 1992, pp. 76-79.

Henneuse, H., "Surface Detection of Vibrations and Drilling Optimization: Field Experience," IADC/SPE Paper No. 23888, presented at the 1992 IADC/SPE Drilling Conference, New Orleans, Louisiana, Feb. 18-21, 1992.

Hudson, J.A., *The Excitation and Propagation of Elastic Waves*, Cambridge University Press, Cambridge, 1980.

Kalsi, M.S., Wang, J.K., and Chandra, U., "Transient Dynamic Analysis of the Drill String Under Jarring Operation Using Finite Element Method," SPE/IADC Paper No. 13446, presented at the 1985 SPE/IADC Drilling Conference, New Orleans, Louisiana, March 6-8, 1985.

Koch, W.M., "Analysis of the First Field Measurements of Downhole Drill Bit Motions and Forces," presented at the Twenty-Seventh Annual Meeting of the American Association of Oilwell Drilling Contractors, Oklahoma City, Oklahoma, Oct. 11-13, 1967.

Langeveld, C.J., "PDC Bit Dynamics," IADC/SPE Paper No. 23867, presented at the 1992 IADC/SPE Drilling Conference, New Orleans, Louisiana, Feb. 18-21, 1992.

Lutz, J., Raynaud, M., Gstalder, S., Quichaud, C., Raynal, J., and Muckleroy, J.A., "Instantaneous Logging Based on a Dynamic Theory of Drilling," Journal of Petroleum Technology, June 1972, pp. 750-758.

Ma, D., and Azar, J.J., "Dynamics of Roller Cone Bits," Journal of Energy Resources Technology, Transactions of the ASME, Dec. 1985, pp. 543-548.

Ma, D., and Azar, J.J., "A Study of Rock-Bit Interaction and Wellbore Deviation," Journal of Energy Resources Technology, Transactions of the ASME, Sept. 1986, pp. 228-233.

Maurer, W.C., "The 'Perfect-Cleaning' Theory of Rotary Drilling," Journal of Petroleum Technology, Nov. 1962, pp. 1270-1274.

Maurer, W.C., "Bit-Tooth Penetration Under Simulated Borehole Conditions," Journal of Petroleum Technology, December 1965, pp. 1433-1442.

Morse, P.M., and Feshbach, H., *Methods of Theoretical Physics, Parts I and II*, McGraw-Hill, New York, 1953.

Outmans, H.D., "The Effect of Some Drilling Variables On the Instantaneous Rate of Penetration," 34th Annual Fall Meeting of the SPE, Dallas, Texas, Oct. 4-7, 1959.

Paslay, P.R., and Bogy, D.B., "Drill String Vibrations Due to Intermittent Contact of Bit Teeth," Journal of Engineering for Industry, Transactions of the ASME, May 1963, pp. 187-194.

Podio, A., and Gray, K.E., "Single-Blow Bit-Tooth Impact Tests on Saturated Rocks Under Confining Pressure, I: Zero Pore Pressure," Society of Petroleum Engineers Journal, September 1965, pp. 211-224.

Rappold, K., "Drilling Optimized With Surface Measurement of Downhole Vibrations," Oil & Gas Journal, Feb. 15, 1993, pp. 58-62.

Rappold, K., "Drillstring Vibration Measurements Detect Bit Stick-Slip," Oil & Gas Journal, Mar.1, 1993, pp. 66-70.

Rector, J.W., and Marion, B.P., "MWD VSP and Checkshot Surveys Using the Drill Bit as a Downhole Energy Source," OTC Paper No. 6024, presented at the 21st Annual Offshore Technology Conference, Houston, Texas, May 1-4, 1989.

Rector, J.W., and Marion, B.P., "Extending VSP to 3-D and MWD: Using the Drill Bit as a Downhole Seismic Source," Oil & Gas Journal, June 19, 1989, pp. 55-58.

Rowley, D.S., Howe, R.J., and Deily, F.H., "Laboratory Drilling Performance of the Full-Scale Rock Bit," presented at the 35th Annual Fall Meeting of the Society of Petroleum Engineers, Denver, Co., Oct. 2-5, 1960.

Wang, J.K., Kalsi, M.S., Chapelle, R.A., and Beasley, T.R., "A Practical Approach to Jarring Analysis," SPE/IADC Paper No. 16155, 1987 SPE/IADC Drilling Conference, New Orleans, March 15-18, 1987.

Warren, T.M., "Drilling Model for Soft Formation Bits," Journal of Petroleum Technology, June 1981, pp. 963-970.

Warren, T.M., "Factors Affecting Torque for a Tricone Bit," SPE Paper No. 11994, presented at the 58th Annual Technical Confrence and Exhibition, San Francisco, Ca., Oct. 5-8, 1983.

Warren, T.M., "Penetration Rate Performance of Roller Cone Bits," SPE Paper No. 13259, presented at the 59th Annual Technical Conference and Exhibition, Houston, Texas, Sept. 16-19, 1984.

Yang, J.H., and Gray, K.E., "Single-Blow Bit-Tooth Impact Tests on Saturated Rocks Under Confining Pressure, II: Elevated Pore Pressure," Society of Petroleum Engineers Journal, December 1967, pp. 389-408.

6
Lateral Bending Vibrations

High-amplitude bending, transverse, or lateral *downhole* vibrations are widely acknowledged to be responsible for *twist-offs* and expensive drillstring failures. The axial and torsional oscillations discussed in this book, which can be dangerous, play only secondary roles in this severe failure mode.

Yet, it is well known that even large, lateral bending vibrations *cannot* be detected uphole at the surface and corrected, even in vertical wells where borehole wall contact and damping are known to be weak. This paradox has eluded practitioners and researchers alike for years, and a deterministic model of this mysterious phenomenon has remained high on R&D priorities.

Why explain this drilling paradox? For one, the requirement for a predictive model is importantly needed in drilling operations, where cost and safety issues are forever looming. Second, the subject of drillstring vibrations cannot be recognized as useful and credible unless the physical explanations for such fundamentally significant observations are identified and understood.

And third, any analytical and numerical formalism for drillstring vibrations, such as that offered in this book, cannot claim to be applicable generally and capable of extension (e.g., to highly deviated drillstrings) unless this first hurdle is overcome through clearly stated engineering principles.

In this chapter, a deterministic and predictive explanation is given. The paradox is simply and straightforwardly explained using new ideas based on *kinematic wave theory* and *group velocity* interpretations of the conventional beam equation. These modern ideas were introduced in Chapter 2.

The analytical model shows how bending disturbances generated both at the drillbit downhole, and uphole by drillstring and borehole interaction, *focus* and *trap* at the *neutral point*, thus leading to large local accumulations of vibration energy which are undetectable uphole. Leaky waves that do escape to the surface will stretch beyond recognition, even when dissipation is unimportant, due to refractive effects. Of course, the presence of any borehole attenuation will hasten the disappearance of these lateral waves.

Having identified the dominant physical mechanisms, we then derive a general partial differential equation model for further study and modeling. More finite difference essentials are introduced, above and beyond those discussed in Chapter 5, and numerical formulations and solution approaches are developed in a step-by-step manner.

Lateral vibrations in deepwater operations. While the emphasis of lateral vibrations in this book is in drilling, such vibrations also figure prominently in *deepwater operations*. Here, we discuss the basic problems.

A *marine riser* is a conductor that links a floating vessel to the *manifold* on the sea bottom, with the sea depth ranging from hundreds to thousands of feet. The riser can be a single rigid pipeline for the *drilling system*, or a bundle of flowlines assembled as an integral unit for a *production system* (the production riser system can be installed on a *tension leg platform* or on a *floating production and storage unit*).

To avoid the collapse and buckling of a riser by its own weight, and to reduce excess load on the manifold in the deep sea, a constant tension is generally required. This is applied at the top end of the riser. The conductor acts like a stiffened cable as far as the end tension is concerned. Also, syntactic foam or foamed aluminum is strapped around production risers at intervals to provide additional buoyancy and to reduce top end tension.

The drilling or production fluid that flows *internally* from one end to the other, of course, experiences centrifugal forces along the inside curved path, where the curved path may be time dependent. The *external* flowing fluid, on the other hand, may induce vibration; this is a function of flow velocity, and unsteady loads arising from vortex shedding will be important.

Chen and Gair (1992), for example, consider a marine riser carrying fluid flow internally, and study instances where internal fluid density and composition, and wave and current loads may vary.

Bending vibrations in directional control. Dynamic lateral bottomhole forces can aid drilling by providing improved directional control and increasing penetration rate. Kane (1984) notes that drillers have little control over lithology, so that bits must be designed to behave well all-around.

He notes that many formations react to peak and not just average force. Thus, it may be possible to use dynamic side forces (which arise only when the bit is rotated, as opposed to the static side force of a pendulum assembly) successfully when the bit begins to break more formation at one side of the hole than another - therefore applying imbalances and eccentrically weighted collars in the drillstring to guide the bit.

Remaining plan for Part II. Chapter 7 temporarily returns to a simpler problem, in particular, a classical wave equation model dealing with torsional oscillations. These are coupled to the axial vibrations of Chapter 5, by way of boundary conditions at the drill bit. Finally, in Chapter 8, all three vibrations modes, that is, axial, torsional, and (two) lateral, will be dynamically coupled for the first time, and numerically solved for a realistic test problem.

In Chapter 8, the individual numerical algorithms introduced separately in Chapters 5, 6 and 7 are integrated. Example calculations show how smooth and rough drilling can both be simulated, and in the process, we demonstrate modern concepts such as *bit bounce* and *torque reversal*.

A DOWNHOLE PARADOX: THE CASE OF THE VANISHING WAVES

The shock and vibration loads generated during drilling are very intense. Typical mile-long, million-pound drillstrings will often break, the result of large static forces and moments, and high cyclic fatigue loadings. Different modes of vibration are excited by rock-bit interaction and drillstring-borehole contact. Unfortunately, exact dynamical analysis appears to be difficult because it is not always possible to characterize these excitation sources accurately.

So much depends on the particular tricone and polycrystalline drillbit configurations used and on their "wobble signatures" as they enter unknown rock formations. Transient drillstring and borehole contacts are equally difficult to describe accurately. The combined pipe motion due to whirling, whipping, axial, torsional and lateral vibration, strongly dependent on the details of the bottomhole assembly used, is far from simple.

The uncertainties are especially the case in the highly deviated holes drilled offshore. However, the practical consequences of vibration are significant. Many twist-offs occur downhole, particularly near the *neutral point*, and expensive, time-consuming *fishing jobs* are needed at times to keep the drilling program on schedule.

Physical features observed at failure. The above unknowns would suggest that each drilling situation means a uniquely different problem requiring a customized description. Despite the apparent differences, however, drillers have long recognized two general features common to rotary drilling:

(i) Far downhole, lateral bending vibrations are the most severe; they are responsible for the greatest frequency of twistoffs and operational failures. And many of these occur at or near the neutral point, which we define by the vanishing of static axial loading.

(ii) Uphole, at the surface, only axial and torsional oscillations can be detected, despite the predominance of severe bending vibrations downhole. This almost contradictory, but universally accepted observation, is true even in vertical wells, where damping due to pipe and borehole wall contact is minimal. Thus, the most damaging downhole vibrations cannot be anticipated from surface measurements and corrected.

Field evidence widely available. The first observation is supported by post-mortem examinations of many fished assemblies, and by direct measurement in field experiments, where recording instruments placed close to the drillbit sampled all vibration modes simultaneously.

Further evidence is found in the computational experiments of Mitchell and Allen (1985), using a six degree-of-freedom dynamical model. These authors noted that axial and torsional vibration analysis of three bottomhole assemblies was unable to identify the cause of repeated washouts and twist-offs. After detailed study, finite element analysis established lateral vibrations as the most likely culprit.

The second observation - *vanishing lateral disturbances at the surface* - almost contradictory but universally accepted, is true even in vertical wells, where damping due to pipe and borehole wall contact is minimal. Wolf, Zacksenhouse, and Arian (1985), for example, noted from their experiments that downhole bending moments *cannot* be observed at the surface. Also, they indicated that the nature of downhole lateral vibrations is not well understood; for example, bending moments estimates based on static models were an order of magnitude smaller than actual values.

Their closing remarks were emphatic, stressing that high bending moments can be observed downhole, with no indication of their presence seen at the surface. Their detailed MWD measurements were made using a high resolution, high-data-rate wireline system for a straight hole.

More recent empirical evidence supports the foregoing observations. Vandiver, Nicholson and Shyu (1989) likewise note that, unlike longitudinal and torsional vibrations, those in bending are not transmitted to the surface by the drillstring. They do suggest, optimistically, that it may be possible to see evidence of downhole bending vibration from surface axial and torsional vibration signals; however, the relationships are not yet understood well enough to be useful for diagnostic purposes.

The authors importantly observed from their case studies that *linear* coupling of axial and bending vibrations of the bottomhole assembly does occur, a physical mechanism developed in detail later in this chapter, expanding on the early model of Chin (1988a,b).

Aldred and Sheppard (1992) recently summarized some industry-wide observations and frustrations. They noted that once damaging transverse vibrations are initiated, they will persist over a wide range of rotary speeds.

That is, a *continuous* range of dangerous frequencies exists, in contrast to simplistic harmonic predictions that pinpoint a discrete set of modal resonant frequencies. These authors, like others, further suggest that models based on the harmonic behaviour of the drillstring do not explain the observed response satisfactorily, possibly because strong nonlinearities are at fault.

Interestingly, some early comments of Nicholson (1975), on the subject of "critical rotary speeds" cautions that "critical" often implies a resonant vibration condition - *but that critical vibrations are not necessarily associated with resonance.* But much confusion exists. Nicholson also notes that resonance has been blamed for serious drillpipe and hole damage, and yet, others intentionally drill "at resonance" without success.

Henneuse (1992), working from an extensive database of results based on Elf Aquitaine's surface vibrations monitor, agrees consistently with Wolf, Zacksenhouse, and Arian (1985) that lateral movements do not propagate easily along the drillstring.

Rewcastle and Burgess (1992) observe that shocks due to high lateral vibrations contain the most energy, also noting that significant mode coupling appears to be present. They conclude, consistently with past field observations, that surface vibration measurements are a poor indicator of the downhole environment.

The destructive force behind downhole lateral vibrations is well documented by real-time downhole measurement. Burgess, McDaniel, and Das (1987) specifically conclude that transverse vibration is a source of downhole MWD tool failure; Cheatham and Comeaux (1992), recognizing this industry problem, provide general guidelines for predicting fatigue life of MWD tools.

Close, Owens, and MacPherson (1988), using downhole MWD data, reiterate the fact that lateral vibrations can be extremely damaging. They concluded that bending accelerations were an order of magnitude greater than axial accelerations. In one case study where destructive vibrations were uncovered, sudden lateral accelerations exceeded axial values by up to twenty-three times.

These authors also observed extreme bending problems while running with a diamond bit, apparently induced by torsional oscillations. Their conclusions cast doubt on long accepted industry standards. For example, critical rotary speed formulas provided in *API RP 7G* since 1970, to prevent bit-excited lateral vibrations in the drill stem, appear to be limited in usefulness.

With respect to the applicability of classical harmonic analysis, field results indicated the presence of severe downhole shock levels even when modeled critical rotary speeds and real-time surface vibrations had been avoided. Thus, there appears to be a different physical mechanism at work. Close *et al* importantly suggest that shock behaviour is perhaps more stable than a purely resonant response.

The recurrent themes cited in our observations (i) and (ii) pervade the literature. From an operational standpoint, it is unacceptable and dangerous that the most damaging downhole vibrations cannot be anticipated from direct surface measurements and corrected. Drillers have long recognized this unfortunate problem; its impact on drilling efficiency, cost and safety, is obviously important, and a good explanation is needed.

Wave trapping, a simple analogy. Catastrophic drillstring failures are often attributed to various linear and nonlinear instabilities, while failure to detect their occurrence from the surface is blamed on attenuation due to borehole wall contact.

However, conventional approaches have yet to produce any results, even qualitative ones, that completely explain the observations indicated above by means of a single self-consistent model, let alone highlight the mysterious role of the neutral point.

One might even suppose that catastrophic events such as twist-offs cannot be described within any linear framework, and that special studies would be needed on a case-by-case basis. Fortunately, this is not true.

A new wave instability theory, proposed at the Massachusetts Institute of Technology in the early 1970s by M.T. Landahl, this author, and others, draws upon the *kinematic wave theory* developed at the California Institute of Technology (Whitham, 1974). The basic elements of kinematic wave theory were introduced in Chapter 2.

The *wave trapping model* was successfully applied to several catastrophic failures in nature, such as vortex breakdown over aircraft delta wings, laminar boundary layer transition to turbulence, and hydraulic jump formation in water.

Landahl (1972) gave the first exposition of the ideas to be developed below, for dynamically steady, and also for fully transient continuum systems. The mathematical formalism for general continuous media is developed in Chin (1976, 1980a). Chin (1979, 1980b, 1981) discusses applications to water waves, whereas Chin (1988a,b) extends these ideas to drillstring failure.

We will develop the fundamental concepts from elementary notions. For simplicity, consider the classical vibrating string, i.e., a violin or guitar string, executing small transverse displacements at some fixed frequency. These strings do not support bending as "drillstrings" do, but the qualitative ideas introduced here can and will be generalized without difficulty.

It is not necessary to solve any differential equations in order to illustrate the physical ideas. We only need to recognize the fact that disturbances propagate at wave speeds of $(T/\rho)^{1/2}$, where T represents tension and ρ is the mass density (see Chapter 1).

Now suppose that $\rho(x)$ increases in the direction of propagation. If the rate of increase is slow enough, it is clear that the foregoing expression for speed still applies locally; more formal asymptotic *WKB* analysis can be used to find explicit high-order corrections.

Eventually, as ρ becomes large, this $(T/\rho)^{1/2}$ expression shows that the disturbance slows down to the point where all wave motion effectively ceases. Physically, the wave cannot overcome the inertia of the system. Because it cannot propagate beyond the *focus*, and since the system is essentially nondissipative, the wave must remain there, and vibrate in place.

The number of waves per unit length, or *wavenumber* k, therefore *piles up* and increases, while significant local accumulations of energy build. This leads to violent *trapped* oscillations which cannot be detected downstream. Drillstring bending vibrations similarly focus, with the focus located close to the neutral point, defined as the transition point between positive and negative axial stress. This will be demonstrated analytically and explained in physical terms.

Extension to general systems. Boundary and initial conditions, typically used in solving differential equations, were not used in our explanation. It was only necessary to consider *wave speed.* When this vanishes due to *refractive effects* which effect slow changes (as opposed to reflections that arise from sudden changes) to wave speed in the propagation medium, energy can *pile up* the way cars pile up near narrow necks of highways.

This explanation is appealing because it is simple and physically intuitive. And, it is easily generalized for more complicated vibrations in arbitrary continuous media. Typically, a wave having a *frequency* ω and a *wavenumber* k will satisfy a *dispersion relationship* $\omega = \omega(k)$ determined by the host vibrations model. These ideas were covered in detail in Chapter 1.

As a brief review, the Fourier assumption $v(x,t) = \sin(kx-\omega t)$ when applied to the transverse string equation $\rho v_{tt} - Tv_{xx} = 0$, leads to $\omega = (T/\rho)^{1/2}\, k$. Thus, every wave of wavenumber k has a well-defined frequency ω that depends on the properties of the medium. The dispersion relationship can be obtained from differential equation analysis, or it can be determined empirically in the laboratory.

Two wave speeds, a *phase velocity* $C_p = \omega/k$, and a *group velocity* $C_g = \partial\omega/\partial k$, are usually defined. These ideas originated with the classical physicists of the nineteenth century, who were disturbed with the fact that the "obvious" ω/k speed inferred from $\sin(kx-\omega t)$ or $\sin k\{x - (\omega/k)t\}$ exceeded the speed of light (ω/k, it turns out, is *not* the speed with which energy propagates).

The applications of these speeds are now familiar to communications and electrical engineers, and to acoustical designers. The velocity used in wave trapping analyses should, naturally, be the one with which energy physically propagates, that is, *we need to consider the manner in which group velocity behaves in waveguides.* For vibrating strings, of course, both velocities *happen* to be identical.

The ideas developed at the Massachusetts Institute of Technology can be summarized in several succinct statements:

(1) Wave trapping represents a strong instability mechanism with a sound physical basis; it may or may not coexist along with the resonant modal ones well known in eigenfunction analysis.

(2) In dynamically steady systems, trapping is implied by zeros or minimums in the expression for group velocity.

(3) In more general transient systems, with higher-order spatial dimensions, the *critical group velocity* turns out to be a complicated (but well defined) function of the *local phase velocity* (e.g., see Landahl (1972)).

WHY DRILLSTRINGS FAIL AT THE NEUTRAL POINT

For the transversely vibrating strings described above, the density *nonuniformities* responsible for vanishing group velocities are easily visualized. In the case of uniform vertical drillstrings, with the exception of unimportant area changes at collar/pipe junctions responsible for reflections only, the required *inhomogeneities* are less apparent.

One might reasonably ask, "What nonuniformity or heterogeneity can possibly induce violent bending motions far downhole, together with the mysterious disappearance of lateral vibrations at the surface?"

The relevant nonuniformities turn out to be those "seen" by the linear equation governing bending motions (for example, refer to Timoshenko and Goodier (1934), Love (1944), Den Hartog (1952) or Abramson, Plass, and Ripperger (1958), Clough and Penzien (1975)).

That is, terms representing the varying axial load due to weight appear as variable coefficients. The loading, compressive at the bit and tensile at the surface, vanishes at the neutral point. We will show that the group velocities of dangerous waves do, in fact, vanish near the neutral point, thus trapping the waves there, leading to large, undetectable local accumulations of vibration energy.

General downhole vibrations consist of waves with many component frequencies. Whitham (1974) describes, as we will below in greater detail, how group velocity can be calculated for monochromatic waves of fixed frequency. Thus, it is possible to identify those waves and frequency ranges most likely to inflict damage.

Exact integrals for steady wave motion can be obtained, and general qualitative and quantitative formulas will be derived. Detailed numerical results for all pertinent vibration parameters, based on the general dimensionless equations, will be tabulated and plotted in order to provide qualitative measures of danger at various points along the drillstring.

Typical applications, for example, may include bottomhole assembly design, the optimal placement and design of downhole MWD tools, and the location of sensors within MWD subs. We will also discuss means to detect violent bending vibrations from the surface. Finally, the use of tapered drillstrings with varying material and geometrical properties (e.g., constructed by using different drill collar and pipe sections with the required variations in cross-sectional properties) is discussed.

Beam equation analysis. Let ρ represent the mass density per unit volume, A the cross-sectional area, E the Young's modulus, I the moment of inertia, k the wave number, ω_0 the vibration frequency, and N the variable axial force distributed along the drillstring.

In the absence of torsion, the *dual bending modes* v and w that generally coexist will decouple, so that it suffices to consider each of the lateral modes individually. Here, we let v(x,t) denote transverse displacements satisfying

$$EI\ \partial^4 v/\partial x^4 + \partial(N\ \partial v/\partial x)/\partial x + \rho A\ \partial^2 v/\partial t^2 = 0 \tag{6.1}$$

The axial force distribution N(x,t) in Equation 6-1 satisfies $N < 0$ for those portions in tension, and $N > 0$ for those portions in compression. The transition $N = 0$ defines the *neutral point* of the drillstring. Once the solution for v(x,t) is available, the *bending moment* is known from $M = -EI\ \partial^2 v/\partial x^2$, while the *shear force* satisfies $V = \partial M/\partial x = -EI\ \partial^3 v/\partial x^3$.

If the *stiffness* EI vanishes identically, and a constant axial force $N < 0$ remains in tension, Equation 6-1 reduces to the classical wave equation. In this limit, the resultant $\partial^2 v/\partial t^2 - c^2\ \partial^2 v/\partial x^2 = 0$ would describe the transverse vibrations of a string, and the speed of sound c would satisfy $c^2 = -N/\rho A$; and since $N = - EA < 0$, it follows that $c^2 = E/\rho > 0$, as is well known from classical wave theory.

Equation 6-1, applicable to straight beams, is discussed in Graff (1975). The $\partial(N\ \partial v/\partial x)/\partial x$ term demonstrates the coupling between N and v(x,t). The assertion in Vandiver, Nicholson, and Shyu (1989) that linear coupling will *not* occur on a perfectly straight beam excited by an axial load which is less that the critical buckling load is therefore incorrect. However, the coupling mechanism that they investigate, namely that due to initial BHA curvature, represents a valid physical problem.

In Equation 6-1, we have not included viscous dissipation, although the extension is trivial; we will discuss the qualitative effects of internal viscous damping later. But an effective damping, however, *is* introduced by the effects of heterogeneity alone, as we will show; in the case of lateral vibrations, it is completely due to axial variations in the longitudinal force N(x,t).

Kinematic wave modeling. Let us consider, as we had in our example for the transverse vibrations of a string, a sinusoidal Fourier wave component having a wavenumber k and a frequency ω. Since Equation 6-1 is linear, this usage involves no further assumptions than those underlying elastic small displacement theory. Thus, it suffices to take

$$v(x,t) = e^{i(kx-\omega t)} \tag{6.2}$$

Substitution of Equation 6-2 in Equation 6-1 leads to the *complex dispersion relation*

$$\rho A\omega^2 = EI\,k^4 - Nk^2 + ikN_x \tag{6.3}$$

where $i = \sqrt{-1}$.

Since Equation 6-3 involves terms with and without "i", ω must necessarily be *complex*. In order to obtain useful results, we rewrite the *complex frequency* ω explicitly as the sum of a *real* part ω_r and an *imaginary* part $i\,\omega_i$, where both ω_r and ω_i are real. Thus, we assert that

$$\omega = \omega_r + i\,\omega_i \tag{6.4}$$

Equations 6-2 and 6-4 allow us to express v(x,t) as the product of two quantities, a *propagating* sinusoidal "$\exp\{i(kx-\omega_r t)\}$" wave, and an exponential "$\exp(\omega_i t)$" *damping*, that is,

$$v(x,t) = \exp(\omega_i t)\exp\{i(kx-\omega_r t)\} \tag{6.5}$$

We now consider only those waves that are not critically damped, that is, only propagating waves with weakly damped oscillations satisfying $|\omega_r| >> |\omega_i|$.

In this limit, substitution of Equation 6-4 in Equation 6-3, and equating real and imaginary parts leads to

$$\rho A\omega_r^2 = EI\,k^4 - Nk^2 \tag{6.6}$$

and

$$\omega_i = kN_x/(2\omega_r\rho A) \tag{6.7}$$

Equation 6-6, describing ω_r, is known as the *real dispersion relation* because it describes the *dispersive characteristics* of a propagating sinusoidal wave $\exp\{i(kx-\omega_r t)\}$. If we solve for the *phase velocity* $C_p = \omega_r(k)/k$, it is clear that waves having different wavelengths travel or *disperse* at different speeds.

This is in contrast to our experience with the second-order wave equation, where the propagation speed is a fixed constant for all waves regardless of form. A wave is *nondispersive* if all waves travel with the same identical speed; otherwise, it is *dispersive*.

Equation 6-7 describes the *imaginary frequency* ω_i. It is important to note that, unlike many examples in elementary physics, a nonzero ω_i function *can* appear even when purely attenuative effects such as viscous dissipation are ignored.

In the present case, Equation 6-7 expresses the energy gained or lost due to interactions with (variable properties in) the medium, namely, N(x,t). Such interactions, consistently with the literature, occur through *radiation stresses.* Sloping beaches, for example, introduce heterogeneities that affect the dynamics of shallow water waves; they play important roles in wave breaking.

Now, since the drillstring is compressive at the bit and tensile at the surface, it follows that $\partial N/\partial x < 0$, as a wave propagates up the drillstring in the direction of increasing x. Thus, the imaginary frequency ω_i in Equation 6-7 is negative, indicating that any wave that propagates to the surface will damp on account of interaction with the background strain field alone. This damping will, of course, be enhanced by the effects of true viscous dissipation.

To determine the *kinematics* of the wave propagation, let us take partial derivatives of Equation 6-6 with respect to k, leading to

$$2\,\rho A\omega_r\partial\omega_r/\partial k = 4\,EI\,k^3 - 2Nk \tag{6.8}$$

and hence

$$C_g = \partial\omega_r/\partial k = (4EI\,k^3 - 2Nk)/2\rho A\omega_r = (2EI\,k^3 - Nk)/\rho A\omega_r \tag{6.9}$$

Again, in a linear system, the *group velocity* C_g is the velocity with which energy and momentum both propagate; in nonlinear systems, energy, momentum, and wavenumber will generally travel at different speeds.

Does C_g ever vanish identically? The answer is a definitive *no*: this would require that $k^2 = N/(2EI)$, so that the right hand side of Equation 6-6, namely, $EI\, k^4 - Nk^2 = -N^2/(4EI)$ would be negative, which is impossible, since the left side requires the opposite, with $\rho A\omega_r^2 > 0$.

To prove that C_g is always nonzero, although *the group velocity turns out to be extremely small near the neutral point,* let us consider (without loss of generality) a dynamically steady, plane, monochromatic wave having the fixed frequency

$$\omega_r = \omega_0 > 0 \qquad (6.10)$$

Thus, we in effect adopt the phase velocity convention $c_p = \omega_0/k > 0$ for upgoing waves. Since $\rho A\omega_r^2 = EI\, k^4 - Nk^2$, it follows that $\rho A\omega_0^2 = EI\, k^4 - Nk^2 > 0$, and hence, that $(EI\, k^3 - Nk) > 0$. Therefore,

$$\begin{aligned} C_g &= \partial\omega_r/\partial k = (2EI\, k^3 - Nk)/\rho A\omega_0 \\ &= (EI\, k^3 - Nk)/\rho A\omega_0 + EI\, k^3/\rho A\omega_0 > 0 \end{aligned} \qquad (6.11)$$

since each of the constituent terms on the right side is greater than zero. In other words, *a wave that is moving up toward the surface in the positive direction will never reflect. Thus, a wave encountering a focus along its direction of travel must remain there, since it cannot reverse direction.*

Now, let us combine Equations 6-6 and 6-10, and solve for the wavenumber distribution k(x). We have, straightforwardly, using the quadratic formula,

$$k^2 = \{N + \sqrt{(N^2 + 4EI\rho A\omega_0^2)}\}/(2EI) \qquad (6.12)$$

and hence,

$$k = \sqrt{[\{N + \sqrt{(N^2 + 4EI\rho A\omega_0^2)}\}/(2EI)]} \qquad (6.13)$$

Our results can be expressed in a more concise and physically revealing form if we introduce the nondimensional (starred) quantities,

$$N^* = N/\{2\omega_0\sqrt{(EI\rho A)}\} \qquad (6.14)$$

$$k^* = \{EI/(\rho A\omega_0^2)\}^{1/4}\, k \qquad (6.15)$$

$$C_g^* = 1/2\, \{\rho A/(EI\omega_0^2)\}^{1/4}\, C_g \qquad (6.16)$$

Then, straightforward algebraic manipulations lead to the polynomial formulas

$$C_g^* = \{N^* + \sqrt{(N^{*2}+1)}\}^{1/2}\, \sqrt{(N^{*2}+1)} \qquad (6.17)$$

$$k^* = \{N^* + \sqrt{(N^{*2}+1)}\}^{1/2} \qquad (6.18)$$

which can be readily programmed and evaluated.

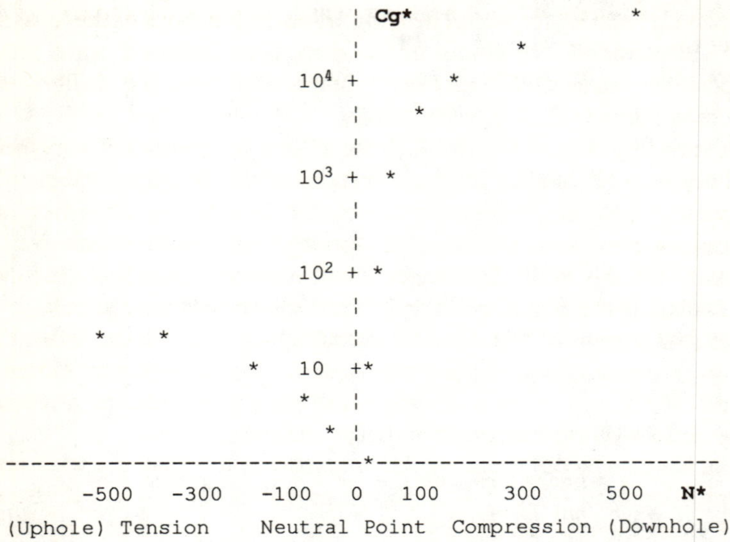

Figure 6-1. Group velocity versus axial force.

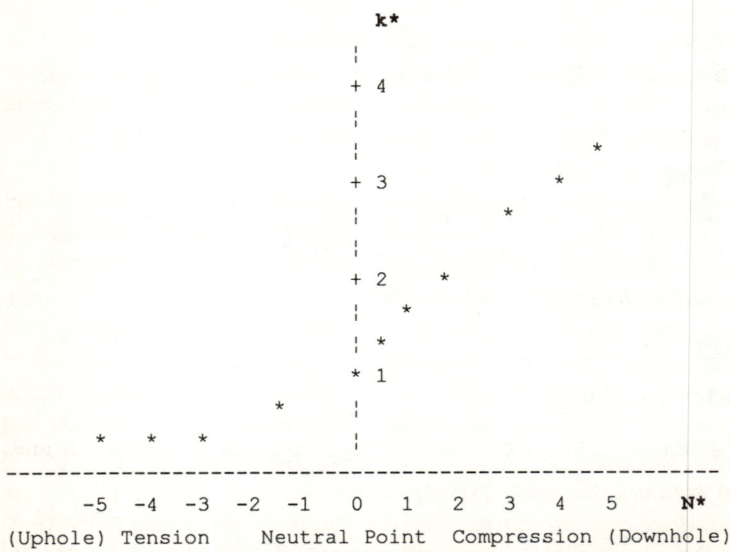

Figure 6-2. Wavenumber versus axial force.

The consequences of these analytical results are intriguing. Figures 6-1 and 6-2 plot C_g* and k* for lateral bending waves as functions of the local axial force N*. Decreasing N* points in the direction of upward wave motion. Again, positive values of N* represent compression near the drillbit, while negative values denote tension at the surface.

It is seen from Figure 6-1 that C_g^*, for all practical purposes, vanishes for zero and negative N*, that is, for those portions of the drillstring uphole of the neutral point. Those small amplitude waves that do "leak" past this *focal point* should dampen eventually, because of drillstring length and borehole wall and mud contact. On this basis, disturbance waves generated downhole near the bit will accumulate at the neutral point and should not be detectable uphole.

Note that regions of low nondimensional group velocity are defined by a *continuous spectrum of values* (consistently with the speculation in Aldred and Sheppard (1992)), and *not* by a discrete set of natural frequencies, as obtained in simple eigenvalue-type, resonant vibration analyses.

What happens to the lateral vibration amplitude *a*? Since the total energy flux (proportional to $a^2 C_g^*$) is expected to be locally conserved, bending amplitudes should grow indefinitely at the focus; they are, in fact, singular like $1/(C_g^*)^{1/2}$ as C_g^* goes to zero (Landahl, 1972).

Figure 6-2 shows how the wavenumber decreases significantly for those waves that do move uphole to the surface; they stretch beyond recognition, owing to refractive (that is, variable N(x)) effects, so that they will be undetectable, even if they have not fully dissipated.

The quantities plotted in Figures 6-1 and 6-2 are tabulated, respectively, in Tables 6-1 and 6-2; these general, nondimensional results are provided to furnish numbers for comparative use along the drillstring in practical application.

Bending vibrations maintain their observable character only for positive N*, that is, only along the section of the drill collar beneath the neutral point. Above the neutral point, observable motions are not possible because energy densities and wavenumbers are both small.

Those disturbances originating at the bit will travel upward to the neutral point, where they may remain; as discussed, they may not reflect downward toward the bit from the neutral point. Bending vibrations are therefore confined to this lowest part of the bottomhole assembly. And because material drill collar damping is weak, these strong vibrations persist, providing a potential source of damaging high-cycle fatigue.

This prediction is substantiated by recent MWD measurements, which track the frequency of high amplitude stresses; by and large, these are greatest in the lowest fifty feet of the drill collar beneath the neutral point.

N*	Cg*	N*	Cg*
-499	15.836	1	2.
-479	15.429	21	136.
-459	15.214	41	371.
-439	14.752	61	673.
-419	14.455	81	1031.
-399	14.114	101	1435.
-379	13.729	121	1882.
-359	13.451	141	2367.
-339	12.975	161	2889.
-319	12.585	181	3443.
-299	12.250	201	4030.
-279	11.839	221	4646.
-259	11.357	241	5291.
-239	10.928	261	5963.
-219	10.477	281	6661.
-199	9.9853	301	7385.
-179	9.4590	321	8133.
-159	8.9145	341	8905.
-139	8.3415	361	9700.
-119	7.7158	381	10517.
- 79	6.2870	421	12216.
- 59	5.4315	441	13097.
- 39	4.4168	461	13998.
- 19	3.0855	481	14919.

Table 6-1a. Cg* versus axial force.

N*	Cg*	N*	Cg*
-5.0	1.6045	0.0	1.000
-4.8	1.5740	0.2	1.126
-4.6	1.5430	0.4	1.309
-4.4	1.5115	0.6	1.549
-4.2	1.4793	0.8	1.847
-4.0	1.4466	1.0	2.197
-3.8	1.4134	1.2	2.596
-3.6	1.3794	1.4	3.039
-3.4	1.3449	1.6	3.523
-3.2	1.3097	1.8	4.045
-3.0	1.2739	2.0	4.602
-2.8	1.2374	2.2	5.192
-2.6	1.2004	2.4	5.813
-2.4	1.1628	2.6	6.464
-2.2	1.1247	2.8	7.143
-2.0	1.0864	3.0	7.850
-1.8	1.0482	3.2	8.582
-1.6	1.0104	3.4	9.339
-1.4	0.9739	3.6	10.120
-1.2	0.9398	3.8	10.924
-1.0	0.9101	4.0	11.751
-0.8	0.8878	4.2	12.600
-0.6	0.8775	4.4	13.470
-0.4	0.8862	4.6	14.362
-0.2	0.9233	4.8	15.273

Table 6-1b. Cg* versus axial force near focus.

N*	k*	N*	k*
-5.0	0.31467	0.0	1.0000
-4.8	0.32103	0.2	1.1044
-4.6	0.32778	0.4	1.2153
-4.4	0.33497	0.6	1.3290
-4.2	0.34265	0.8	1.4424
-4.0	0.35086	1.0	1.5538
-3.8	0.35969	1.2	1.6619
-3.6	0.36920	1.4	1.7665
-3.4	0.37949	1.6	1.8673
-3.2	0.39065	1.8	1.9645
-3.0	0.40284	2.0	2.0582
-2.8	0.41619	2.2	2.1486
-2.6	0.43090	2.4	2.2361
-2.4	0.44721	2.6	2.3207
-2.2	0.46541	2.8	2.4028
-2.0	0.48587	3.0	2.4824
-1.8	0.50904	3.2	2.5598
-1.6	0.53553	3.4	2.6351
-1.4	0.56610	3.6	2.7086
-1.2	0.60171	3.8	2.7802
-1.0	0.64359	4.0	2.8501
-0.8	0.69327	4.2	2.9185
-0.6	0.75246	4.4	2.9853
-0.4	0.82282	4.6	3.0508
-0.2	0.90543	4.8	3.1150

Table 6-2. Wavenumber versus axial force.

In our analysis, we modeled harmonic bending disturbances as they propagated on a time-stationary axial stress field, with N(x) depending only on x and not t. The resulting dynamically steady problem, greatly simplified, then yielded to closed form analytical solutions that indicated strong instabilities due to energy accumulation and wave trapping.

This new mechanism is consistent with Wolf, Zacksenhouse, and Arian (1985). In their evaluation of downhole lateral and longitudinal interactions, the authors observed that WOB fluctuations were relatively low, or were in the process of decaying, when high bending moments developed. When high WOB fluctuations developed, interestingly, bending moments turned out to be low. These limited observations suggest that the root cause of large bending fluctuations may have to do with static distributions of N(x) rather than fluctuating ones.

Bending amplitude distribution in space. In this section, we will expand upon the physical arguments offered above, and provide more rigorous and exact results. It can be shown (Whitham, 1974; Chin, 1976, 1980a) that the *wave momentum density* M and the *wave energy density* E satisfy

$$\partial M/\partial t + \partial(C_g M)/\partial x = M\{2\omega_i - \omega_{r,x}(k,x,t)/k\} \quad (6.19)$$

$$\partial E/\partial t + \partial(C_g E)/\partial x = E\{2\omega_i + \omega_{r,t}(k,x,t)/\omega_r\} \quad (6.20)$$

If the kinematic *wavenumber* field k(x,t) is available, the general initial value problems for these amplitude-like equations can be solved in closed form using the method of characteristics (see Chapter 2). In the foregoing equations, $\omega_{r,t}(k,x,t)$ and $\omega_{r,x}(k,x,t)$ denote partial derivatives of the real frequency function with all other variables *including k* held fixed.

Again, we did not explicitly consider the attenuative effects of internal viscous dissipation, since these were not germane to our arguments. We wished to show that an ω_i damping arose from the effects of heterogeneity alone; the effects of irreversible friction would, in any case, produce additive exponential terms with respect to the imaginary frequency.

Earlier we determined that the real dispersion relation satisfied $\rho A\omega_r^2 = EI\, k^4 - Nk^2$. Hence, upon partial differentiation with respect to x, we obtain $2\rho A\omega_r\omega_{r,x} = -N_x k^2$, thus implying that

$$-\omega_{r,x}(k,x,t)/k = (N_x k)/(2\rho A\omega_0) \tag{6.21}$$

From the imaginary part of the dispersion relation, we have

$$2\,\omega_i = 2N_x k/(2\rho A\omega_0) \tag{6.22}$$

Thus, we have, on combining Equations 6-19, 6-21, and 6-22,

$$\begin{aligned}\partial M/\partial t + \partial(C_g M)/\partial x &= M\{2\omega_i - \omega_{r,x}(k,x,t)/k\}\\ &= M\,[3kN_x/(2\rho A\omega_0)]\end{aligned} \tag{6.23}$$

As we are considering dynamically steady wave systems oscillating with a real frequency ω_0 unchanged with time, the partial derivative with respect to "t" vanishes, and we have

$$C_g dM/dx + M\, dC_g/dx = M\,[3kN_x/(2\rho A\omega_0)] \tag{6.24}$$

Note that only ordinary derivatives with respect to space appear in the foregoing equation. Division by MC_g throughout leads to

$$(1/M)\, dM/dx + (1/C_g)\, dC_g/dx = [3kN_x/(2\rho A\omega_0)]\,/C_g \tag{6.25}$$

Now, we observe that a *wave group trajectory* is kinematically defined by

$$dx/dt = C_g \tag{6.26}$$

Thus, it follows that

$$\begin{aligned}&d(\log M)/dx + d(\log C_g)/dx = d\{\log (MC_g)\}/dx\\ &\quad = [3kN_x/(2\rho A\omega_0)]\,/C_g = [3kN_x/(2\rho A\omega_0)]\, dt/dx\end{aligned} \tag{6.27}$$

In turn, we therefore have

$$d\{\log_e (MC_g)\} = [3kN_x/(2\rho A\omega_0)]\, dt \tag{6.28}$$

$$\log_e (MC_g) = \int [3kN_x/(2\rho A\omega_0)]\, dt \tag{6.29}$$

and finally, *the closed form spatial distribution for the wave momentum*

$$M(x) = \{1/C_g(x)\} \ \exp \{ \int [3kN_x/(2\rho A\omega_0)]\, dt \} \tag{6.30}$$

Now, since $\omega_0 > 0$ by requirement, and because $N'(x) < 0$ as the wave makes its way uphole to the surface, it follows that the second factor above vanishes exponentially far away at the surface, even without explicitly considering internal viscous dissipation. But near the neutral point, the effects of this factor are locally weak; the dominant term is the near-zero group velocity C_g.

Similar considerations apply to energy density using Equation 6-20. Since explicit dependences are disallowed because N is (by assumption) solely dependent upon x and not t, Equation 6-20 reduces to

$$\partial E/\partial t + \partial(C_g E)/\partial x = 2\omega_i E \tag{6.31}$$

because the term $\omega_{r,t}(k,x,t)$ vanishes identically. Thus, the integral corresponding to this simplified energy equation, following a derivation similar to that for the momentum derivation above, is

$$E(x) = \{1/C_g(x)\} \ \exp \{ \int [2kN_x/(2\rho A\omega_0)]\, dt \} \tag{6.32}$$

and physical interpretations analogous to those for M(x) apply. The functions $C_g(x)$ and k(x) are given in our earlier graphs and tables.

Close to the neutral point, the wave momentum density M(x) and wave energy density E(x) are both singular like $1/C_g(x)$ as $C_g(x)$ approaches zero. This indicates that the singularity is locally strong. In fact, as earlier tabulated and plotted formulas for the group velocity show, the singularity is strongly algebraic. Note that N(x) varies linearly with x along the drillstring, so that any explicit N dependence can be replaced by a constant multiple of x.

Our model explains why high concentrations of bending amplitude are more the rule than the exception, and why their mysterious absence at the surface is really quite natural when axial load variations are accounted for along the length of the drillstring.

Analyses for axial and torsional vibrations can be similarly performed. But because analogous sources of spatial inhomogeneity do not exist in the governing equations, their motions are not dynamically restricted or physically altered by *kinematic barriers* such as those introduced here. Thus, instabilities associated with axial and torsional vibrations can and should be modeled using conventional *eigenfunction, natural frequency, or resonant modal* techniques.

Incidentally, surface measurements of these non-lateral modes, as we shall see, properly deconvolved, should provide useful information on downhole vibrations and lithology. This possibility will be pursued later.

We are not suggesting fault with existing approaches, e.g., conventional modal theories, nonlinear models, buckling simulations, or parametric instabilities. In fact, they most likely will apply to other aspects of the vibration process; however, the observed failures occurring at the neutral point can only be described by bending models that specifically allow for axial load variations.

Importantly, we have provided a physically plausible explanation without recourse to nonlinear models. The model is consistent with observed phenomena, and supports the view points expressed by recent authors.

We emphasize that *the frequencies associated with large vibration amplitudes are not necessarily the natural frequencies of the system.* Linear system resonances always lead to instabilities. However, they are not the only source of violent fluctuations, and in the problem at hand, they do not represent the most dangerous form of downhole vibration.

It is also interesting that our general breakdown criterion does not depend on the details (that is, the boundary or initial conditions) of the bottomhole assembly or drillstring configuration used. Thus, our wave trapping instability model supports the speculation behind Rewcastle and Burgess (1992), that shock behaviour is more stable than a purely resonant response.

Of course, the details will be important to the extent that they *do* determine the exact conditions under which dangerous waves are produced and the resultant magnitudes; in the final analysis, detailed boundary value problem solutions are required.

The above results lead us to speculate that impending failures might be avoided by substantially increasing or decreasing weight-on-bit (WOB), at least temporarily, so that the drillstring is completely in compression or in tension, in order to remove the neutral point singularity.

This would relieve the intense, localized, dynamic stress concentrations formed as a result of wave trapping. But the possibility that another more conventional type of instability will appear in response to this change in WOB must be explored for the bottomhole assembly of interest.

Designing safe drill collars. It is clear that the equations derived above can be used to design (un)tapered drillstrings (with continuously varying A, E, I, and ρ) that are less susceptible to the undesirable effects associated with wave trapping and energy accumulation.

For a given target frequency ω_0, the wavenumber solution $k = [\{N+\sqrt{(N^2+4EI\rho A\omega_0^2)}\}/(2EI)]^{1/2}$ shown in Equation 6-13 can be used to simplify the group velocity formula $C_g = (2EIk^3 - Nk)/\rho A\omega_0$ given by Equation 6-6. The resulting expression

$$C_g = (2EI\,[\{N+\sqrt{(N^2+4EI\rho A\omega_0{}^2)}\}/(2EI)]^{3/2} - N[\{N+\sqrt{(N^2+4EI\rho A\omega_0{}^2)}\}/(2EI)]^{1/2})/\rho A\omega_0 \qquad (6.33)$$

is an algebraic quantity that is variable along its length.

Since N(x) is a known variable, the cross-sectional geometry and materials should be selected so that the group velocity remains more or less constant and never vanishing. Clearly, it is impractical to redesign the entire drillstring in this manner; most of the danger, perhaps, may be removed just by improving the transmission properties of the drill collar.

Viscous dissipation. We developed our model for wave trapping in attenuation-free media in order to obtain simple closed form solutions. This strategy also allowed us to circumvent the problems which arise in dealing with dimensionless groups in addition to those of Equations 6-14 to 6-16.

The more complete beam equation accounting for internal viscous dissipation, which we do solve numerically later, takes the form

$$EI\ \partial^4 v/\partial x^4 + \partial(N\ \partial v/\partial x)/\partial x + \beta\ \partial v/\partial t + \rho A\ \partial^2 v/\partial t^2 = 0 \qquad (6.34)$$

where the damping factor satisfies $\beta > 0$. If now we substitute Equations 6-2 and 6-4, we obtain on taking real parts,

$$\rho A(\omega_r{}^2 - \omega_i{}^2) - \beta\omega_i = EI\ k^4 - Nk^2 \qquad (6.35)$$

whereas by taking imaginary parts, we have

$$\omega_i = kN_x/(2\omega_r\rho A) - \beta/2\rho A \qquad (6.36)$$

From Equation 6-36 above, we observe that bending vibrations dissipate like $\exp\{kN_x/(2\omega_r\rho A) - \beta/2\rho A\}t$. Not only does the inclusion of the $\beta\ \partial v/\partial t$ term introduce an *attenuative* $-\ \beta t/2\rho A$ in the exponential factor, but viscous dissipation imparts the phase change evident from Equation 6-35. Such phase effects are discussed in detail in Chapter 1. However, since $|\omega_r| >> |\omega_i|$, modifications for dissipation should not affect our results qualitatively.

Later in this chapter, the qualitative ideas developed in this section will be quantified, and combined with the results of Chapter 5 to produce a numerical and analytical model that simulates coupled axial and bending vibrations only. Again, torsion is introduced in Chapter 7, while coupled axial, torsional, and bending vibrations are considered in Chapter 8.

SURFACE DETECTION OF DOWNHOLE BENDING DISTURBANCES

That the violent lateral vibrations occurring downhole cannot be detected from the surface as high amplitude bending waves, of course, does not mean that such catastrophic events cannot be monitored using *other* related measurements.

Detecting lateral vibrations. Vandiver, Nicholson and Shyu (1989) note that unlike longitudinal and torsional vibrations, lateral vibrations are not transmitted to the surface by the drillstring. However, they express cautious optimism, suggesting that it may be possible to see evidence of downhole bending vibration in the surface axial and torsional vibration signals although the relationships are not understood well enough to be used diagnostically.

Direct downhole observation remains a possibility, for example, by real-time monitoring of the timewise density of transverse shocks exceeding a pre-programmed threshold amplitude, and transmitting the result to the surface using MWD transmission channels.

Such a technique, in fact, is reported in Rewcastle and Burgess (1992). However, the low data rates characteristic of existing MWD systems mean that any such transmissions are likely to provide no more than highly compressed information of limited value, e.g., a *rms* value indicative of dynamic stresses at some point in the bottom hole assembly. Also, MWD vibrations measurements are necessarily taken at the expense of directional and formation data.

Paslay, Jan, Kingman and Macpherson (1992) attempt to detect large downhole lateral vibrations from *surface* longitudinal and torsional measurements. Their analysis, based on simplified geometrical beam deflection arguments, predicted motions at top of string due to BHA forward synchronous whirl. We will discuss their field and calculated results in more detail later.

Here, we will develop a different conceptual framework, and draw upon formal results in nonlinear elasticity. From the theory of nonlinear vibrations, *bending motions will act as distributed sources of axial disturbances,* which *are* detectable at the surface as longitudinal oscillations, and possibly, additionally through their effect on torsional oscillations.

This effect mathematically manifests itself as an external *forcing function* to the linear longitudinal wave operator; again, the modeling of forcing functions is discussed in Chapter 1, and in more specialized contexts, in Chapter 5. Thus, *the spectral content obtained by monitoring surface axial vibrations should change noticeably and predictably when large bending motions are encountered or when drillstring failure is impending.*

Nonlinear axial equation. To illustrate the basic ideas, we will ignore the effects of static and dynamic torsion, concentrating instead on longitudinal and lateral mode interactions only. Earlier we indicated that undamped longitudinal oscillations for the *axial displacement* variable u(x,t) satisfy the classical wave equation

$$\rho\, \partial^2 u/\partial t^2 - E\, \partial^2 u/\partial x^2 = 0 \qquad (6.37)$$

where ρ is the mass density per unit volume and E is Young's modulus.

Equation 6-37 holds when all other vibration modes are weak in amplitude; it shows that u(x,t) acts independently of other types of vibration, except to the extent that there may exist some form of coupling through rock-bit boundary conditions. If we revisit the more complete system of coupled governing equations that apply *before* the linearizing assumption behind Equation 6-37 was made, it turns out that

$$\rho A\, \partial^2 u/\partial t^2 - EA\, \partial^2 u/\partial x^2 = EI\, \partial\{(\partial^3 v(x,t)/\partial x^3)\, (\partial v/\partial x)\}/\partial x \qquad (6.38)$$

applies at the next hierarchical level of approximation, where v(x,t) represents the *lateral* bending displacement (Graff, 1975).

The right side term describes the nonlinear coupling between longitudinal and bending modes, one that will in general be weak; it is usually neglected because it is quadratic in a small dimensionless lateral displacement, thus leading to classical wave equation for u(x,t), which is independent of axial contraction and expansion due to bending.

When lateral vibrations are large, though, the right side of Equation 6-38 will not be small. But the simplicity behind the classical wave operator *can* be retained: insofar as the wave-like left side of Equation 6-38 is concerned, the right side plays the role of an *external forcing function*, literally providing a *source* of distributed axial force.

This book-keeping is not *mere* accounting: it is consistent with actual observation. Since up and down-going axial waves *do* exist, whether or not downhole lateral vibrations are small, our interpretation of Equation 6-38 as a type of *almost linear* axial wave equation is physically justified.

For large lateral vibrations, we can therefore view the left side of Equation 6-38 as we did in Chapter 1, but interpret the right side of Equation 6-38 as a forced excitation. Additional simplifications are possible. Since the bending singularity is only sizable near the neutral point, its localized nature allows us to view this *compact source* of axial stress as a *concentrated load*, particularly amenable to *delta function* modeling as described in Chapter 1.

Detecting lateral vibrations from the surface. We will accept this physical interpretation. Thus, if ω_b is the dominant frequency characterizing downhole *lateral* excitation, then the quadratic nature of the right-side source term indicates that the *axial vibration frequency spectrum measured at the surface* should contain a spectral line having frequency $2\omega_b$ - *this exists only if downhole lateral vibrations are significant!*

In other words, a drillstring that is undergoing large-amplitude trapped *transverse* vibrations downhole at ω_b will create *longitudinal* oscillations near the neutral point having frequency $2\omega_b$, which will, in turn, be freely transmitted to the surface.

Continuous monitoring of the axial spectrum for sudden unexpected changes, therefore, should provide a clue as to the likelihood of impending danger. Furthermore, if drill collar lateral vibration information can be ascertained from real-time, downhole MWD measurements, and transmitted from time to time to the surface, more controlled surface monitoring is possible.

This lateral-to-axial mode of energy transfer was first noted in Chin (1988a,b). It was later suggested independently by Rewcastle and Burgess (1992), who stated explicitly that bending vibration can induce axial vibration at twice the frequency. However, these authors did not publish details related to this statement.

The literature provides little evidence regarding this speculation. Monitoring surface axial vibrations for downhole lateral instability, needless to say, is neither obvious nor ordinary. An interesting clue is offered by Paslay, Jan, Kingman and Macpherson (1992), who monitor the surface axial vibration spectrum due to a drillstring in a vertical well.

Their experiments were conducted with a drill bit rotation rate in the 2.3 Hz (or 138 rpm) range, and its corresponding spectral line prominently appears in their measurements. For example, their Figure 3 clearly shows a strong 2.3 Hz line, while their Figure 5 similarly shows a well-defined 2.1 Hz line.

Also noted on these respective figures, though, are unexpectedly strong 4.6 and 4.2 Hz lines. An examination of their Figure 1 shows that the lateral resonant frequency *happens* to be 2.2 Hz, which is extremely close to their 2.1 and 2.3 Hz rotation rates. If BHA lateral resonance was, in fact, ongoing at the time axial data was collected, the 4 Hz axial signals may well represent twice the resonant lateral frequency and not so much a confusing twice-rotation-rate.

LINEAR BOUNDARY VALUE PROBLEM FORMULATION

So far we have discussed fundamental physical ideas only: in particular, why lateral vibrations tend to be violent downhole and undetectable at the surface, plus the possibility of detecting their downhole occurrence by monitoring the axial wave spectrum at the surface.

Here we formulate the linear boundary value problem, to compute solutions satisfying prescribed conditions. By *linear*, we refer to lateral displacement solutions when the axial strain field is specified, without determining the *back-interaction* of lateral displacements on axial stresses. We also give a stable numerical finite difference scheme that leads to easily inverted algebraic equations requiring efficient *banded matrix solvers* only.

General linear equation. Our beam equation for the transverse displacement v(x,t), Equation 6-1, or $EI\ \partial^4 v/\partial x^4 + \partial(N\ \partial v/\partial x)/\partial x + \rho A\ \partial^2 v/\partial t^2 = 0$, included only the essential terms needed to model wave focusing and trapping. In general, the complete equation takes the form

$$EI\ \partial^4 v/\partial x^4 + \partial(N\ \partial v/\partial x)/\partial x + kv + \beta\ \partial v/\partial t + \rho A\ \partial^2 v/\partial t^2 = q^{(v)}(x,t) \tag{6.40}$$

where we have introduced a general loading $q^{(v)}(x,t)$ to account for additional sources of external excitation, and a *resistive force* $kv + \beta\ \partial v/\partial t$ consisting of *elastic* (kv) and *dissipative* ($\beta\ \partial v/\partial t$) contributions.

In Equation 6-40, N(x,t) represents the axial force, positive for compression, and negative for tension; in the limit of vanishing stiffness EI and dissipation β, we recover the classical wave equation for transverse vibrations. The reader should perform a global energy analysis similar to that of Chapter 5, constructing energy inequalities, in order to understand the exact physical meaning behind each term in Equation 6-40.

Now recall from Chapter 5 how our sign convention assumed that the axial force satisfies $AE\ u_x < 0$ for compression, while we have $AE\ u_x > 0$ for tension. Thus, we consistently set

$$N(x,t) = -AE\ \partial u/\partial x \tag{6.41}$$

so that the bending equation takes the explicit form

$$EI\ \partial^4 v/\partial x^4 - \partial\{(AE\ \partial u/\partial x)\ \partial v/\partial x\}/\partial x + kv + \beta\ \partial v/\partial t + \rho A\ \partial^2 v/\partial t^2 = q^{(v)} \tag{6.42}$$

The following review is not intended to replace a complete study of strength of materials and elasticity. For the transverse deflection v(x,t), the slope is simply $\partial v(x,t)/\partial x$. The moment M(x,t) and the shear V(x,t) satisfy

$$M = -\ EI\ \partial^2 v(x,t)/\partial x^2 \tag{6.43}$$

and

$$V = -\ EI\ \partial^3 v(0,t)/\partial x^3 \tag{6.44}$$

even when axial forces are present.

Auxiliary conditions. Equation 6-42, a partial differential equation for v(x,t), must be solved with auxiliary boundary and initial conditions. Let us first review the fundamental *boundary conditions* applicable to beam bending formulations. In short, we have,

Pinned end:

$$v(0,t) = 0 \tag{6.45}$$

$$\partial^2 v(0,t)/\partial x^2 = 0 \tag{6.46}$$

Fixed end:

$$v(0,t) = 0 \tag{6.47}$$

$$\partial v(0,t)/\partial x = 0 \tag{6.48}$$

Free end:

$$\partial^2 v(0,t)/\partial x^2 = 0 \tag{6.49}$$

$$\partial^3 v(0,t)/\partial x^3 = 0 \tag{6.50}$$

Deflected spring:

$$\partial^2 v(0,t)/\partial x^2 = 0 \tag{6.51}$$

$$EI\ \partial^3 v(0,t)/\partial x^3 = -\ k_d\ v(0,t) \tag{6.52}$$

Torsional spring:

$$v(0,t) = 0 \tag{6.53}$$

$$EI\ \partial^2 v(0,t)/\partial x^2 = k_T\ \partial v(0,t)/\partial x \tag{6.54}$$

Mass loaded:

$$\partial^2 v(0,t)/\partial x^2 = 0 \tag{6.55}$$

$$-\ EI\ \partial^3 v(0,t)/\partial x^3 = m\ \ \partial^2 v(0,t)/\partial t^2 \tag{6.56}$$

Dashpot dampened:

$$\partial^2 v(0,t)/\partial x^2 = 0 \tag{6.57}$$

$$EI\ \partial^3 v(0,t)/\partial x^3 = -\beta\ \ \partial v(0,t)/\partial t \tag{6.58}$$

Although we have chosen to emphasize the rock-bit contact point x = 0, the above conditions also apply at the surface x = L as necessary. In the foregoing boundary conditions, m, k_T, k_d, and β are suitable constants.

Finally, suitable *initial conditions* must be used to start any analytical or numerical integrations. If we assume that the initial drillstring is straight and starts from rest, we can write

$$v(x,0) = 0 \tag{6.59}$$

$$\partial v(x,0)/\partial t = 0 \tag{6.60}$$

We are now in a position to simulate lateral bending vibrations. We emphasize that while numerous analytical solutions are available in the mechanical engineering literature, they do not contain variable N(x,t) functions that hold particular relevance to drillstring vibrations. Thus, we will refrain from presenting these, preferring instead to consider more powerful computational methods that can be ultimately extended to handle general coupled vibrations.

FINITE DIFFERENCE MODELING

Because lateral vibrations play significant roles in downhole mechanical failure and borehole instability, the ability to model and simulate their behaviour holds practical importance. Spanos and Payne (1992), for example, have developed a numerical *finite element method* based on *Euler-Bernoulli beam theory* to simulate *harmonic* events occurring downhole.

We have already identified the predominant physical mechanisms controlling downhole instability, using kinematic wave theory as applied to individual Fourier wave components, so we will now concentrate on fully transient theory capable of modeling shocks and events induced by general longitudinal transients. Such a formulation, of course, will lead to harmonic results if such results in fact exist.

Pentadiagonal difference equations. Unlike the axial wave equation in Chapter 5, which contains *second-order* spatial derivatives, the beam equation is *fourth-order*. This means that the simple *tridiagonal equations* we obtained earlier for longitudinal vibrations no longer apply.

However, the idea behind *central differencing*, which preserves left-and-right symmetry, still applies, since a symmetrically constrained beam acting under symmetric axial stresses, must respond symmetrically about its midpoint. If we apply our earlier central difference formulas recursively, it is possible to show that

$$\partial^4 u/\partial x^4 = (U_{i-2} - 4U_{i-1} + 6U_i - 4U_{i+1} + U_{i+2})/(\Delta x)^4 + O\{(\Delta x)^2\} \tag{6.61}$$

applies at the point x_i. Thus, Equation 6-40 can be discretized or finite differenced as follows,

$$\begin{aligned}
&EI\,(v_{i-2,n} - 4v_{i-1,n} + 6v_{i,n} - 4v_{i+1,n} + v_{i+2,n})/(\Delta x)^4 \\
&+ N_{i,n}\,(v_{i-1,n} - 2v_{i,n} + v_{i+1,n})/(\Delta x)^2 \\
&+ \{(N_{i+1,n} - N_{i-1,n})/2\Delta x\}\{(v_{i+1,n} - v_{i-1,n})/2\Delta x\} \\
&+ kv_{i,n} + \beta\,(v_{i,n} - v_{i,n-2})/2\Delta t \\
&+ \rho A(v_{i,n} - 2v_{i,n-1} + v_{i,n-2})/(\Delta t)^2 = q^{(v)}{}_{i,n-1}
\end{aligned} \tag{6.62}$$

with

$$N_{i,n} = - AE \{u_{i+1,n} - u_{i-1,n}\}/2\Delta x \tag{6.63}$$

where, as in Chapter 5, the time derivatives are approximated using backward differences by virtue of *causality* requirements.

Now, we can rewrite Equation 6-62 in a more enlightening form which is immediately suggestive of a stable computational algorithm or recipe. We recast our difference approximation in the form

$$\begin{aligned}
&\{EI/(\Delta x)^4\}\, v_{i-2,n} \\
&+ \{-4EI/(\Delta x)^4 + N_{i,n}/(\Delta x)^2 - (N_{i+1,n}-N_{i-1,n})/(4(\Delta x)^2)\}\, v_{i-1,n} \\
&+ \{6EI/(\Delta x)^4 - 2N_{i,n}/(\Delta x)^2 + k + \beta/(2\Delta t) + \rho A/(\Delta t)^2\}\, v_{i,n} \\
&+ \{-4EI/(\Delta x)^4 + N_{i,n}/(\Delta x)^2 + (N_{i+1,n}-N_{i-1,n})/(4(\Delta x)^2)\}\, v_{i+1,n} \\
&+ \{EI/(\Delta x)^4\}\, v_{i+2,n} \\
&\quad = \quad q^{(v)}{}_{i,n-1} \\
&\qquad + (\beta/(2\Delta t))v_{i,n-2} + \{2\rho A/(\Delta t)^2\}v_{i,n-1} - \{\rho A/(\Delta t)^2\}v_{i,n-2}
\end{aligned} \tag{6.64}$$

Finite difference beam recipe. Following the procedure of Chapter 5, we write Equation 6-64 for a subset of *internal* drillstring nodes. This results, of course, in more unknowns than equations; the problem is remedied by imposing the required number of boundary conditions.

In particular, we will need to write Equation 6-64 for the drillstring indexes i = 3, 4, 5, ... , imax-3, imax-2. The resulting equations involve unknowns ranging from $v_{1,n}$ to $v_{imax,n}$ at any time step n. Thus, four additional equations are needed to completely specify the algebraic problem.

The equations corresponding to the i = 1 and 2 and i = imax-1 and imax, of course, are reserved for downhole and surface boundary conditions, respectively. Once these are written, we have in matrix form, the *pentadiagonal* governing equations,

$$\begin{vmatrix}
B_1 & C_1 & E_1 & & & & & & \\
A_2 & B_2 & C_2 & E_2 & & & & & \\
D_3 & A_3 & B_3 & C_3 & E_3 & & & & \\
 & D_4 & A_4 & B_4 & C_4 & E_4 & & & \\
 & & D_5 & A_5 & B_5 & C_5 & E_5 & & \\
 & & & & \cdots & & & & \\
 & & & & & \cdots & & & \\
 & & & & D_{imax-2} & A_{imax-2} & B_{imax-2} & C_{imax-2} & E_{imax-2} \\
 & & & & & D_{imax-1} & A_{imax-1} & B_{imax-1} & C_{imax-1} \\
 & & & & & & D_{imax} & A_{imax} & B_{imax}
\end{vmatrix}$$

$$\times \begin{vmatrix} v_1 \\ v_2 \\ v_3 \\ v_4 \\ v_5 \\ \ldots \\ \ldots \\ v_{imax-2} \\ v_{imax-1} \\ v_{imax} \end{vmatrix} = \begin{vmatrix} W_1 \\ W_2 \\ W_3 \\ W_4 \\ W_5 \\ \ldots \\ \ldots \\ W_{imax-2} \\ W_{imax-1} \\ W_{imax-1} \end{vmatrix} \qquad (6.65)$$

This pentadiagonal (or, five diagonal) system, with right-hand-sides determined from earlier time steps, can be *reduced* using standard operations from *linear algebra* to simple *tridiagonal form*. By removing the two outermost left and right bands, the resulting matrix can be inverted using the tridiagonal Fortran algorithm given in Chapter 5, to yield solutions for the lateral displacements v_i.

Although the computational procedure is similar to that of Chapter 5, the physics requires a somewhat different perspective. For axial vibrations, disturbances propagate at a constant speed "c", theoretically anyway, and time step sizes are chosen consistently with the *Courant-Friedrichs-Lewy* (or, *CFL*) condition $\Delta t < (\Delta x)/c$. In the case of lateral vibrations, we have shown that propagation speeds depend upon the wavenumber, the disturbance waves being *dispersive*; thus, *the choice of Δt will automatically eliminate a range of lateral waves from consideration.*

Additional modeling considerations. We will briefly address several subjects that are important in modeling simple bending vibrations, which apply also to the dual torque-coupled lateral modes considered in Chapter 8.

Borehole wall contacts. Closed form transverse solutions for *static* drillstrings in boreholes are difficult to obtain because wall constraints due to finite hole radii lead, possibly, to multiple contact points, or none at all. This is compounded by *static indeterminacy*, thus forcing the use of more complicated solution methods. Nor are static computational approaches any less troublesome. At contact points, if any, our difference model does not apply directly, and modifications to include local *shear force effects* (at positions unknown *a priori*) in the overall force balance are needed.

But treating static problems as large-time limits of transient processes completely circumvents the problem. Finite hole radii with formation contact is simply modeled with Fortran *if-then* logic that invokes a spring-like restoring force if the borehole radius is exceeded at any time step (the distributed force model can be used in conjunction with elastic, inelastic, or plastic spring responses to model different formations).

This transient approach is useful in three dimensions, when torque couples both lateral bending modes, and spiral etch patterns have been computed. Moreover, this device is well grounded physically, because it is exactly the same process or path that nature takes. Suitable initial conditions are needed, e.g., a starting deflected or undeflected state, with or without initial velocity. The end displacement pattern, and number and position of contact points, incidentally, may vary with initial condition.

Aldred and Sheppard (1992), interestingly, propose a mechanism by which transverse drillstring motions couple to rotary ones by way of borehole wall contact. They suggest that harmonic resonant behaviour is not primarily responsible for observed high shock levels, and that these shock levels can lead to an increase in surface torque and a decrease in the rate-of-penetration. Their ideas can be tested and refined using the algorithm presented here.

Modeling steady state oscillations. Dynamically steady oscillations were discussed in Chapter 5 for axial vibrations, and the same separation of variables techniques developed there apply to Equation 6-42. Naturally, because the beam equation is significantly more complicated than the classical wave equation, the resulting complex variables based results will be more unwieldy.

The early paper of Huang and Dareing (1968) considers critical buckling loads and natural frequencies of lateral vibration modes for long vertical pipes, suspended in fluid, simply supported at the top and vertically guided at the bottom. As discussed, wall contacts cannot be modeled, thus limiting the usefulness of the results. The authors also restricted the drillstring analysis to uniform cross-sectional areas.

Simulating area changes. In our discussion thus far, we have ignored the fact that real drillstrings contain changes in cross-sectional properties, particularly at the interface separating the drillpipe and the drill collar. At these discontinuities, the differential equation breaks down because not all derivatives exist, and recourse to separate matching conditions is necessary. For transverse vibrations, these include continuity of displacement, moment, and shear; slope, of course, may or may not be conserved through the transition. These matching conditions can be developed into a diagonally dominant finite difference model, as in Chapter 5, to ensure stable time calculations.

EXAMPLE FORTRAN IMPLEMENTATION

In this section, we list the Fortran statements needed to describe transverse vibrations, but limit ourselves to uniform drillstrings for simplicity. Therefore, we will concentrate on the finite difference representation in Equation 6-64, and describe its matrix implementation as defined by the pentadiagonal system in Equation 6-65.

Again, we will refer to statements in the general coupled program for axial, torsional, and (two) lateral vibrations developed in Chapter 8. Since portions of this code exist in Chapters 5, 6, and 7, only those sections directly applicable to the subject of this chapter will be presented. The relevant Fortran statements are listed in Figure 6-3 below.

```
C     .
C     INITIALIZE AXIAL, ANGULAR AND BENDING DISPLACEMENTS
      DO 130  I=1,IMAX
      X = XS*(I-1)/(IMAX-1)
      UNM1(I) = RHO*G*(X**2)/(2.*ELAST) +WZERO*X/AE
C     .
      VNM1(I) = 0.
      VNM2(I) = 0.
      WNM1(I) = 0.
      WNM2(I) = 0.
 130  CONTINUE
C     BEGIN TIMEWISE INTEGRATION
      T = 0.
      DO 900  N=1,NMAX
      T = N*DT
C     AXIAL VIBRATIONS
      DO 200  I=2,IMAXM1
      A(I) = 1.
      C(I) = 1.
      B(I) = -2.-RHO*DX*DX/(ELAST*DT*DT)-GAMA*DX*DX/(2.*ELAST*DT)
      W(I) = -RHO*DX*DX*(2.*UNM1(I)-UNM2(I))/(ELAST*DT*DT)
     1         -GAMA*DX*DX*UNM2(I)/(2.*ELAST*DT)+RHO*G*(DX**2)/ELAST
 200  CONTINUE
      A(IMAX) = -AE/DX
      B(IMAX) = TBMASS/(DT**2) +BETA/(2.*DT) +AE/DX +SPRING
      C(IMAX) = 99.
      W(IMAX) = 2.*TBMASS*UNM1(IMAX)/(DT**2)
     1          +(BETA/(2.*DT) -TBMASS/(DT**2))*UNM2(IMAX)-TBMASS*G
      A(1) = 99.
      VEL = (UNM1(1)-UNM2(1))/DT
      IF(VEL.LE.0.) GO TO 205
C     .
C     TORSIONAL VIBRATIONS
C     .
C     .
```

Figure 6-3. Fortran listing, lateral vibrations (to be continued).

```
C     .
C     .
C     COUPLED BENDING VIBRATIONS
C     V LATERAL MODE
      DO 600  I=3,IMAXM2
      D(I) = EI/DX4
      E(I) = EI/DX4
      A(I) = -4.*EI/DX4 -AE*DUDX(I)/(DX*DX)
     1        +AE*(DUDX(I+1)-DUDX(I-1))/(4.*DX*DX)
      C(I) = -4.*EI/DX4 -AE*DUDX(I)/(DX*DX)
     1        -AE*(DUDX(I+1)-DUDX(I-1))/(4.*DX*DX)
      B(I) = 6.*EI/DX4 +2.*AE*DUDX(I)/(DX*DX)
     1        +BEAMK +BEAMB/(2.*DT) +RHO*AREA/(DT**2)
      W(I) = Q -GJ*DTHDX(I)*
     1        (-WNM1(I-2)+2.*WNM1(I-1)
     2        -2.*WNM1(I+1)+WNM1(I+2))/(2.*DX**3)
     3        -GJ*(DTHDX(I+1)-DTHDX(I-1))*(1./(2.*DX))*
     4        (WNM1(I-1)-2.*WNM1(I)+WNM1(I+1))/(DX**2)
     5        +BEAMB*VNM2(I)/(2.*DT)
     6        +2.*RHO*AREA*VNM1(I)/(DT**2)
     7        -RHO*AREA*VNM2(I)/(DT**2)
 600  CONTINUE
C     .
C     .
C     Define downhole and surface beam boundary conditions next,
C     reduce pentadiagonal equations to tridiagonal form and solve.
C     .
C     .
      CALL REDUCE(A,B,C,D,E,W,IMAX)
      CALL TRIDI(A,B,C,VECTOR,W,IMAX)
      DO 650  I=1,IMAX
      VBENDN(I) = VECTOR(I)
 650  CONTINUE
C     W LATERAL MODE
C     .
C     .
  900  CONTINUE
C     .
C     .
      STOP
      END
```

Figure 6-3. Fortran listing, lateral vibrations (continued).

Figure 6-3 deserves elaboration. If we compare Equations 6-64 and 6-65, the matrix coefficients **D**, **A**, **B**, **C**, **E**, and **W** (for the right-hand-side of Equation 6-65) are easily identified; these appear in the "VBEND" logic shown in the 600 *do-loop*.

Also shown in this loop are right-side effects that deal with coupled torque and W-mode bending, that the reader should ignore for the purposes of this chapter. Once Loop 600 is executed, matrix coefficients related to i indexes 1, 2, imax-1, and imax must be defined, subject to the boundary conditions assumed at the bit and the surface. This process is similar to that discussed in detail in Chapter 5.

The implementation described here appears to be unconditionally stable on a von Neumann basis, and transient calculations at this writing have consistently remained stable. Again, examples for smooth and rough drilling will be given in Chapter 8.

NONLINEAR INTERACTION BETWEEN AXIAL AND LATERAL BENDING VIBRATIONS

Our closed form solution shows that the static axial stress field imposed along a drillstring can lead to the trapping of lateral waves, and hence, localized high cycle fatigue. But as lateral amplitudes become singular, they can alter the axial stress state that induced the severe bending, and theoretically anyway, act as to remove the cause of the instability.

But this is probably unlikely. Severe lateral vibrations will remain an issue, as unfortunate field experiences constantly remind us; thus, the axial-lateral mode coupling to be discussed is likely to provide only minor changes to the linear model discussed so far. This statement represents, of course, opinion and not fact.

For formulation purposes, it is worthwhile to return to two previous equations, namely,

$$EI\ \partial^4 v/\partial x^4 - \partial\{(AE\ \partial u/\partial x)\ \partial v/\partial x\}/\partial x + kv + \beta\ \partial v/\partial t + \rho A\ \partial^2 v/\partial t^2 = q^{(v)} \tag{6.42}$$

$$\rho A\ \partial^2 u/\partial t^2 - EA\ \partial^2 u/\partial x^2 = EI\ \partial\{(\partial^3 v(x,t)/\partial x^3)\ (\partial v/\partial x)\}/\partial x \tag{6.38}$$

Again, note that the second term in Equation 6-42 represents the effects of variable axial loading, with the force $AE\ u_x < 0$ for compression and $AE\ u_x > 0$ for tension; in terms of our earlier notation, we can write $N(x,t) = -EA\partial u/\partial x$.

In elementary *beam* theory, Equation 6-42 is solved with all coefficients uniquely defined; for example, A, E, I, k, β, and ρ are typically constants, and $\partial u/\partial x$ and $q^{(v)}$ are prescribed functions of x. While complicated, the beam formulation is still a linear one.

Equation 6-38, if the right-side is set to zero, is no more than the classical wave equation discussed at length in Chapters 1 and 5. This model is a linear one. However, when lateral vibrations are significant downhole, $v(x,t)$ and its spatial derivatives may no longer be small, and the right-side of Equation 6-38 cannot in general be neglected.

Thus, Equations 6-42 and 6-38 couple nonlinearly, and together describe a *nonlinear model* in which axial and lateral bending modes continuously exchange energy in time. The coupled system in general requires numerical solution methods, which at first sight may appear overwhelming.

However, the only visible structural change is the nonzero right-side of Equation 6-38. This suggests that all of the recipes we have developed so far still apply, provided we update our axial displacement results at each time step using latest available bending solutions. The nonlinear boundary value problem in this manner is rendered amenable to numerical solution.

REFERENCES

Abramson, H.N., Plass, H.J., and Ripperger, E.A., "Stress Wave Propagation in Rods and Beams," in Advances in Applied Mechanics, Vol. V, edited by Hugh Dryden and Theodore von Karman, Academic Press, New York, 1958.

Aldred, W.D., and Sheppard, M.C., "Drillstring Vibrations: A New Generation Mechanism and Control Strategies," SPE Paper No. 24582, presented at the 67th Annual Technical Conference and Exhibition of the Society of Petroleum Engineers, Washington, D.C., Oct. 4-7, 1992.

Burgess, T.M., McDaniel, G.L., and Das, P.K., "Improving BHA Tool Reliability with Drillstring Vibration Models: Field Experience and Limitations," SPE/IADC Paper No. 16109, 1987 SPE/IADC Drilling Conference, New Orleans, March 15-18, 1987.

Cheatham, C.A., and Comeaux, B.C., "General Guidelines for Predicting Fatigue Life of MWD Tools," IADC/SPE Paper No. 23906, presented at the 1992 IADC/SPE Drilling Conference, New Orleans, Louisiana, Feb. 18-21, 1992.

Chen, B.C.M., and Gair, R., "Flowing Fluid Dynamics Can No Longer Be Ignored in Deepwater Operations," Offshore Magazine, April 1992, pp. 54-59.

Chin, W.C., *Physics of Slowly Varying Wavetrains in Continuum Systems, Ph.D. Thesis,* Massachusetts Institute of Technology, Cambridge, Massachusetts, 1976.

Chin, W.C., "Kinematic Wave Approach to Hydraulic Jumps with Waves," Journal of Hydronautics, Oct. 1979.

Chin, W.C., "Wave Focusing and Hydraulic Jump Formation," Journal of Hydronautics, July 1980.

Chin, W.C., "Effect of Dissipation and Dispersion on Slowly Varying Wavetrains," AIAA Journal, Feb. 1980.

Chin, W.C., "Kinematic Barrier for Gravity Waves on Variable Currents, Journal of Hydronautics," Jan. 1981.

Chin, W.C., "Fatal Tubular Bending Motions Difficult to Detect Uphole," Offshore Magazine, OTC Preview Issue, April 1988.

Chin, W.C., "Why Drill Strings Fail at the Neutral Point," Petroleum Engineer International, Downhole Tools and Technology Issue, May 1988.

Close, D.A., Owens, S.C., and MacPherson, J.D., "Measurement of BHA Vibration Using MWD," IADC/SPE Paper No. 17273, presented at the 1988 IADC/SPE Drilling Conference, Dallas, Texas, Feb. 28 - Mar. 2, 1988.

Clough, R.W., and Penzien, J., *Dynamics of Structures*, McGraw-Hill, New York, 1975.

Dareing, D.W., "Rotary Speed, Drill Collars Control Drill String Bounce," Oil & Gas Journal, June 6, 1983, pp. 63-68.

Den Hartog, J.P., *Advanced Strength of Materials*, McGraw-Hill, New York, 1952.

Graff, K.F., *Wave Motion in Elastic Solids*, Ohio State University Press, 1975.

Henneuse, H., "Surface Detection of Vibrations and Drilling Optimization: Field Experience," IADC/SPE Paper No. 23888, presented at the 1992 IADC/SPE Drilling Conference, New Orleans, Louisiana, Feb. 18-21, 1992.

Huang, T., and Dareing, D.W., "Buckling and Lateral Vibration of Drill Pipe," Journal of Engineering for Industry, Transactions of the ASME, Nov. 1968, pp. 613-619.

Kane, J.E., "Dynamic Lateral Bottom Hole Forces Aid Drilling," Drilling, Aug. 1984, pp. 43-48.

Landahl, M.T., "Wave Mechanics of Breakdown," Journal of Fluid Mechanics, Vol. 56, Part 4, 1972, pp. 775-802.

Love, A.E.H., *A Treatise on the Mathematical Theory of Elasticity*, Dover Publications, New York, 1944.

Mitchell, R.F., and Allen, M.B., "Lateral Vibration: The Key to BHA Failure Analysis," World Oil, March 1985, pp. 101-106.

Morse, P.M., and Feshbach, H., *Methods of Theoretical Physics, Parts I and II*, McGraw-Hill, New York, 1953.

Nicholson, R.W., "Critical Rotary Speeds," presented at the 1975 Rotary Drilling Conference of the International Association of Drilling Contractors (IADC), March 5-7, 1975.

Paslay, P.R., Jan, Y.M., Kingman, J.E.E., and Macpherson, J.D., "Detection of BHA Lateral Resonances While Drilling With Surface Longitudinal and Torsional Sensors," SPE Paper No. 24583, presented at the 67th Annual Technical Conference and Exhibition of the Society of Petroleum Engineers, Washington, D.C., Oct. 4-7, 1992.

Rewcastle, S.C., and Burgess, T.M., "Real-Time Downhole Shock Measurements Increase Drilling Efficiency and Improve MWD Reliability," IADC/SPE Paper No. 23890, presented at the 1992 IADC/SPE Drilling Conference, New Orleans, Louisiana, Feb. 18-21, 1992.

Spanos, P.D., and Payne, M.L., "Advances in Dynamic Bottomhole Assembly Modeling and Dynamic Response Determination," IADC/SPE Paper No. 23905, presented at the 1992 IADC/SPE Drilling Conference, New Orleans, Louisiana, Feb. 18-21, 1992.

Timoshenko, S.P., and Goodier, J.N., *Theory of Elasticity*, McGraw-Hill, New York, 1934.

Vandiver, J.K., Nicholson, J.W., and Shyu, R.J., "Case Studies of the Bending Vibration and Whirling Motion of Drill Collars," SPE/IADC Paper No. 18652, presented at the 1989 SPE/IADC Drilling Conference, New Orleans, Louisiana, February 28-March 3, 1989.

Whitham, G.B., *Linear and Nonlinear Waves*, John Wiley and Sons, New York, 1974.

Wolf, S.F., Zacksenhouse, M., and Arian, A., "Field Measurements of Downhole Drilling Vibrations," SPE Paper No. 14330, 60th Annual Technical Conference and Exhibition of the Society of Petroleum Engineers, Las Vegas, Nevada, Sept. 22-25, 1985.

7
Torsional and Whirling Vibrations

There are dangers associated with generalization, but it is often true that *axial vibrations are associated with tricone bits*, which are vertically excited by displacement sources, while *torsional vibrations are associated with PDC bits having flatter faces* that are characterized by high levels of friction.

Torsional oscillations represent twisting actions about the axis of the drillstring, while whirling oscillations consist of drillstring motions about the central borehole axis. Whirling motion, a subset of lateral bending vibrations, is *not* torsional; however, whirl is included in this chapter because it does not exist without some rotational driving mechanism.

Why study torsional vibrations? These dynamic oscillations are associated with *stick-slip* phenomena, that is, a winding and unwinding of the drillstring that proves detrimental to bit life. Stick-slip conditions are discussed in detail, and simple mathematical models are proposed. Then, there are more obvious problems, e.g., over-tightening of drillpipe connections over a period which can lead to equipment fatigue and premature wear.

Why study whirling? The *backward whirling* of PDC bits is detrimental to cutter life. When a bit whirls, its instantaneous center moves about the bit face rather than around its true center, as in smooth rotation. The path taken by any single cutter is not circular: cutters impact the formation at off-design angles, leading to damage in form of spalling, chipping, and premature bit failure (Bobrosky and Osmak, 1993). Whirling will be discussed in detail and modeled mathematically here and in Chapter 8.

TORSIONAL VIBRATIONS

In this section, we present the differential equation describing torsional waves, and show how solutions for *drillstring wind-up* and *solid body rotation* appear as special limits. Then, the dynamic process consisting of static *torque-up, torsional wave generation*, and *winding and unwinding* due to momentary bit bounce or bottom slippage is outlined. This is followed by a broad discussion on *stick-slip* oscillations and its modeling.

Torsional wave equation. The partial differential equation governing *torsional displacements* is similar to that describing longitudinal displacements along a drillstring (see Chapters 1 and 5). Because of these similarities, we will not rehash the fundamentals in any significant detail; analogous global energy balances, static-to-dynamic energy mode transfers, finite difference schemes, and so on, apply with obvious changes in nomenclature.

If $\theta(x,t)$ represents the *angle of twist* along our *elastic line* at any point x at a given time t, the *torque* $T(x,t)$ satisfies

$$T = C\, \partial\theta/\partial x \tag{7.1}$$

where C is known as the *torsional rigidity* or the *torsional stiffness.* For circular cross-sections,

$$C = GJ \tag{7.2}$$

where G is the *shear modulus* and J is the *polar moment of inertia* (again, for circular cross-sections, $J = \pi a^4/2$, *a* being the radius). For such problems, Equation 7-1 becomes

$$T = GJ\, \partial\theta/\partial x \tag{7.3}$$

In the simplest case, angular momentum considerations show that the angle of twist satisfies the classical wave equation

$$\partial^2\theta/\partial t^2 - (C/\rho J)\, \partial^2\theta/\partial x^2 = 0 \tag{7.4}$$

where we have assumed a nondissipative system without external excitation. In general, Equations 7-2 and 7-4 suggest the definition

$$c_s^2 = G/\rho \tag{7.5}$$

At the same time, following Chapters 1 and 5, the obvious extension of Equation 7-4 is

$$\partial^2\theta/\partial t^2 - \xi\, \partial\theta/\partial t - c_s^2\, \partial^2\theta/\partial x^2 = \Theta_e \tag{7.6}$$

Note that c_s is the *shear wave propagation velocity*, and that it is different from the *axial speed of sound* $c = (E/\rho)^{1/2}$ for longitudinal disturbances. In Equation 7-6, we have introduced a right side Θ_e term to represent the effect of external excitation due to borehole friction and contact.

The fact that different distinct characteristic speeds exist along a drillstring is important to future MWD telemetry methods, especially with respect to *multiplexing*, that is, transmitting multiple signals along the same communications channel.

For example, downhole information may be conveyed to the surface using axial pipe waves propagating at a speed $(E/\rho)^{1/2}$, torsional pipe waves at a speed $(G/\rho)^{1/2}$, *and* acoustic mud pulse signals at a speed $(B_{mud}/\rho_{mud})^{1/2}$, where B_{mud} and ρ_{mud} represent the bulk modulus and density of the mud. Lateral bending waves, as discussed in Chapter 6, cannot be used for MWD telemetry because they are associated with low signal-to-noise ratios at the surface.

We have also introduced an attenuation term $\xi\ \partial\theta/\partial t$ along the lines noted in Chapter 5. The damping factor ξ may in general be different from the ζ appearing in $\rho\ \partial^2u/\partial t^2 + \zeta\ \partial u/\partial t - E\ \partial^2u/\partial x^2 = 0$ for axial vibrations. For detailed expositions on torsional mechanics, the reader is referred to classic books by Timoshenko and Goodier (1934), Love (1944), Den Hartog (1952), and Timoshenko (1958), and the more vibrations textbook by Graff (1975).

Although Equation 7-6 is suggestive of transient torsional waves, it is important to understand that it applies as well to the *combined* static and dynamic angle of twist. In Chapter 5, we emphasized the significance of the combined static and dynamic axial displacement field, since energy transfer between the two in general exists through coupling at the bit. This, as we will find, is similarly the case with torsional oscillations.

In a purely *static* problem, the simplification $\partial/\partial t = 0$ leads to the ordinary differential equation $\partial^2\theta/\partial x^2 = 0$, which integrates to $\theta_s(x) = Ax + B$, with A and B being constants. This static solution shows that successive cross-sections along the drillstring displace azimuthally with a linear variation.

In a purely *transient* limit, the assumption $\partial/\partial x = 0$ leads to $\partial^2\theta/\partial t^2 = 0$, so that we obtain the *solid body rotation* $\theta(t) = \omega t$ (plus an arbitrary integration constant) where ω is a constant drillstring rotation speed. This type of motion, ideally anyway, exists when all other kinds of compressible transients have decayed. In early drilling publications, this unrealistic solid body assumption was actually quite popular.

Wave-like torsional disturbances ride on these solid body rotations, the same way, as in Chapter 5, dynamic displacements ride on the static stress field due to weight. In the simplest dynamical problems, the static plus dynamic book-keeping $\theta(x,t) = \theta_s(x) + \theta_d(x,t)$, by substitution in Equation 7-6, shows that the two component strain fields can be treated independently via *linear superposition*, at least when partial differential equations alone are considered.

This approach, natural in mechanical engineering, is assumed in the early drilling literature. However, it is not representative of actual drilling operations. In drillstring vibrations, the static mean is never clearly defined. At the start of drilling, an unstressed drillstring will not rotate until some threshold *torque-up* level that overcomes static frictional effects at the face of the formation is achieved. The exact threshold depends on the bit type, the formation, and the static weight-on-bit. Only when it is sufficiently wound is the drillstring ready to rotate; once rotating, dynamic rock-bit interaction will excite the drillstring with torsional waves that propagate to the surface or to areal changes, and reflect.

The complete picture does not end here. As we have seen in Chapter 5, the dynamic interaction between the rock and the bit can lead to transient bit bouncing, which implies a free torsional end. Low dynamic weight-on-bit will also imply free torsional ends, since greater azimuthal slippage is allowed. Both effects will momentarily remove the source of rock-bit excitation and allow the drillstring to unwind. Then, the wind-up process repeats itself.

Thus, the transfer between mean potential energy and wave potential and kinetic energy, due to axial induced boundary condition coupling at the bit, may be quite active. Consequently, as in Chapter 5, it is more natural to study the complete transient angular displacement $\theta(x,t)$ rather than separate $\theta_s(x)$ and $\theta_d(x,t)$ components. This is the approach undertaken in this book.

Stick-slip oscillations. Here we will describe in detail the complicated dynamical process that is known as *stick-slip* oscillation. When rotation commences, the drillstring twists in place and stores torsional strain (or, *spring potential*) energy until the applied torque exceeds the total static frictional torque acting on the bottomhole assembly. The BHA now starts to rotate in a steady state mode.

This steady motion may not be permanent. If the static friction exceeds the dynamic friction acting at the bit, the previously stored potential energy may be transferred into kinetic energy, so that acceleration to a speed faster than steady state rotation is possible. Sometimes this process can result in self-sustaining oscillations that cause drillpipe fatigue and damage to the bit.

Equally dangerous *drillstring unwinding* can occur, as described, through axial-torsional mode coupling. When longitudinal vibrations induce momentary bit bounce or bottom bit slippage, severe pendulum-like unwinding and winding is possible. These effects have been experimentally observed in numerous field tests.

Brett (1991) notes that relatively moderate and continuous reductions in dynamic rotating torque (obtained as PDC bit rotary speeds increase, while held at constant weight-on-bit) may create similar vibrations. That is, it is not necessary for pure static friction to exist before self-excited torsional oscillations occur.

Brett (1992) presents lab and field torsional measurements, comparing these with modeling results, to show that the observed vibrations can be explained by a reduction in PDC bit torque as rotary speed increases. Apparently, the reduction in torque during rotational acceleration releases static strain energy for transfer into dynamic component modes. This so-called *negative damping* is a manifestation of the interaction between static and dynamic torsional states.

Brett's laboratory results for PDC bits show that the reduction in torque with rotary speed is greatest for a dull bit in hard rock, and that as weight-on-bit increases, the absolute value of the reduction in torque also increases. From Brett (1991), we might tentatively conclude that torsional vibrations that arise from bit torque characteristics are more common with PDC bits than with tricones; also, they appear to be more severe at higher applied weight-on-bits, at lower rotary speeds, and with duller bits.

Dufeyte and Henneuse (1991) and Henneuse (1992) describe a number of Elf Aquitaine experiments in which downhole vibrations were continuously monitored at the surface. Their results, obtained over numerous operating hours and a variety of well types, showed that *stick-slip* phenomena represent very frequent occurrences. In particular, stick-slip oscillation appeared as a result of a natural tendency for PDC bits to develop substantial torsional waves. The exact severity of any damaging backward spin depended on the bits' cutting or abrasion operating mode.

Close, Owens, and MacPherson (1988) found, upon detailed analysis of MWD data, that PDC diamond bits may be responsible for high levels of lateral vibration as well. Correlation between surface measured WOB and RPM, and downhole lateral vibrations, was obtained.

Torque and bending coupling may also arise when borehole sticking due to large lateral displacements cause the drillstring to twist in place. In Chapter 8, we will show exactly how this and the above dynamic coupling arises, and present the governing partial differential equations, numerical formulation, and typical solutions.

Again, it is significant to note that the quasi-periodic transfer between mean torsional strain energy and dynamic rotational energy, responsible for stick-slip oscillations, is mathematically analogous to the energy transfer that causes changes in axial mean load and fluctuating weight-on-bit in longitudinal vibrations.

Rewcastle and Burgess (1992) also noted stick-slip motion, in cases where the bit or stabilizers stall in contact with the formation. String torque builds up until sufficient torque is available to overcome the resisting force, and the energy is suddenly released.

Stick-slip behavior is usually seen as cyclical signatures in surface measured torque; real-time surface torque measurements, therefore, are valuable in detecting rough drilling conditions. Stick-slip oscillations can cause the bit to spin backwards and destroy a PDC bit, unscrew the bottomhole assembly, induce lateral vibrations which are detrimental to the BHA and to downhole MWD instruments. Thus, there is substantial interest in active control over torsional vibrations.

For example, Sananikone, Kamoshima, and White (1992) have investigated the use of *active feedback and control systems* which appear to be operationally simple and effective. Lin and Wang (1990) gave a preliminary explanation that attempts to define the relationship between stick-slip and observed beating phenomena using a simplified one-degree-of-freedom mathematical model.

Energy considerations. The above observations on stick-slip oscillations indicate that significant energy transfer occurs between static and dynamic torsional states. Thus, as in the case for axial vibrations, although for different reasons, there is little incentive or rationale to consider separate formulations for the static and dynamic problems.

It is of operational interest to determine the critical rotational speed beyond which stick-slip stops, but this critical speed may be unique to each operation, and may be overly optimistic. It may well be that such a parameter cannot be determined from purely torsional considerations alone; from our discussions, the role of important axial vibrations certainly cannot be ignored.

The arguments behind the energy transfer mechanisms described are quantified by considering the *energy density* $1/2\ \{\rho J\ (\partial\theta/\partial t)^2 + C\ (\partial\theta/\partial x)^2\}$. The first term represents the system kinetic energy, and applies to the dynamic solution only; the second represents the potential strain energy due to torsion, and applies to both static and dynamic displacement modes.

If we now integrate this energy density over the entire length of the drillstring, the net system energy can be described by the energy integral E(t), where

$$E(t) = 1/2 \int \{\rho J\ (\partial\theta/\partial t)^2 + C\ (\partial\theta/\partial x)^2\}\ dx \tag{7.7}$$

In Equation 7-7, we have omitted the limits of integration $x = 0$ to L for brevity. Now, let us resolve the angle of twist into steady and time-dependent parts, taking $\theta(x,t) = \theta_s(x) + \theta_d(x,t)$.

We follow an integration by parts procedure similar to that in Chapter 5 to show that changes to E(t) are due to internal dissipation (e.g., arising from the $\xi\ \partial\theta/\partial t$ term in Equation 7-6) and power input at $x = 0$ and $x = L$.

The power contribution depends on the product $(\partial\theta/\partial t)(\partial\theta/\partial x)$ or $(\partial\theta_d/\partial t)(\partial\theta_s/\partial x + \partial\theta_d/\partial x)$ at upper and lower ends. This contribution is zero, positive, or negative, depending on the complete dynamics of the system, which in turn depends on the *if-then* boundary condition at the bit, as conceptualized in Chapter 5.

As we have indicated that, when the bit bounces, the boundary condition downhole is the torsional free end $\partial\theta/\partial x = 0$. But when it drills ahead and makes hole, the appropriate prescription is an indicator of total bit torque, say $T(0,t) = GJ\, \partial\theta(0,t)/\partial x = f\, AE\, \{\partial u(0,t)/\partial x\}$, where f is an empirical *Coulomb friction* factor. In this physically oriented model, the instantaneous value of the axial strain $\partial u(0,t)/\partial x$ is obtained from the complementary axial vibrations problem.

These ideas are consistent with the results of Clayer, Vandiver, and Lee (1990). They found that the natural frequencies of bottomhole assemblies in torsion are best modeled by free stiffness boundary conditions at the bit, in particular, descriptions involving $\partial\theta(0,t)/\partial x$ in conjunction with large bit damping.

The Θ_e external excitation shown in Equation 7-6 plays a role analogous to the F_e we encountered in modeling axial vibrations. If borehole wall friction or substantial borehole contact is significant, Θ_e cannot be neglected along the (contact) length of the drillstring. This will also affect the energy balance, as indicated in Chapter 5.

Static torque effects on bending. Earlier we emphasized the need to study a single *static plus transient* torque, rather than separate static and transient combinations. This is all the more important for lateral vibrations: the existence of the slightest torque will dynamically couple both orthogonal lateral bending modes, as discussed in the "eraser experiment" of Chapter 4, and again in Chapter 8.

Thus, even when dynamic torsional oscillations are completely absent, static torque must be accounted for in the overall dynamical picture, if bending effects such as *precessional instabilities* are to be modeled. Again, drillstrings unwind when they are off-bottom; the amount of unwinding is in part set by the initial torque-up or static torque. Thus, modeling the *complete* torque variable is of paramount importance.

Finite difference modeling. Equation 7-6 is identical to the axial wave equation discussed in Chapter 5. Thus, the same tridiagonal finite difference equations, implementation recipes, and diagonal dominance considerations apply, with obvious changes in nomenclature, and identical matrix inversion processes can be used.

Simplifications, of course, are found in the auxiliary conditions. For instance, the mass-spring-damper surface model used to simulate traveling block oscillations is now replaced by the simpler time-dependent displacement condition $\theta = \omega t$, where ω is the *applied* rotation rate. For example, we might have an ω of 60 *rpm*, written in terms of radians; note that we do *not* use 3 x 60 *rpm*, which applies to tricone-induced axial vibrations (also, the displacement source model used to simulate vertical tricone excitation does not apply here).

In our discussion so far, we have considered a purely uniform drillstring, ignoring cross-sectional discontinuities along drillstring lengths, particularly at the interfaces separating drillpipe from drill collar. In Chapter 5, we showed that *continuity of force and of longitudinal displacement apply to axial motions*, and derived stable finite difference formulas for solution matching at impedance change junctions.

The required matching conditions for torsional vibrations are analogous, that is, we assume *continuity of torque and of angular displacement*. Like axial strain, the torsional strain is double-valued at the impedance junction: it suddenly jumps in value. The analytical and numerical results of Chapter 5 apply directly, with only obvious changes in nomenclature; for example, observe the similarity between *force* $\propto$ AE $\partial u(x,t)/\partial x$ for axial vibrations, whereas *torque* $\propto$ GJ $\partial\theta(x,t)/\partial x$ for torsional vibrations.

WOB/TOB. The ratio of average *weight-on-bit* to *torque-on-bit* provides useful information for bit bearing failure detection and formation identification. This parameter, when available in real-time, is therefore useful in anticipating drilling problems, and in providing mechanical-based "second opinions" in support of lithology predictions founded on gamma ray, resistivity, and porosity measurements.

Applications to MWD telemetry. *Mud pulse telemetry*, in which acoustic pulses are transmitted through the mud column inside the drillstring, represents the most popular form of Measurement-While-Drilling communications at the present time. However, other telemetry methods are possible, and are under continual industry evaluation and research.

Bradbury and Wilhoit (1962) importantly considered the effect of tool joints on the passage of plane longitudinal and torsional waves along a drillpipe. Approximate solutions to the governing partial differential equations for an idealized tool joint showed that tool joints had negligible effect for exciting frequencies of the same order as common rotary speeds, and that the drillpipe can be taken as a uniform pipe with negligible error.

Barnes and Kirkwood (1972) also considered longitudinal and torsional vibrations as a means for MWD communication. For their analysis, only zeroth mode vibrations (i.e., plane wave) were considered, the assumption being that higher modes would be damped out over short distances. The basic ideas on modal eigenfunctions and evanescent waves were discussed in Chapters 1 and 3.

In their idealized model, where the drillstring is taken as an infinitely long cylindrical steel pipe loaded at equal intervals with identical couplings, the authors show that most of the *passbands* for compressional waves are about twice the width of those for torsional waves (passbands are frequencies for which waves propagate without loss). Refer to the paper for further details on the exact geometry used and the numerical results obtained.

Torsional wave propagation along drillstrings, as a means of MWD acoustic telemetry, was reconsidered by Squire and Whitehouse (1979). They importantly studied attenuative effects posed by the drilling mud environment, with and without gelling, and results of field experiments were reported. The conclusions cited in the foregoing papers, which have not been independently verified by this author, are given for completeness only.

EXAMPLE FORTRAN IMPLEMENTATION

Here, as in Chapters 5 and 6, we will give portions of a Fortran source code listing, showing how the torsional finite difference equations are introduced by defining the required **A**, **B**, **C**, and **W** coefficients, and how they are solved using the TRIDI subroutine call. Because the similarities to the axial case are strong, we will provide this listing with a minimum of elaboration.

```
C     .
C     .
C     INITIALIZE AXIAL, ANGULAR AND BENDING DISPLACEMENTS
      DO 130  I=1,IMAX
      X = XS*(I-1)/(IMAX-1)
      UNM1(I) = RHO*G*(X**2)/(2.*ELAST) +WZERO*X/AE
      UNM2(I) = UNM1(I)
      TNM1(I) = 0.
      TNM2(I) = 0.
C     .
C     .
C     BEGIN TIMEWISE INTEGRATION
      T = 0.
      DO 900  N=1,NMAX
      T = N*DT
C     .
C     .
```

Figure 7-1. Torsional vibrations, Fortran listing (to be continued).

```
C       .
C       AXIAL VIBRATIONS
        DO 200  I=2,IMAXM1
        A(I) = 1.
        C(I) = 1.
        B(I) = -2.-RHO*DX*DX/(ELAST*DT*DT)-GAMA*DX*DX/(2.*ELAST*DT)
        W(I) = -RHO*DX*DX*(2.*UNM1(I)-UNM2(I))/(ELAST*DT*DT)
      1        -GAMA*DX*DX*UNM2(I)/(2.*ELAST*DT)+RHO*G*(DX**2)/ELAST
 200    CONTINUE
C       .
C       TORSIONAL VIBRATIONS
        DO 400  I=2,IMAXM1
        A(I) = 1.
        C(I) = 1.
        B(I) = -2.-RHO*DX*DX/(SHRMOD*DT*DT)-
               GAMA2*DX*DX/(2.*SHRMOD*DT)
        W(I) = -RHO*DX*DX*(2.*TNM1(I)-TNM2(I))/(SHRMOD*DT*DT)
      1        -GAMA2*DX*DX*TNM2(I)/(2.*SHRMOD*DT)
 400    CONTINUE
        A(IMAX) = 0.
        B(IMAX) = 1.
        C(IMAX) = 99.
        W(IMAX) = RPM*2.*PI*T/60.
        A(1) = 99.
        B(1) = -1.
        C(1) = 1.
        W(1) = (FCOEF*AE/GJ)*(UN(2)-UN(1))
        CALL TRIDI(A,B,C,VECTOR,W,IMAX)
        DO 450  I=1,IMAX
        THETAN(I) = VECTOR(I)
 450    CONTINUE
        DTHDX(1) = (THETAN(2)-THETAN(1))/DX
        TORBIT = GJ*DTHDX(1)
        DTHDX(IMAX) = (THETAN(IMAX)-THETAN(IMAXM1))/DX
        DO 505  I=2,IMAXM1
        DTHDX(I) = (THETAN(I+1)-THETAN(I-1))/(2.*DX)
 505    CONTINUE
C       .
 900    CONTINUE
```

Figure 7-1. Torsional vibrations, Fortran listing (continued).

WHIRLING MOTIONS

Initially straight machine shafts, carrying a torque between bearings, with or without pulleys or similar loads, will at certain rotation speeds *whirl* or rotate in a bent configuration about the original axis (Bickley and Talbot, 1961).

When the shaft rotation rate equals the natural frequency of the shaft in bending, highly destructive motions are possible in rotating machinery. The mechanical engineering literature deals with the vibration and balancing of flexible rotating shafts, and a considerable body of literature is available (e.g., see Bishop (1959), or Bishop and Gladwell (1959)).

Whirling motions are solutions of the dynamic beam equation governing lateral vibrations. However, they are introduced in this chapter on torsional vibrations, if only because such motions cannot exist unless rotation is present to initiate the whirling. From Chapter 3, the elementary bending equation is

$$EI\ \partial^4 v/\partial x^4 = -\ \rho A\ \partial^2 v/\partial t^2 \tag{7.8}$$

where E is Young's modulus, I is the moment of inertia, A is the cross-sectional area, and ρ is the mass per unit volume.

If $v(x,t)$ is the shaft centerline deflection from its equilibrium straight position at any point x, the equivalent elastic restoring force must equal that necessary to produce the centripetal acceleration.

The exact form of the centripetal acceleration term can be obtained by examining the cross-plane perpendicular to the shaft, where the linear coordinate s of the circular motion and the angle θ are related by $s = R\theta$, with R being the radius. The linear speed is therefore $ds/dt = R\ d\theta/dt$.

From particle dynamics, the centripetal acceleration per unit length of shaft is $\rho A(ds/dt)^2/R$ or $\rho A(d\theta/dt)^2 R$. If ω denotes the constant angular velocity $d\theta/dt$ of the rotating shaft and $v(x)$ is the local radius R, we have the expression $\rho A\omega^2 v$. Thus, whirling motions satisfy

$$EI\ \partial^4 v/\partial x^4 = \rho A\omega^2 v \tag{7.9}$$

where the explicit dependence on time disappears in this *dynamically steady* limit (in real drilling, *transient whirling* is typically the rule in the presence of arbitrary axial and torsional static and transient fields).

Example 7-1: Machine shaft example. Let us rewrite Equation 7-9 as $d^4v/dx^4 = \alpha^4 v$ where $\alpha^4 = \rho A\omega^2/EI$. The general solution to this fourth-order ordinary differential equation is $v(x) = A \cosh \alpha x + B \sinh \alpha x + C \cos \alpha x + D \sin \alpha x$. For the simple *pinned-pinned* machine shaft problem defined on $0 < x < L$, $v(x)$ and the moment $EI\ d^2v/dx^2$ both vanish at $x = 0$ and $x = L$. The sinusoidal mode shape $v(x) = \sin n\pi x/L$ is obtained, where n = 1, 2, 3, ... with the corresponding natural frequencies ω_n satisfying $\omega_n^2 = n^4\pi^4 EI/\rho AL^4$.

Note that the frequencies do not form a complete harmonic sequence, although they are all integral multiples of the fundamental. All of the mode configurations have constrained nodes at the ends of the shaft, as assumed, and the *n*th mode also has (n-1) internal nodes, equally spaced along its length.

Example 7-2: Generalized whirl. We have given, in the above analysis, a self-contained derivation for whirling oscillations that is independent of the lateral vibration equations to be covered in Chapter 8. However, we emphasize that circular whirl can be viewed as the resultant of two perpendicular lateral vibrations v(x,t) and w(x,t), which combine to form a local radial displacement $\sqrt{v^2(x,t) + w^2(x,t)}$ from an equilibrium x = 0 state.

Thus, the results obtained here, and their extensions under more practical boundary conditions, should emerge as subsets of the general coupled model to be presented later. A combined lateral displacement variable similar to this is used by Dunayevsky and Judzis (1983), Dunayevsky, Judzis, and Mills (1984), and Dunayevsky, Judzis, Mills, and Griffin (1985) in their studies of precessional motions. Thus, it is not necessary to consider special differential equations for whirling motions; if they exist for specific non-simple sets of boundary conditions, the transient coupled axial, torsional and lateral equations derived in Chapter 8 will model them.

Causes of whirling motions. In drillstring vibrations, whirling represents the centrifugally induced bowing of drill collars resulting from rotation. If the center of gravity of a drill collar is not located precisely along the center line of the hole, then the collar rotates, and a centrifugal force acts at the center of gravity causing the collar to bend. Eccentricity may arise, for example, from initially bent drill collars, or from collar sag due to gravity and high compressive loads at the bit.

Motions of drill collars may vary from simple whirling motion, like that of an unbalanced centrifuge, to highly irregular motion caused by nonlinear effects of fluid forces, stabilizer clearance, and borehole wall contact. While whirling does not generally cause precipitous drill collar failures, because lateral amplitudes limited by wall contact, constant wall contact does result in high levels of abrasion.

Practical questions such as stabilizer modeling arise in the course of vibration simulation. Stabilizers have larger diameters than drillpipes, but small length. Thus, they can be viewed as simple *lumped masses* on beams, at least in the first approximation; where stabilizer friction due to rotation and borehole contact with the formation are important, nonzero right-side "q" terms or special elastic responses may be invoked and defined as necessary.

An excellent study on whirling appears in Vandiver, Nicholson, and Shyu (1989), providing qualitative descriptions of the vibration process based on extensive high-data-rate field measurements. Jansen (1992) investigated the effects of eccentricity of the center of mass of the drill collar, fluid damping, stabilizer clearance, and friction on forward and backward whirl.

In Chapter 8, a transient coupled formulation for axial, torsional, and (two) lateral vibrations is given, which builds upon the ideas developed in Chapters 4-7. This model is solved numerically, and computed results corresponding to "smooth" and "rough drilling" are discussed.

REFERENCES

Barnes, T.G., and Kirkwood, B.R., "Passbands for Acoustic Transmission in an Idealized Drill String, Vol. 51, No. 5, Journal of the Acoustical Society of America, 1972, pp. 1606-1608.

Bickley, W.G., and Talbot, A., *An Introduction to the Theory of Vibrating Systems*, Oxford University Press, London, 1961.

Bishop, R.E.D., "The Vibration of Rotating Shafts," Journal of Mechanical Engineering Science, Vol. 1, No. 1, 1959, pp. 50-65.

Bishop, R.E.D., and Gladwell, G.M.L., "The Vibration and Balancing of an Unbalanced Flexible Rotor," Journal of Mechanical Engineering Science, Vol. 1, No. 1, 1959, pp. 66-77.

Bobrosky, D., and Osmak, G., "Antiwhirl PDC Bits Increased Penetration Rates in Alberta Drilling," Oil & Gas Journal, July 5, 1993, pp. 41-46.

Bradbury, R.E., and Wilhoit, J.C., "Effect of Tool Joints on Passage of Plane Longitudinal and Torsional Waves Along a Drill Pipe," Paper No. 62-PET-32, presented at the Petroleum Mechanical Engineering Conference, American Society of Mechanical Engineers, Dallas, Tx., Sept. 23-26, 1962.

Brett, J.F., "The Genesis of Bit-Induced Torsional Drillstring Vibrations," SPE/IADC Paper No. 21943, presented at the 1991 SPE/IADC Drilling Conference, Amsterdam, Mar. 11-14, 1991.

Brett, J.F., "The Genesis of Torsional Drillstring Vibrations," SPE Drilling Engineering Journal, Sept. 1992, pp. 168-174.

Clayer, F., Vandiver, J.K., and Lee, H.Y., "The Effect of Surface and Downhole Boundary Conditions on the Vibration of Drillstrings," SPE Paper No. 20447, presented at the 65th Annual Technical Conference and Exhibition of the Society of Petroleum Engineers, New Orleans, La., Sept. 23-26, 1990.

Close, D.A., Owens, S.C., and MacPherson, J.D., "Measurement of BHA Vibration Using MWD," IADC/SPE Paper No. 17273, presented at the 1988 IADC/SPE Drilling Conference, Dallas, Texas, Feb. 28 - Mar. 2, 1988.

Den Hartog, J.P., *Advanced Strength of Materials*, McGraw-Hill, New York, 1952.

Dufeyte, M.P., and Henneuse, H., "Detection and Monitoring of the Slip-Stick Motion: Field Experiments," SPE/IADC Paper No. 21945, presented at the SPE/IADC Drilling Conference, Amsterdam, Mar. 11-14, 1991.

Dunayevsky, V.A., and Judzis, A., "Conservative and Nonconservative Buckling of Drillpipe," SPE Paper No. 11991, presented at the 58th Annual Technical Conference and Exhibition of the Society of Petroleum Engineers, San Francisco, Ca., Oct. 5-8, 1983.

Dunayevsky, V.A., Judzis, A., and Mills, W.H., "Onset of Drillstring Precession in a Directional Borehole," SPE Paper No. 13027, presented at the 59th Annual Technical Conference and Exhibition of the Society of Petroleum Engineers, Houston, Texas, Sept. 16-19, 1984.

Dunayevsky, V.A., Judzis, A., Mills, W.H., and Griffin, G., "Dynamic Stability of Drillstrings Under Fluctuating Weights-on-Bit," SPE Paper No. 14329, 60th Annual Technical Conference and Exhibition of the Society of Petroleum Engineers, Las Vegas, Nevada, Sept. 22-25, 1985.

Graff, K.F., *Wave Motion in Elastic Solids*, Ohio State University Press, 1975.

Henneuse, H., "Surface Detection of Vibrations and Drilling Optimization: Field Experience," IADC/SPE Paper No. 23888, presented at the 1992 IADC/SPE Drilling Conference, New Orleans, Louisiana, Feb. 18-21, 1992.

Jansen, J.D., "Whirl and Chaotic Motion of Stabilized Drill Collars," SPE Drilling Engineering Journal, June 1992, pp. 107-114.

Lin, Y.Q., and Wang, Y.H., "New Mechanism in Drillstring Vibration," OTC Paper No. 6225, presented at the 22nd Annual Offshore Technology Conference, Houston, Texas, May 7-10, 1990.

Love, A.E.H., *A Treatise on the Mathematical Theory of Elasticity*, Dover Publications, New York, 1944.

Rewcastle, S.C., and Burgess, T.M., "Real-Time Downhole Shock Measurements Increase Drilling Efficiency and Improve MWD Reliability," IADC/SPE Paper No. 23890, presented at the 1992 IADC/SPE Drilling Conference, New Orleans, Louisiana, Feb. 18-21, 1992.

Sananikone, P., Kamoshima, O., and White, D.B., "A Field Method for Controlling Drillstring Torsional Vibrations," IADC/SPE Paper No. 23891, presented at the 1992 IADC/SPE Drilling Conference, New Orleans, Louisiana, Feb. 18-21, 1992.

Squire, W.D., and Whitehouse, H.J., "A New Approach to Drill-String Acoustical Telemetry," SPE Paper No. 8340, presented at the 54th Annual Fall

Technical Conference and Exhibition of the Society of Petroleum Engineers of the AIME, Las Vegas, Nevada, Sept. 23-26, 1979.

Timoshenko, S.P., *Strength of Materials, Part II: Advanced Theory and Problems*, Van Nostrand Reinhold Company, New York, 1958.

Timoshenko, S.P., and Goodier, J.N., *Theory of Elasticity*, McGraw-Hill, New York, 1934.

Vandiver, J.K., Nicholson, J.W., and Shyu, R.J., "Case Studies of the Bending Vibration and Whirling Motion of Drill Collars," SPE/IADC Paper No. 18652, presented at the 1989 SPE/IADC Drilling Conference, New Orleans, Louisiana, February 28-March 3, 1989.

8
Coupled Axial, Torsional, and Lateral Bending Vibrations

We have discussed the fundamentals behind axial, torsional, and lateral vibrations, and the physics of modal interaction through drillbit boundary conditions. Here, we integrate all of our previous models into a single one that allows axial, torsional, and two lateral vibration modes to dynamically interact.

We warn against the dangers of accepting the Fortran listing at face value. Like all research products, its underlying implicit assumptions are numerous. For example, the back-interaction of large lateral vibrations on the axial stress field as discussed in Chapter 6, is not included in the printed listing, although simple modifications will extend the program to handle such nonlinear effects.

Also, the code may not be directly useful to certain applications. For instance, there are essential differences between *rotary drilling* where the entire drillstring rotates, and *turbodrilling*, where the drillstring does not rotate, although the drillbit does. On the other hand, vertical, deviated, and horizontal drillstrings see different kinds of gravity effects and formation reactions. *The emphasis of this book is therefore tutorial, and not, software oriented.*

In the drilling literature, the meaning of linearity is sometimes confused with the presence of dynamic mode coupling. We emphasize that mode coupling does *not* imply nonlinearity, when simple classical boundary conditions are used. This is not to say that the resulting problem is simple!

But our use of *if-then* rock-bit boundary conditions, which clearly relate to changes triggered by transient threshold *amplitude* levels, renders the complete boundary value problem nonlinear, despite any apparent linearity in the governing partial differential equations. This observation is crucial: computed results may appear similar, but calculated results from one run will generally not scale into those for another.

Importance to PDC bit dynamics. At the present time, a significant commercial driver for understanding coupled vibrations is costly PDC bit failure. The failure of such bits in hard formations is often blamed on uneven thermal expansion rates between the binder and the diamonds during rapid frictional heating (e.g., see Cohen, Maurer, and Westcott (1993)).

However, part of the blame rests with bit-induced coupled axial, torsional and lateral vibrations in hard formations: conventional PDC bits tend to whirl backward, resulting in reduced penetration rates and drillbit life. When a bit whirls, it walks around the well bore backwards, increasing hole size and reducing penetration rate, and decreasing directional control. Bit whirl shows how a "simple boundary condition" can control the entire drillstring dynamics.

PDC bits generate high dynamic loads under normal rotary drilling conditions. Because of their cutting action, such bits induce high-amplitude torsional and axial drillstring vibrations. This is further aggravated by the tendency of PDC bits to whirl backwards. Langeveld (1992) describes recent research efforts directed at numerical simulation.

New developments in *antiwhirl PDC bits* were recently reported in the *Oil & Gas Journal* (Staff, 1992). This article discusses newly designed antiwhirl bits which exhibit strongly reduced tendencies to whirl backward, although these bits do become unstable in backward whirl above a critical rotary speed that depends on the hole size and weight-on-bit.

Coupled axial, torsional and bending vibrations. In this chapter, the titled coupling is studied primarily for vertical wells, and computations which test the formulations posed are directed at validating the underlying physical ideas. At its conclusion, when the objectives of Part II of this book have been met, extensions of the models to handle highly deviated and horizontal wells are outlined to stimulate further research.

The fundamental equations governing coupled drillstring axial, torsional and lateral vibrations can be inferred from those for rotating machinery shafts in mechanical engineering. These basic equations are given in classical elasticity and will not be rederived, although we will consider the proper form for the torque-dependent terms later. However, the subtle bit boundary conditions described qualitatively in Chapters 4-7 are new, and their combined consequences within the framework of a broad integrated model will be the subject of this chapter.

Before we consider dynamic coupling in any detail, it is worthwhile to explore the role of torque under simple static conditions. We need not understand the methods of mathematical elasticity in order to appreciate the profound aspects of dual lateral mode coupling, since a simple desktop experiment suffices.

Example 8-1: Simple desktop experiment. There are key physical differences between the *in-plane* lateral deflections of Chapter 6, and the dual, *out-of-plane*, lateral mode coupling present whenever torque exists. For the purposes of experiment, you should obtain one of the long, flexible, rubber erasers used in draftsmen's electrical rotating erasers.

First, without applying any external torque or twist, introduce a compressive axial load, and observe how the eraser executes lateral in-plane deflections. Next, repeat this experiment, but first impose a static torque by twisting the eraser. Then, apply the same compressive force. The resulting transverse deflections are seen to occur in two orthogonal planes.

Whereas in torque-free problems we considered a $v(x,t)$ deflection mode only, we need now consider an additional $w(x,t)$ transverse deflection coupled through torque, with the coupling disappearing as torque vanishes. When axial and torque loads are dynamic, drillstring vibrations can be quite complicated.

Vandiver, Nicholson, and Shyu (1989), for example, note that axial, torsional and lateral vibrations are all present and strongly coupled. Bit bounce, stick-slip, forward and backward whirl, linear and parametric coupling between axial and lateral vibration all occur during the drilling process.

Notes on the coupled model. Before presenting mathematical details, we wish to emphasize those aspects of the physics that are incorporated in our formulation. Importantly, the analytical solutions given in Chapter 6 show that a bending equation that takes account of the static axial stress field will successfully predict catastrophic lateral vibrations at the neutral point, and simultaneously explain the disappearance of violent bending motions uphole.

Thus, the coupled equations presented below contain two important physical characteristics. First, torque is included to couple both lateral modes; and second, axial forces are included in both bending equations to account for extensions of the wave trapping model given in Chapter 6.

The axial formulation, even by itself, is nontrivial. Recall from Chapter 5 that this model was designed to simulate bit bounce, rate of penetration, and rock-bit interaction. Likewise, our torsional wave model was designed to use information available from axial calculations, in order to model torque reversal and stick-slip motions. How all of these models worked in concert was the main research objective of this chapter.

Drillstring vibration analyses can support field operations in several ways. They can identify rotation rates that are likely to induce fatigue, and define safe operating envelopes for damage avoidance. Also, properly designed simulators can help drillers penetrate more efficiently, by harnessing the vibration energy available within the drillstring. The model developed below, hopefully, represents a first step in this direction.

COUPLED AXIAL, TORSIONAL, AND BENDING EQUATIONS

We now present the partial differential equations for a drillstring vibrations model that embodies coupled axial, torsional and (two) lateral bending modes. The coupled equations in their entirety have not been given explicitly, either in the petroleum industry, or in related mechanical and aerospace applications.

However, their general form can be inferred from Timoshenko and Goodier (1934), Love (1944), Den Hartog (1952), and Clough and Penzien (1975) on beams and shafts. In addition, Nordgren (1974) formulated a 3D nonlinear, large-amplitude, transient problem, and solved it by finite differences; Garrett (1982) considered the nonlinear dynamics of slender rods.

Recent clues related to the form of the coupled equations can be found in Dunayevsky and Judzis (1983), Dunayevsky, Judzis, and Mills (1984), Dunayevsky, Judzis, Mills, and Griffin (1985), and Dunayevsky, Judzis, and Abbassian (1989), dealing with drillstring buckling and precessional motions.

Partial differential equations. To begin our discussion, we refer to Figures 8-1 and 8-2, bearing in mind that four vibration modes dynamically interact. Let $v(x,t)$ and $w(x,t)$ represent the two orthogonal lateral bending displacement modes, lying in planes perpendicular to the drillstring axis, that are coupled by the torque $T(x,t)$. These modes can be shown to satisfy

$$EI\ \partial^4 v/\partial x^4 + \partial(T\partial^2 w/\partial x^2)/\partial x + \partial(N\ \partial v/\partial x)/\partial x \\ + kv + \beta\ \partial v/\partial t + \rho A\partial^2 v/\partial t^2 = q^{(v)}(x,t) \tag{8.1}$$

$$EI\ \partial^4 w/\partial x^4 - \partial(T\partial^2 v/\partial x^2)/\partial x + \partial(N\ \partial w/\partial x)/\partial x \\ + kw + \beta\ \partial w/\partial t + \rho A\partial^2 w/\partial t^2 = q^{(w)}(x,t) \tag{8.2}$$

where the $q^{(v)}(x,t)$ and $q^{(w)}(x,t)$ indicate that the right-hand-side "q" terms need not be the same for both v and w bending modes. These equations will be derived and justified at the end of this chapter.

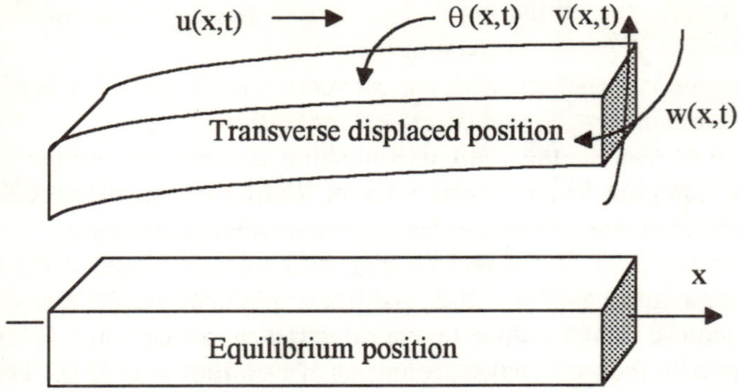

Figure 8-1. Coupled axial, torsional, and lateral displacements.

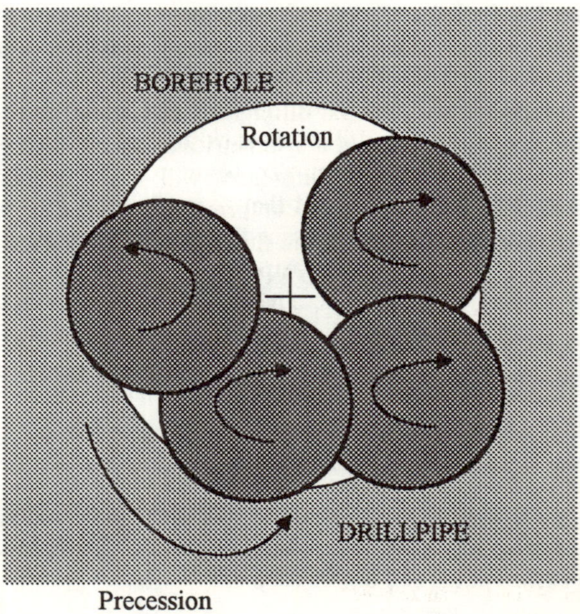

Figure 8-2. Drillstring precession, rotation, torque reversal.

Sign conventions. Observe that the axial force N(x,t) in Equations 8-1 and 8-2 satisfies $N > 0$ for compression and $N < 0$ for tension, and that

$$N = - AE\, \partial u/\partial x \qquad (8.3)$$

This expression is consistent with our convention in Chapter 5, which assumes $AE\, u_x < 0$ for compression and $AE\, u_x > 0$ for tension.

Also note that in the limit of vanishing stiffness EI, torque T, spring constant k, damping factor β, and external loading q, Equations 8-1 and 8-2 reduce to classical wave equations for transverse string vibrations.

We further point to the obvious sign differences in front of the torsional terms in Equations 8-1 and 8-2 and their physical implications. These equations require that any slope in one direction causes curvature in the other, consistently with the well known arguments of Den Hartog (1952). The sign of the axial force N, of course, remains the same regardless of lateral mode.

Finite differencing the coupled bending equations. As in Chapter 6 for lateral vibrations without torque, our coupled equations describing v(x,t) and w(x,t) are each characterized by *fourth-order* spatial derivative terms. This means that the approximating *difference equations* will again take on a *banded pentadiagonal structure*. The equations are no longer *tridiagonal*, as in Chapters 5 and 7, for axial and torsional vibrations.

In addition, we now have *third-order* spatial derivative terms related to nonvanishing torque, e.g., $\partial(T\partial^2 w/\partial x^2)/\partial x = T\partial^3 w/\partial x^3 + (\partial T/\partial x)(\partial^2 w/\partial x^2)$. These must be modeled with central differences throughout in order to ensure midpoint symmetry, in the limit of a symmetrically constrained uniform beam (the use of backward or forward differences will lead to incorrectly biased asymmetric results for symmetric problem formulations).

Algebraic manipulations similar to those in Chapter 5 show that $\partial^3 u/\partial x^3 = (- U_{i-2} + 2U_{i-1} - 2U_{i+1} + U_{i+2})/\{2(\Delta x)^3\} + O(\Delta x)^2$ and, as in Chapter 6, $\partial^4 u/\partial x^4 = (U_{i-2} - 4U_{i-1} + 6U_i - 4U_{i+1} + U_{i+2})/(\Delta x)^4 + O(\Delta x)^2$. These formulas provide the required spatial symmetry about the index i, and offer second-order accuracy in addition to stability benefits. Equations 8-1 to 8-3 become, respectively,

$$\begin{aligned}
&EI\,(v_{i-2,n} - 4v_{i-1,n} + 6v_{i,n} - 4v_{i+1,n} + v_{i+2,n})/(\Delta x)^4 \\
&+ N_{i,n}\,(v_{i-1,n} - 2v_{i,n} + v_{i+1,n})/(\Delta x)^2 \\
&+ \{(N_{i+1,n} - N_{i-1,n})/2\Delta x\}\{(v_{i+1,n} - v_{i-1,n})/2\Delta x\} \\
&+ kv_{i,n} + \beta\,(v_{i,n} - v_{i,n-2})/2\Delta t \\
&+ \rho A(v_{i,n} - 2v_{i,n-1} + v_{i,n-2})/(\Delta t)^2 \\
&\quad = \quad q^{(v)}{}_{i,n-1} \\
&\qquad - T_{i,n}(-w_{i-2,n-1} + 2w_{i-1,n-1} - 2w_{i+1,n-1} + w_{i+2,n-1})/2(\Delta x)^3 \\
&\qquad - \{(T_{i+1,n} - T_{i-1,n})/2\Delta x\}(w_{i-1,n-1} - 2w_{i,n-1} + w_{i+1,n-1})/(\Delta x)^2 \qquad (8.4)
\end{aligned}$$

$$\begin{aligned}
&EI\,(w_{i-2,n} - 4w_{i-1,n} + 6w_{i,n} - 4w_{i+1,n} + w_{i+2,n})/(\Delta x)^4 \\
&+ N_{i,n}\,(w_{i-1,n} - 2w_{i,n} + w_{i+1,n})/(\Delta x)^2 \\
&+ \{(N_{i+1,n} - N_{i-1,n})/2\Delta x\}\{(w_{i+1,n} - w_{i-1,n})/2\Delta x\} \\
&+ kw_{i,n} + \beta\,(w_{i,n} - w_{i,n-2})/2\Delta t \\
&+ \rho A(w_{i,n} - 2w_{i,n-1} + w_{i,n-2})/(\Delta t)^2 \\
&\quad = \quad q^{(w)}{}_{i,n-1} \\
&\qquad + T_{i,n}(-v_{i-2,n-1} + 2v_{i-1,n-1} - 2v_{i+1,n-1} + v_{i+2,n-1})/2(\Delta x)^3 \\
&\qquad + \{(T_{i+1,n} - T_{i-1,n})/2\Delta x\}(v_{i-1,n-1} - 2v_{i,n-1} + v_{i+1,n-1})/(\Delta x)^2 \qquad (8.5)
\end{aligned}$$

and

$$N_{i,n} = -\,AE\,\{u_{i+1,n} - u_{i-1,n}\}/2\Delta x \qquad (8.6)$$

Note $q^{(v)}$ and $q^{(w)}$ need not be identical, and likewise, similar considerations apply to the spring constant k and damping factor β (since v and w are displaced by 90°, they "see" different sides of the hole). Also note that the Greek letters used in Chapter 6 and here, should not be confused with those used in Chapters 5 and 7. Now, Equations 8-4 and 8-5 can be rearranged to highlight their pentadiagonal structure as follows,

$$\begin{aligned}
&\{EI/(\Delta x)^4\}\, v_{i-2,n} \\
&+ \{-4EI/(\Delta x)^4 + N_{i,n}/(\Delta x)^2 - (N_{i+1,n} - N_{i-1,n})/(4(\Delta x)^2)\}\, v_{i-1,n} \\
&+ \{6EI/(\Delta x)^4 - 2N_{i,n}/(\Delta x)^2 + k + \beta/(2\Delta t) + \rho A/(\Delta t)^2\}\, v_{i,n} \\
&+ \{-4EI/(\Delta x)^4 + N_{i,n}/(\Delta x)^2 + (N_{i+1,n} - N_{i-1,n})/(4(\Delta x)^2)\}\, v_{i+1,n} \\
&+ \{EI/(\Delta x)^4\}\, v_{i+2,n} \\
&\quad = \quad q^{(v)}{}_{i,n-1} \\
&\qquad - T_{i,n}\,(-w_{i-2,n-1} + 2w_{i-1,n-1} - 2w_{i+1,n-1} + w_{i+2,n-1})/\{2(\Delta x)^3\} \\
&\qquad - \{(T_{i+1,n} - T_{i-1,n})/(2\Delta x)\}\{(w_{i-1,n-1} - 2w_{i,n-1} + w_{i+1,n-1})/(\Delta x)^2\} \\
&\qquad + (\beta/(2\Delta t))v_{i,n-2} + \{2\rho A/(\Delta t)^2\}v_{i,n-1} - \{\rho A/(\Delta t)^2\}v_{i,n-2} \qquad (8.7)
\end{aligned}$$

$$\begin{aligned}
&\{EI/(\Delta x)^4\}\, w_{i-2,n} \\
&+ \{-4EI/(\Delta x)^4 + N_{i,n}/(\Delta x)^2 - (N_{i+1,n} - N_{i-1,n})/(4(\Delta x)^2)\}\, w_{i-1,n} \\
&+ \{6EI/(\Delta x)^4 - 2N_{i,n}/(\Delta x)^2 + k + \beta/(2\Delta t) + \rho A/(\Delta t)^2\}\, w_{i,n} \\
&+ \{-4EI/(\Delta x)^4 + N_{i,n}/(\Delta x)^2 + (N_{i+1,n} - N_{i-1,n})/(4(\Delta x)^2)\}\, w_{i+1,n} \\
&+ \{EI/(\Delta x)^4\}\, w_{i+2,n} \\
&\quad = \quad q^{(w)}{}_{i,n-1} \\
&\qquad + T_{i,n}\,(-v_{i-2,n-1} + 2v_{i-1,n-1} - 2v_{i+1,n-1} + v_{i+2,n-1})/\{2(\Delta x)^3\} \\
&\qquad + \{(T_{i+1,n} - T_{i-1,n})/(2\Delta x)\}\{(v_{i-1,n-1} - 2v_{i,n-1} + v_{i+1,n-1})/(\Delta x)^2\} \\
&\qquad + (\beta/(2\Delta t))w_{i,n-2} + \{2\rho A/(\Delta t)^2\}w_{i,n-1} - \{\rho A/(\Delta t)^2\}w_{i,n-2} \qquad (8.8)
\end{aligned}$$

Computational recipe. As in Chapter 6, the required finite difference equations are obtained by writing Equations 8-7 and 8-8 for the internal nodes i = 3, 4, ... , imax-3, imax-2. This process yields, for each of v and w, *imax-4* equations in *imax* unknowns at the indexes i = 1 to imax. For each dependent variable, the additional four equations are obtained from boundary conditions. The two corresponding to the indexes i = 1 and 2 represent downhole bit conditions, whereas those at i = imax-1 and imax are surface constraints.

The right sides of Equations 8-7 and 8-8 represent known functions from previous time steps that have been stored in memory. The variable coefficients related to N(x,t) and T(x,t) on the left sides of these equations are computed at the same time step using the algorithms given in Chapters 5 and 7. Once the coefficients of pentadiagonal matrixes have been defined, the two outermost diagonals in the five-banded matrix are reduced (using standard operations from linear algebra) to tridiagonal form. Then, the tridiagonal matrix solver given in Chapter 5 can be used to solve for v_i and w_i for i = 1 to imax.

Modes of coupling. Dynamic coupling appears at several distinct points in the analysis. We have emphasized, in Chapters 5 and 7, axial and torsional coupling through boundary conditions at the bit. The foregoing equations emphasize that v(x,t) and w(x,t) coupling also occurs through the explicit presence of T(x,t) and N(x,t) in both bending partial differential equations; these two variable fields represent known quantities. They are first determined by solving

$$\rho\, \partial^2 u/\partial t^2 + \zeta\, \partial u/\partial t - E\, \partial^2 u/\partial x^2 = 0 \tag{8.9}$$

for axial displacement vibrations, and

$$\partial^2\theta/\partial t^2 - \xi\, \partial\theta/\partial t - c_s^2\, \partial^2\theta/\partial x^2 = \Theta_e \tag{8.10}$$

for torsional vibrations, using numerical methods already described. As discussed earlier, the complete boundary value problems associated with Equations 8-9 and 8-10 provide the capability of modeling bit-bounce, rate-of-penetration, and rock-bit interaction in the case of axial vibrations, and stick-slip oscillations and torque reversals for torsional vibrations.

Should lateral mode back-interaction on the axial strain field become important, an equation of the form $\rho A\, \partial^2 u/\partial t^2 - EA\, \partial^2 u/\partial x^2 = \partial\{EI\, (\partial^3 v/\partial x^3)\, (\partial v/\partial x)\}/\partial x$ cited in Chapter 6, extended to handle dual bending modes with torsion, would replace Equation 8-9 above.

Numerical considerations. Our numerical model was implemented using time-dependent, implicit, finite difference schemes, initialized by static, start-up conditions. Transient terms were approximated using three-point time difference formulas to minimize storage requirements; thus, only three time levels of axial, torsional and lateral wave data are required in computer memory at any given time. Bit displacement, stress and velocity histories are written to output graphics files and tables immediately upon solution.

All spatial derivatives were modeled using central difference approximations. These lead to computationally efficient tridiagonal matrixes in the case of axial and torsional vibrations, and only slightly more complicated pentadiagonal banded matrixes for both coupled lateral bending modes. All of the component modules, that is, axial, torsional, v and w mode bending, are believed to be extremely stable. Thus, *practical PC-based simulations can be performed in minutes for typical bottomhole assemblies without difficulty.*

Again, as explained in Chapter 1, numerical models are often the final resort in modeling reality. Morse and Feshbach (1953) and Graff (1975) clearly state that *classical normal mode analyses are often inappropriate, when certain complicated boundary conditions are used,* because *completeness* requirements are not fulfilled by the corresponding mathematical eigenvalue problem. However, solution can be obtained by solving (numerical) transient formulations; these also yield dynamically steady solutions, if they exist.

Numerical approaches offer other advantages. For example, the problem of *borehole wall contact,* in the static limit, is ripe with computational complexity because *static indeterminacies* arise at contact points that are unknown *a priori*. One particularly simple device that avoids these conceptual difficulties requires us to imagine that a continuous transverse spring loading exists, with a spring "constant" k acting on the beam. In the Fortran logic, this spring constant may be assumed to be vanishingly weak. Only if the net local transverse displacement $\sqrt{\{v^2(x,t) + w^2(x,t)\}}$ equals or exceeds the wellbore radius, at any instant during the course of the computations, at the required angle, is k(x) set to an arbitrarily large or possibly finite number.

Computed results are only as accurate as the assumed boundary conditions, and these, we emphasize, are continually under development. Cook, Nicholson, Sheppard, and Westlake (1989), following exhaustive study of data obtained in the Gulf of Mexico, conclude that considerable uncertainty in actual downhole boundary conditions make accurate BHA simulation difficult, indicating the value of real-time downhole measurements.

This, of course, reinforces the role of real-time MWD measurements in refining our understanding of downhole vibrations. The integrated model proposed in this chapter, therefore, should be viewed as a vehicle by which hypotheses can be tested, validated, or refuted. It is also clear that boundary condition types must change as the drillbit enters different layers of earth strata. Thus, numerical simulators and real-time measurements are best used in environments where their synergies can both be exploited.

In numerical simulation and field implementation, care must be exercised to ensure that any and all mesh widths used are significantly smaller than the wavelengths of the disturbances modeled. The following question is crucial: *What do we mean by disturbance?* In general, a disturbance will consist of axial, torsional, and lateral bending modes, and each of these will be characterized by different length scales, wavelengths, and periods.

The disturbances must be resolvable on spatial scales much smaller than a wavelength, and on time scales much smaller than a typical period; otherwise, the details of the vibration process will be lost, and truncation errors will dominate the solutions. In addition, it is important to satisfy all required *CFL* conditions, as explained in Chapter 5, to ensure that causality constraints and stability conditions are fulfilled.

The importance of *numerical stability* in an integration algorithm cannot be de-emphasized. This ensures that *real instabilities*, when they do arise, for example, from wave trapping at the neutral point or from conventional resonance, are physical and not computational anomalies. A vibrations algorithm should *not* be designed to suppress all high-amplitude phenomena, since a major practical objective is the prediction of real high-amplitude events.

GENERAL FORTRAN IMPLEMENTATION

In this section, we give a "bird's eye view" of the Fortran source code listing for coupled axial, torsional, and (dual) bending mode vibrations. Lower order details are specifically deleted in order to convey "the big picture", but these details are re-introduced later in separate expanded discussion.

```
C     .
C     .
C     Fortran dimension and common statements, parameter
C     input queries, specific constants, etc., not shown.
C     .
C     .
C     INITIALIZE AXIAL, ANGULAR AND BENDING DISPLACEMENTS
      DO 130  I=1,IMAX
      X = XS*(I-1)/(IMAX-1)
      UNM1(I) = RHO*G*(X**2)/(2.*ELAST) +WZERO*X/AE
      UNM2(I) = UNM1(I)
      TNM1(I) = 0.
      TNM2(I) = 0.
      VNM1(I) = 0.
      VNM2(I) = 0.
      WNM1(I) = 0.
      WNM2(I) = 0.
 130  CONTINUE
```

Figure 8-3. Coupled vibrations, Fortran listing (to be continued).

```
C      BEGIN TIMEWISE INTEGRATION
       T = 0.
       DO 900  N=1,NMAX
       T = N*DT

C      AXIAL VIBRATIONS
       DO 200  I=2,IMAXM1
       A(I) = 1.
       C(I) = 1.
       B(I) = -2.-RHO*DX*DX/(ELAST*DT*DT)-GAMA*DX*DX/(2.*ELAST*DT)
       W(I) = -RHO*DX*DX*(2.*UNM1(I)-UNM2(I))/(ELAST*DT*DT)
      1       -GAMA*DX*DX*UNM2(I)/(2.*ELAST*DT)+RHO*G*(DX**2)/ELAST
  200  CONTINUE
C      .
C      See Chapter 5 for axial vibration code listing with boundary
C      condition models for rate-of-penetration and bit bounce, and
C      displacement source model, from drill bit kinematic modeling.
C      .
C      .
C      TORSIONAL VIBRATIONS
       DO 400  I=2,IMAXM1
       A(I) = 1.
       C(I) = 1.
       B(I) = -2.-RHO*DX*DX/(SHRMOD*DT*DT)-
              GAMA2*DX*DX/(2.*SHRMOD*DT)
       W(I) = -RHO*DX*DX*(2.*TNM1(I)-TNM2(I))/(SHRMOD*DT*DT)
      1       -GAMA2*DX*DX*TNM2(I)/(2.*SHRMOD*DT)
  400  CONTINUE
C      .
C      .
C      See Chapters 5 and 7 for finite difference equation
C      definition, setup, and  boundary condition modeling.
C      Immediately below we prescribe constant surface rpm
C      via the angular displacement THETAN = RPM*2*PI*T/60.
C
       A(IMAX) = 0.
       B(IMAX) = 1.
       C(IMAX) = 99.
       W(IMAX) = RPM*2.*PI*T/60.
C
C      The following Fortran indicates that the drillbit torque
C      is proportional to the product of a friction coefficient
C      FCOEF and the dynamic weight-on-bit as obtained from the
C      axial calculations above.
C
       A(1) = 99.
       B(1) = -1.
       C(1) = 1.
       W(1) = (FCOEF*AE/GJ)*(UN(2)-UN(1))
C
       CALL TRIDI(A,B,C,VECTOR,W,IMAX)
       DO 450  I=1,IMAX
       THETAN(I) = VECTOR(I)
  450  CONTINUE
```

Figure 8-3. Coupled vibrations, Fortran listing (to be continued).

```
      DTHDX(1) = (THETAN(2)-THETAN(1))/DX
      TORBIT = GJ*DTHDX(1)
      DTHDX(IMAX) = (THETAN(IMAX)-THETAN(IMAXM1))/DX
      DO 505  I=2,IMAXM1
      DTHDX(I) = (THETAN(I+1)-THETAN(I-1))/(2.*DX)
 505  CONTINUE

C     COUPLED BENDING VIBRATIONS
C     V LATERAL MODE
C     This section on coupled vibrations implements Equations
C     8-7 and 8-8.  We annotate the listing for Equation 8-7
C     describing v(x,t) in detail.  The coefficients, D, E, A,
C     C, B and W are determined by one-to-one comparison from
C     the pentadiagonal form in Equation 8-7.
C
      DO 600  I=3,IMAXM2
      D(I) = EI/DX4
      E(I) = EI/DX4
      A(I) = -4.*EI/DX4 -AE*DUDX(I)/(DX*DX)
     1       +AE*(DUDX(I+1)-DUDX(I-1))/(4.*DX*DX)
      C(I) = -4.*EI/DX4 -AE*DUDX(I)/(DX*DX)
     1       -AE*(DUDX(I+1)-DUDX(I-1))/(4.*DX*DX)
      B(I) = 6.*EI/DX4 +2.*AE*DUDX(I)/(DX*DX)
     1       +BEAMK +BEAMB/(2.*DT) +RHO*AREA/(DT**2)
      W(I) = Q -GJ*DTHDX(I)*
     1       (-WNM1(I-2)+2.*WNM1(I-1)
     2       -2.*WNM1(I+1)+WNM1(I+2))/(2.*DX**3)
     3       -GJ*(DTHDX(I+1)-DTHDX(I-1))*(1./(2.*DX))*
     4       (WNM1(I-1)-2.*WNM1(I)+WNM1(I+1))/(DX**2)
     5       +BEAMB*VNM2(I)/(2.*DT)
     6       +2.*RHO*AREA*VNM1(I)/(DT**2)
     7       -RHO*AREA*VNM2(I)/(DT**2)
 600  CONTINUE
C     .
C     .
C     Next, the coefficients for the downhole indexes i = 1 and
C     2, and for the surface indexes i = imax-1 and imax,  must
C     be given in order to have imax equations in imax unknowns.
C     These, not shown for now, are discussed later in the text.
C     .
C     .
C     Once all  of the required matrix coefficients are defined,
C     the outer-most "bands" of the pentadiagonal matrix can be
C     eliminated, to reduce the matrix to tridiagonal form, and
C     the result can be solved solved using the TRIDI call.
C
      CALL REDUCE(A,B,C,D,E,W,IMAX)
      CALL TRIDI(A,B,C,VECTOR,W,IMAX)
C     .
C     Then, the v(x,t) solution is updated by copying the newly
C     obtained solution from the tridiagonal solver.
C
      DO 650  I=1,IMAX
      VBENDN(I) = VECTOR(I)
 650  CONTINUE
```

Figure 8-3. Coupled vibrations, Fortran listing (to be continued).

```
C       .
C       The sequential process described, axial ---> torsional --->
C       ---> v bending requires one final step,  i.e., the solution
C       " ---> w bending" for the second bending mode.   This logic
C       is listed without comment.
C       .
C       .
C       COUPLED BENDING VIBRATIONS
C       W LATERAL MODE
        DO 800  I=3,IMAXM2
        D(I) = EI/DX4
        E(I) = EI/DX4
        A(I) = -4.*EI/DX4 -AE*DUDX(I)/(DX*DX)
       1       +AE*(DUDX(I+1)-DUDX(I-1))/(4.*DX*DX)
        C(I) = -4.*EI/DX4 -AE*DUDX(I)/(DX*DX)
       1       -AE*(DUDX(I+1)-DUDX(I-1))/(4.*DX*DX)
        B(I) = +6.*EI/DX4 +2.*AE*DUDX(I)/(DX*DX)
       1       +BEAMK +BEAMB/(2.*DT) +RHO*AREA/(DT**2)
        W(I) = +Q+GJ*DTHDX(I)*
       1       (-VNM1(I-2)+2.*VNM1(I-1)
       2       -2.*VNM1(I+1)+VNM1(I+2))/(2.*DX**3)
       3       +GJ*(DTHDX(I+1)-DTHDX(I-1))*(1./(2.*DX))*
       4       (VNM1(I-1)-2.*VNM1(I)+VNM1(I+1))/(DX**2)
       5       +BEAMB*WNM2(I)/(2.*DT)
       6       +2.*RHO*AREA*WNM1(I)/(DT**2)
       7       -RHO*AREA*WNM2(I)/(DT**2)
  800   CONTINUE
C       .
C       .
        CALL REDUCE(A,B,C,D,E,W,IMAX)
        CALL TRIDI(A,B,C,VECTOR,W,IMAX)
        DO 850  I=1,IMAX
        WBENDN(I) = VECTOR(I)
  850   CONTINUE
C       .
C       At this point, all  of  the calculations for a time
C       step N are completed and command returns to the top
C       of the 900 do-loop for the next time integration.
C       .
  900   CONTINUE
C       .
        STOP
        END
```

Figure 8-3. Coupled vibrations, Fortran listing (continued).

Next, we proceed to use the algorithm given in Figure 8-3, to calculate realistic drillstring coupled vibrations. Two detailed examples were selected to demonstrate the range of capabilities behind the algorithm designed in this book, to show that events like *bit bounce, forward rate-of-penetration, stick-slip, torque-reversal, and drillbit precession* can be modeled and simulated straightforwardly in a stable fashion.

EXAMPLE CALCULATIONS: BIT-BOUNCE, STICK-SLIP, RATE-OF-PENETRATION, AND DRILLSTRING PRECESSION

At this writing, our integrated axial, torsional, and bending vibrations formulation has been implemented in a fully debugged Fortran algorithm, at least in the programming sense. *Various ideas and coding details are still in research phase, and are under continual testing; thus, Examples 8-2 and 8-3 represent preliminary results subject to correction and refinement.*

For this reason, the physical details of the drillstring assumed in the calculations reported below are not discussed, since any computed results are qualitative at best. Our integrated vibrations solutions are only intended to provide a flavor of eventual capabilities, and for this reason, the reader should not place undue emphasis on the exact numbers themselves.

But we do wish to emphasize that real drilling vibrations are complicated motions that encompass varied types of observations reported by the myriad of authors cited in this book - and that these observations *can* be simulated rather directly. Modeled phenomena include bit bounce, stick-slip oscillations, forward rate-of-penetration, torque reversal, and drillstring precessional motions. We will demonstrate this with two contrasting calculated examples.

Example 8-2: Smooth drilling and making hole. In this first calculation, the rock-bit interaction model (a.k.a., the *bottomhole impedance boundary condition*) $\alpha u_x + \beta u_t + \lambda u = 0$ assumed in Chapter 5 is executed with the numerical values shown in Figure 8-4, as extracted from the Fortran listing of Figure 8-3 for coupled motions.

```
C     DEFINE UNITS OF ROCK-BIT INTERACTION
      RBALPH = -1.
      RBBETA = +1.
C     RBLAMB = +20 GIVES BIT BOUNCE,-20 GIVES FORWARD DRILL ONLY.
C     RBLAMB = +20.
      RBLAMB = -20.
```

Figure 8-4. Rock-bit impedance model, smooth drilling.

The RBALPH, RBBETA, and RBLAMB parameters shown in the Fortran above represent α, β, and λ respectively. Our computed results for the axial displacement u (at i = 1, the bit position) demonstrate that the drillstring is *making hole* in a well-behaved manner: u descends vertically, with the descent rate or rate-of-penetration du/dt always negative.

The *torque-at-bit*, interestingly, reverses sign at regular intervals, and is therefore indicative of stick-slip and torsional unwinding; in Figure 8-5, we separately identify *four* blocks of computed time data, taken whenever torque reverses in sign.

The strain $\partial u/\partial x$ at the bit, for the parameters used in $\alpha u_x + \beta u_t + \lambda u = 0$, is compressive, but sometimes tensile, depending upon the formation interaction (recall that $u_x < 0$ for compression and $u_x > 0$ for tension). For our *fictitious* rock, both push and pull exist, and descent is due to a good extent to gravity. Bending displacements v and w at the bit appear to be stable computationally over numerous time steps despite rapid changes in axial and torsional solutions.

Start of drilling ...

N	U	DU/DT	DU/DX	TORQ	V	W
1	-.11E-06	.00E+00	-.20E-03	-.22E+05	.14E-01	.40E+01
2	-.20E-06	-.19E-03	-.19E-03	-.20E+05	.27E-01	.40E+01
3	-.30E-06	-.18E-03	-.18E-03	-.19E+05	.41E-01	.40E+01
4	-.38E-06	-.17E-03	-.16E-03	-.17E+05	.54E-01	.40E+01
5	-.46E-06	-.16E-03	-.15E-03	-.15E+05	.68E-01	.40E+01
6	-.53E-06	-.14E-03	-.13E-03	-.14E+05	.82E-01	.40E+01
7	-.59E-06	-.13E-03	-.12E-03	-.12E+05	.95E-01	.40E+01
8	-.65E-06	-.12E-03	-.11E-03	-.11E+05	.11E+00	.40E+01
9	-.70E-06	-.11E-03	-.92E-04	-.91E+04	.12E+00	.40E+01
10	-.74E-06	-.94E-04	-.80E-04	-.77E+04	.14E+00	.40E+01
11	-.78E-06	-.83E-04	-.68E-04	-.64E+04	.15E+00	.40E+01
12	-.81E-06	-.72E-04	-.56E-04	-.51E+04	.16E+00	.40E+01
13	-.84E-06	-.61E-04	-.45E-04	-.39E+04	.18E+00	.40E+01
14	-.86E-06	-.51E-04	-.34E-04	-.27E+04	.19E+00	.40E+01
15	-.88E-06	-.41E-04	-.23E-04	-.15E+04	.20E+00	.40E+01
16	-.89E-06	-.31E-04	-.13E-04	-.33E+03	.22E+00	.40E+01

.

Torque reversal ...

N	U	DU/DT	DU/DX	TORQ	V	W
17	-.90E-06	-.21E-04	-.29E-05	.80E+03	.23E+00	.40E+01
18	-.90E-06	-.11E-04	.70E-05	.19E+04	.24E+00	.40E+01
19	-.89E-06	-.13E-05	.17E-04	.30E+04	.26E+00	.40E+01
20	-.10E-01	.83E-05	.26E-04	.00E+00	.27E+00	.40E+01
.						
.						
225	-.95E-01	-.19E+01	.25E-04	.22E+04	.28E+01	.29E+01
226	-.96E-01	-.19E+01	.19E-04	.15E+04	.28E+01	.29E+01
227	-.97E-01	-.19E+01	.13E-04	.83E+03	.28E+01	.29E+01
228	-.98E-01	-.19E+01	.73E-05	.17E+03	.28E+01	.29E+01

.

Torque reversal ...

N	U	DU/DT	DU/DX	TORQ	V	W
229	-.99E-01	-.20E+01	.15E-05	-.50E+03	.28E+01	.28E+01
230	-.10E+00	-.20E+01	-.44E-05	-.12E+04	.28E+01	.28E+01
231	-.10E+00	-.20E+01	-.10E-04	-.18E+04	.28E+01	.28E+01
232	-.10E+00	-.20E+01	-.16E-04	-.25E+04	.28E+01	.28E+01
233	-.10E+00	-.20E+01	-.22E-04	-.32E+04	.28E+01	.28E+01
234	-.10E+00	-.21E+01	-.28E-04	-.39E+04	.29E+01	.28E+01
235	-.11E+00	-.21E+01	-.34E-04	-.45E+04	.29E+01	.28E+01
236	-.11E+00	-.21E+01	-.40E-04	-.52E+04	.29E+01	.28E+01
237	-.11E+00	-.21E+01	-.46E-04	-.59E+04	.29E+01	.28E+01
238	-.11E+00	-.22E+01	-.52E-04	-.66E+04	.29E+01	.28E+01
239	-.11E+00	-.22E+01	-.57E-04	-.72E+04	.29E+01	.28E+01
240	-.11E+00	-.22E+01	-.63E-04	-.79E+04	.29E+01	.27E+01
241	-.11E+00	-.22E+01	-.69E-04	-.86E+04	.29E+01	.27E+01
242	-.11E+00	-.23E+01	-.75E-04	-.93E+04	.29E+01	.27E+01
243	-.12E+00	-.23E+01	-.81E-04	-.99E+04	.29E+01	.27E+01

Figure 8-5. Coupled vibrations, smooth drilling (to be continued).

Torque reversal ...

```
  N    U          DU/DT     DU/DX      TORQ        V          W
452 -.11E+01 -.22E+02 -.79E-05  .52E+03  .40E+01  .14E+00
453 -.11E+01 -.22E+02  .45E-05  .19E+04  .40E+01  .12E+00
454 -.11E+01 -.23E+02  .17E-04  .34E+04  .40E+01  .11E+00
455 -.12E+01 -.23E+02  .30E-04  .49E+04  .40E+01  .95E-01
987 -.38E+03 -.75E+04  .28E+00  .33E+08 -.85E+00 -.39E+01
988 -.38E+03 -.76E+04  .29E+00  .33E+08 -.86E+00 -.39E+01
989 -.39E+03 -.76E+04  .29E+00  .33E+08 -.88E+00 -.39E+01
990 -.39E+03 -.77E+04  .29E+00  .34E+08 -.89E+00 -.39E+01
991 -.39E+03 -.78E+04  .30E+00  .34E+08 -.90E+00 -.39E+01
992 -.40E+03 -.79E+04  .30E+00  .34E+08 -.92E+00 -.39E+01
```

Figure 8-5. Coupled vibrations, smooth drilling (continued).

Example 8-3: Rough drilling with bit bounce. In our second calculation, the rock-bit interaction model $\alpha u_x + \beta u_t + \lambda u = 0$ takes on the values shown in Figure 8-6, that is,

```
C       DEFINE UNITS OF ROCK-BIT INTERACTION
        RBALPH = -1.
        RBBETA = +1.
C       RBLAMB = +20 GIVES BIT BOUNCE, -20 GIVES FORWARD DRILL ONLY
        RBLAMB = +20.
C       RBLAMB = -20.
```

Figure 8-6. Rock-bit impedance model, rough drilling.

The sign change in the λ term introduces some computationally interesting results, namely, the existence of bit bounce and rough drilling. From the following results, we find torque reversal as before, plus well-behaved bending displacements that confirm numerical stability.

However, we find that the bit speed $\partial u/\partial t$ may also reverse sign, demonstrating that bit-bounce is possible for the $\alpha u_x + \beta u_t + \lambda u = 0$ assumed. When this is so, the Fortran logic enforces zero bit strain, and continues to the next time integration; gravity ultimately pulls the drillstring down, at which point the rock-bit interaction model again applies. The computational results in Figure 8-7 are again conveniently broken out accordingly at significant transition points in the physics.

Start of drilling ...

N	U	DU/DT	DU/DX	TORQ	V	W
1	-.10E-06	.00E+00	-.20E-03	-.22E+05	.14E-01	.40E+01
2	-.20E-06	-.19E-03	-.19E-03	-.20E+05	.27E-01	.40E+01
3	-.28E-06	-.18E-03	-.18E-03	-.19E+05	.41E-01	.40E+01
4	-.36E-06	-.16E-03	-.16E-03	-.17E+05	.54E-01	.40E+01
5	-.43E-06	-.14E-03	-.15E-03	-.15E+05	.68E-01	.40E+01
6	-.49E-06	-.12E-03	-.13E-03	-.14E+05	.82E-01	.40E+01
7	-.54E-06	-.11E-03	-.12E-03	-.12E+05	.95E-01	.40E+01
8	-.58E-06	-.94E-04	-.11E-03	-.11E+05	.11E+00	.40E+01
9	-.62E-06	-.81E-04	-.92E-04	-.91E+04	.12E+00	.40E+01
10	-.65E-06	-.67E-04	-.80E-04	-.77E+04	.14E+00	.40E+01
11	-.67E-06	-.55E-04	-.68E-04	-.64E+04	.15E+00	.40E+01
12	-.69E-06	-.43E-04	-.56E-04	-.51E+04	.16E+00	.40E+01
13	-.70E-06	-.31E-04	-.45E-04	-.39E+04	.18E+00	.40E+01
14	-.70E-06	-.20E-04	-.34E-04	-.27E+04	.19E+00	.40E+01
15	-.70E-06	-.93E-05	-.23E-04	-.15E+04	.20E+00	.40E+01
16	-.13E-01	.10E-05	-.13E-04	.00E+00	.22E+00	.40E+01

.

Torque reversal ...

N	U	DU/DT	DU/DX	TORQ	V	W
279	-.43E+00	.70E+01	.00E+00	.00E+00	.33E+01	.23E+01
280	-.43E+00	-.37E+02	.00E+00	.46E+04	.33E+01	.23E+01
281	-.43E+00	.86E+01	.40E-04	.00E+00	.33E+01	.23E+01
282	-.44E+00	.40E+01	.00E+00	.00E+00	.33E+01	.23E+01
283	-.44E+00	-.34E+02	.00E+00	.33E+04	.33E+01	.23E+01
284	-.44E+00	.88E+01	.29E-04	.00E+00	.33E+01	.23E+01
285	-.46E+00	.98E+00	.00E+00	.00E+00	.33E+01	.23E+01
286	-.45E+00	-.32E+02	.00E+00	.19E+04	.33E+01	.23E+01
287	-.45E+00	.90E+01	.17E-04	.00E+00	.33E+01	.22E+01

.

Rapid torque oscillations and bit bounce ...

N	U	DU/DT	DU/DX	TORQ	V	W
288	-.45E+00	-.21E+01	.00E+00	.16E+04	.33E+01	.22E+01
289	-.45E+00	.90E+01	.14E-04	.00E+00	.33E+01	.22E+01
290	-.45E+00	-.94E+01	.00E+00	-.90E+03	.33E+01	.22E+01
291	-.46E+00	.90E+01	-.79E-05	.00E+00	.33E+01	.22E+01
292	-.45E+00	-.17E+02	.00E+00	-.24E+04	.34E+01	.22E+01
293	-.46E+00	.90E+01	-.21E-04	.00E+00	.34E+01	.22E+01
294	-.46E+00	-.21E+02	.00E+00	-.36E+04	.34E+01	.22E+01
295	-.47E+00	.92E+01	-.32E-04	.00E+00	.34E+01	.22E+01
296	-.47E+00	-.23E+02	.00E+00	-.46E+04	.34E+01	.21E+01
297	-.48E+00	.93E+01	-.40E-04	.00E+00	.34E+01	.21E+01
298	-.47E+00	-.25E+02	.00E+00	-.55E+04	.34E+01	.21E+01
299	-.49E+00	.95E+01	-.48E-04	.00E+00	.34E+01	.21E+01
300	-.48E+00	-.26E+02	.00E+00	-.63E+04	.34E+01	.21E+01

.

Torque stabilization, with bit bounce ...

N	U	DU/DT	DU/DX	TORQ	V	W
625	-.59E+00	.58E+01	.00E+00	.00E+00	.34E+01	-.21E+01
626	-.58E+00	-.46E+01	.00E+00	.20E+04	.34E+01	-.21E+01
627	-.58E+00	.12E+02	.18E-04	.00E+00	.34E+01	-.21E+01
628	-.58E+00	.99E+01	.00E+00	.00E+00	.34E+01	-.21E+01
629	-.58E+00	-.71E+01	.00E+00	.35E+04	.34E+01	-.22E+01
630	-.57E+00	.12E+02	.31E-04	.00E+00	.34E+01	-.22E+01
.						
.						
900	-.71E+00	.14E+02	.12E-04	.00E+00	.33E+00	-.40E+01
901	-.70E+00	-.22E+01	.00E+00	.12E+04	.31E+00	-.40E+01
902	-.71E+00	.14E+02	.11E-04	.00E+00	.30E+00	-.40E+01

Figure 8-7. Coupled vibrations, rough drilling.

Model limitations and extensions. There are various areas for improvement and research, insofar as the modeling presented thus far is concerned. For one, significant research remains in properly defining the parameters in $\alpha u_x + \beta u_t + \lambda u = 0$, to ensure that they are realistic and simulate actual bit behavior. The empirical data cited in Chapter 5 for single tooth impacts represent a first step in this direction.

As coded, our coupled "one way" model (at any given time step) computes axial solutions first, torsional ones next, followed by those in bending. This *cascading effect* does not allow *nonlinear back-interaction* of lateral bending vibrations on the axial and subsequent torsional solutions.

The fix needed is easily implemented: *simply replace our axial wave equation with the corrected one containing nonlinear right-side bending terms,* as given in Chapter 6 (e.g., for a single bending mode, $\rho A\ \partial^2 u/\partial t^2 - EA\ \partial^2 u/\partial x^2 = EI\ \partial\{(\partial^3 v(x,t)/\partial x^3)\ (\partial v/\partial x)\}/\partial x$). This approach to nonlinear back-interaction was first proposed by Chin and Rizzetta (1979) and Chin (1981) for unsteady transonic flows in high speed aerodynamics, where significant interaction between mean and harmonic flows exist.

PRECESSIONAL INSTABILITIES

The study of drillstring buckling under torsion and drillstring precessional instabilities was initiated by Dunayevsky and his co-workers in the 1980s, and reported in Dunayevsky and Judzis (1983), Dunayevsky, Judzis, and Mills (1984), Dunayevsky, Judzis, Mills, and Griffin (1985), Dunayevsky, Judzis, and Abbassian (1989), and Dunayevsky, Abbassian, and Judzis (1993).

Our "twisted eraser" experiment of Example 8-1 clearly shows that a long twisted rod under axial load will bend in two orthogonal planes. Dunayevsky and Judzis (1983), in their analysis of critical loads for 3D drillstring buckling, importantly include in the governing bending equations, terms due to torque and axial force. This early paper dealt with static deflections only, and did not address transient problems; the torque in their governing *elastic line* equations was *constant*, while the axial force varied linearly as discussed in Chapter 5.

Later, Dunayevsky *et al* studied drillstring precession in directional boreholes and drillstring stability, when the rotation is constrained by hole geometry. Under certain conditions, axial vibration energy will be transformed into transverse vibration energy; at the same time, drillstring rotation about the *drillstring axis* becomes unstable, and the transition to rotation about the *borehole axis* takes place. This is *precessional* motion.

Their critical conditions are formulated in terms of rotary speeds, amplitude of weight-on-bit variations, damping, and geometry. A *finite element procedure* reduces the partial differential equation describing drillstring dynamics to a set of *Matthew-Hill* ordinary differential equations. The *stability boundaries* of *parametric resonances* for these equations determines the onset of precessional motion of the drillstring. The results can be used to predict stable performance of drillstrings during *directional drilling*.

We emphasize that the instability studied by Dunayevsky *et al* represents one of many types that can damage the drillstring; like other instabilities considered in the literature, whether or not it exists downhole is subject to empirical verification.

But this is not easy. Rewcastle and Burgess (1992) note that precession of the bottomhole assembly, in reality, is accompanied by high levels of friction and abrasion: the lateral-tangential collar motion causes the BHA to impact the borehole wall and to decelerate rapidly whenever maximum deflections are reached. This will damp out all evidence in measured surface torques.

Comments on Dunayevsky *et al.* The equations governing combined axial, torsional, and lateral bending vibrations used by Dunayevsky *et al* were obtained by specializing the *Kirchoff vector equations* (Bolotin, 1964) for a general *elastic line*, therefore requiring some very unwieldy, error-prone algebra. Thus, several high-order terms in torque and axial force seem to be inconsistently modeled.

Actually, self-consistent results can be obtained more simply, with greater physical insight, using results from *strength of materials*. To illustrate the basic ideas, we start with the standard *moment equations* for bending, governing rods under a torque T and an axial force F (Den Hartog, 1952), and restrict ourselves to static deflections. The required equations are, namely,

$$EI\,v'' = -Fv - Tw' \tag{8.11}$$

$$EIw'' = -Fw + Tv' \tag{8.12}$$

where primes denote x derivatives, following the notation established earlier. Let us differentiate Equation 8-11 for v(x,t) twice with respect to x, so that

$$(EI\,v'')' = -F'v - Fv' - T'w' - T\,w'' \tag{8.13}$$

$$\begin{aligned}(EI\,v'')'' &= -F''v - F'v' - F'v' - Fv'' - T'w'' - T''w' - Tw''' - T'w'' \\ &= -(Fv')' - (Tw'')' - (F'v)' - (T'w')' \\ &= -(Fv')' - (Tw'')' + H.O.T.\end{aligned} \tag{8.14}$$

We next apply the same differentiations to Equation 8-12 for w(x,t), with the result that

$$(EIw'')' = -F'w - Fw' + T'v' + Tv'' \tag{8.15}$$

$$(EI\,w'')'' = -F'w' - F''w - F'w' - Fw'' + T''v' + T'v'' + T'v'' + Tv'''$$
$$= -(Fw')' + (Tv'')' - (F'w)' + (T'v')'$$
$$= -(Fw')' + (Tv'')' + H.O.T. \qquad (8.16)$$

Now in *linear elasticity*, the high-order flux terms (denoted *H.O.T.*) in Equations 8-14 and 8-15 containing the spatial derivatives $F_x(x,t)$ and $T_x(x,t)$ can be neglected, since their corresponding background strain fields are assumed to be slowly varying. This leaves

$$(EI\,v'')'' + (Tw'')' + (Fv')' = 0 \qquad (8.17)$$

$$(EI\,w'')'' - (Tv'')' + (Fw')' = 0 \qquad (8.18)$$

This derivation gives the proper differential forms satisfied by the bending equations when T(x,t) and F(x,t) are nonzero. The above differential operators appear in our Equations 8-1 and 8-2, where inertia, damping, and restoring force terms are appended in the usual way.

In Dunayevsky and Judzis (1983), which deals with 3D *static* buckling, F is taken as a linear function of x, in agreement with Chapter 5, and T is assumed to be a known constant. These consistent requirements implicitly assume that borehole wall friction and dog-legs are absent.

The assumptions behind the early *dynamics* papers are less clear. However, Dunayevsky, Abbassian, and Judzis (1993) restated them in a recent paper. Essentially, the drillstring remains in permanent contact with the wellbore wall along its entire length (this *kinematic constraint* linking v to w also reduces the number of unknown variables).

Several boundary condition models were allowed, namely, (1) simple supports at the surface (*top-drive condition*), or at the bottom (e.g., *bit with slick assembly*), with $v = v_{xx} = 0$ at $x = 0$ and L, and (2) clamped at top (e.g., *kelly rotary table*), or at the bottom (*packed assembly condition*), with $v = v_x = 0$ at $x = 0$ and L.

Importantly, the authors noted that *the effects due to drillstring rotary motion and twisting vibrations are ignored in the precession model.* Thus, the model does *not* include torque, and hence, *it cannot model stick-slip oscillations and their interaction with bending.* Of course, the objective of the authors' model was to examine lateral vibrations as induced by axial excitation; in that regard, the work represents a significant contribution to the literature.

Direct simulation of bit precession. We indicate that the formulation of Dunayevsky *et al* will produce *stability envelopes* only: it will define parameter ranges for which dangerous motions are anticipated, but it will *not* determine forces, moments, displacements, velocities or accelerations.

But direct transient analysis will satisfy both objectives. Since actual amplitude outputs are obtained from the fast difference scheme given earlier, exact magnitudes are available, where the results are to be interpreted along with the user's (subjective) assessment of stability.

The formulation given here *will* allow prescribed circular motions at the bit, and in fact, the computed results given above assumed just such a model. The Fortran listing which implemented this motion is given in Figure 8-8, where the first source block reads `1 x v(1) + 0 x v(2) + 0 x v(3) = PRERAD*SIN(WMGPRE*T)`, or v(1) = PRERAD*SIN(WMGPRE*T). Similarly, the second block reads w(1) = PRERAD*COS(WMGPRE*T).

Here, PRERAD denotes a prescribed precession radius R_{prec}, while WMGPRE is a precession frequency ω_{prec}. Since $\sin^2 \omega_{prec}t + \cos^2 \omega_{prec}t = 1$, it follows that $v(1)^2 + w(1)^2 = R_{prec}{}^2$ applies at the bit index i = 1. Of course, we could have *parameterized* v(1) and w(1) to any other functions of time, in order to model the effects of bit motions other than perfect circles.

Prescribing circular motions at the bit with a radius equal to that of the hole, in our analytical model, in no way constrains the entire drillstring to "wrap around" the borehole wall - unless, of course, that is the outcome of the calculations. Thus, the approach is more general, because it allows a dynamical system to seek out its own stable equilibria, rather than imposing their existence on an *a priori* basis.

```
        .
        .
C       V LATERAL MODE, FIRST BIT BOUNDARY CONDITION
        B(1) = 1.
        C(1) = 0.
        E(1) = 0.
        W(1) = PRERAD*SIN(WMGPRE*T)
        .
        .
C       W LATERAL MODE, FIRST BIT BOUNDARY CONDITION
        B(1) = 1.
        C(1) = 0.
        E(1) = 0.
        W(1) = PRERAD*COS(WMGPRE*T)
        .
        .
```

Figure 8-8. Precession boundary condition logic at the bit.

DRILLSTRING VIBRATIONS IN HORIZONTAL WELLS

Having discussed and motivated our model for axial, torsional and lateral vibrations in detail, we assume at this point that the ideas rest comfortably with the reader, and that he is ready to continue the work outlined here. In this concluding section, we suggest various modifications needed to extend the algorithm to deviated and horizontal wells.

Because we are now familiar with the derivations and their limitations, the suggestions incorporated here are given with a minimum of elaboration. The Fortran listing given earlier may serve as a starting point for a more complete computer program for arbitrary drillstring inclinations and areal discontinuities.

In highly deviated and horizontal wells, the effect of gravity must be properly included. Thus, only the appropriate component of gravity along the drillstring axis should be incorporated in the axial vibrations model. In the vertical drillstring model considered here, gravity played no significant role in lateral vibrations; however, in deviated drillstrings, gravity now acts as a transverse load that must be accounted for in bending modeling.

Deviated drillstrings very often rest on the formation. In addition to gravity, a second transverse loading is provided by formation support, which manifests itself through variable elastic supports and damping factors. Borehole friction will affect torsional oscillations, but only at the lower side of the hole; thus, the Θ_e excitation term in $\partial^2\theta/\partial t^2 - \xi\, \partial\theta/\partial t - c_s^2\, \partial^2\theta/\partial x^2 = \Theta_e$ will in general be a function of θ, rendering the governing equation nonlinear.

Borehole wall contacts can complicated. Earlier, again for vertical holes, we suggested that when computed lateral displacements are large, taking on displacements with $\sqrt{(v^2 + w^2)} > R_{hole}$, Fortran *if-then* logic can be invoked to apply a formation restoring force. Under some conditions, an additional logical test based on the value of θ may be needed, to distinguish upper from lower sides of the hole.

When a drillstring lies on its side, it is possible that, in addition to rotating, the bit "walks up and down in circles." This precession-like spinning can also be modeled using the Fortran logic shown in Figure 8-8. All of the changes suggested in this section can be straightforwardly implemented as extensions to our computational algorithm.

REFERENCES

Bolotin, V.V., *The Dynamic Stability of Elastic Systems*, Holden-Day, Inc., San Francisco, 1964.

Chin, W.C., "Harmonic Analysis of Unsteady Transonic Flow," AIAA Journal, Feb. 1981.

Chin, W.C., and Rizzetta, D., "Effect of Frequency in Unsteady Transonic Flow," AIAA Journal, July 1979.

Clough, R.W., and Penzien, J., *Dynamics of Structures*, McGraw-Hill, New York, 1975.

Cohen, J.H., Maurer, W.C., and Westcott, P.A., "High-Performance TSD Bits Improve Penetration Rate," Oil & Gas Journal, Apr. 12, 1993, pp. 47-51.

Cook, R.L., Nicholson, J.W., Sheppard, M.C., and Westlake, W., "First Real Time Measurements of Downhole Vibrations, Forces, and Pressures Used to Monitor Directional Drilling Operations," SPE/IADC Paper No. 18651, presented at the 1989 SPE/IADC Drilling Conference, New Orleans, Louisiana, February 28 - March 3, 1989.

Den Hartog, J.P., *Advanced Strength of Materials*, McGraw-Hill, New York, 1952.

Dunayevsky, V.A., and Judzis, A., "Conservative and Nonconservative Buckling of Drillpipe," SPE Paper No. 11991, presented at the 58th Annual Technical Conference and Exhibition of the Society of Petroleum Engineers, San Francisco, Ca., Oct. 5-8, 1983.

Dunayevsky, V.A., Judzis, A., and Mills, W.H., "Onset of Drillstring Precession in a Directional Borehole," SPE Paper No. 13027, presented at the 59th Annual Technical Conference and Exhibition of the Society of Petroleum Engineers, Houston, Texas, Sept. 16-19, 1984.

Dunayevsky, V.A., Judzis, A., Mills, W.H., and Griffin, G., "Dynamic Stability of Drillstrings Under Fluctuating Weights-on-Bit," SPE Paper No. 14329, 60th Annual Technical Conference and Exhibition of the Society of Petroleum Engineers, Las Vegas, Nevada, Sept. 22-25, 1985.

Dunayevsky, V.A., Judzis, A., and Abbassian, F., "Use of Dynamic Drillstring Simulator for RPM and Bottomhole Assembly Optimization," OTC Paper 5902, presented at the 21st Annual Offshore Technology Conference, Houston, Texas, May 1-4, 1989.

Dunayevsky, V.A., Abbassian, F., and Judzis, A., "Dynamic Stability of Drillstrings Under Fluctuating Weight on Bit," SPE Drilling and Completion Journal, June 1993, pp. 84-92.

Garrett, D.L., "Dynamic Analysis of Slender Rods," Journal of Energy Resources Technology, Transactions of the ASME, Dec. 1982, pp. 302-306.

Graff, K.F., *Wave Motion in Elastic Solids*, Ohio State University Press, 1975.

Langeveld, C.J., "PDC Bit Dynamics," IADC/SPE Paper No. 23867, presented at the 1992 IADC/SPE Drilling Conference, New Orleans, Louisiana, Feb. 18-21, 1992.

Love, A.E.H., *A Treatise on the Mathematical Theory of Elasticity*, Dover Publications, New York, 1944.

Morse, P.M., and Feshbach, H., *Methods of Theoretical Physics, Parts I and II*, McGraw-Hill, New York, 1953.

Nordgren, R.P., "On Computation of the Motion of Elastic Rods," Journal of Applied Mechanics, Transactions of the ASME, Sept. 1974, pp. 777-780

Rewcastle, S.C., and Burgess, T.M., "Real-Time Downhole Shock Measurements Increase Drilling Efficiency and Improve MWD Reliability," IADC/SPE Paper No. 23890, presented at the 1992 IADC/SPE Drilling Conference, New Orleans, Louisiana, Feb. 18-21, 1992.

Staff, "Antiwhirl PDC Bit Designs Reduce Vibrations," Oil & Gas Journal, Nov. 30, 1992, pp. 59-62.

Timoshenko, S.P., and Goodier, J.N., *Theory of Elasticity*, McGraw-Hill, New York, 1934.

Vandiver, J.K., Nicholson, J.W., and Shyu, R.J., "Case Studies of the Bending Vibration and Whirling Motion of Drill Collars," SPE/IADC Paper No. 18652, presented at the 1989 SPE/IADC Drilling Conference, New Orleans, Louisiana, February 28-March 3, 1989.

Part III
Modern Petroleum Engineering Applications

9
Mud Dynamics in the Borehole Annulus

From a propagation viewpoint, borehole acoustics is synonymous with *traditional acoustics* (i.e., small amplitude problems) in air and underwater sound: *it is nondispersive and linear.* The thermodynamic loss mechanisms are different, but these high-order effects are usually modeled *after-the-fact* using lumped exponential damping (e.g., see Chapter 2). In this chapter, we describe new modeling concepts specific to modern petroleum applications only.

Conventional acoustics involves area modulated horns, transducer design, three-dimensional concert halls, and other short wavelength phenomena. In borehole and drillpipe acoustics, wavelengths are long compared to typical cross-sectional dimensions. Ratios exceeding *five hundred* are not uncommon.

Modeling issues deal with abrupt area discontinuities, and in swab-surge and MWD applications, the monopole or dipole nature of the internal acoustic source. Here, the fundamental equations are derived, and their application to *swab-surge* is discussed. *Mud pulse telemetry* is treated in Chapter 10, along with an introduction to alternative telemetry methods.

Similarities between the longitudinal vibrations in bars and acoustic oscillations in fluid columns are highlighted in elementary books, since similar mathematical formulations apply in many problems. However, these analogies break down whenever area or material discontinuities exist, e.g., at drillpipe and collar junctions in the case in borehole and drillpipe applications. These subtleties are considered here.

GOVERNING LAGRANGIAN EQUATIONS

In mud acoustics, two fluid-dynamical models can be formulated, each with its own strengths and areas of application. These are the *Lagrangian* model, in which momentum book-keeping follows a fixed element of mass, and the *Eulerian* model, in which we focus our attention at specific points in space.

The Lagrangian derivation is unchanged from that of Chapter 5 for drillstring axial vibrations, except to the extent that Young's modulus E is replaced by the *bulk modulus* B. Let us first repeat this derivation for a uniform cross-sectional area, using terminology appropriate to the present chapter.

We emphasize that viscous shear stresses that act on a smaller hydraulic scale can be ignored in the *propagation* formulation. The details of this flow will no doubt affect the *attenuation* model, but these effects can be introduced after-the-fact through an imaginary frequency. In modeling long wave *propagation*, only compressibility effects due to pressure need to be considered.

Hydraulic versus acoustic motion. In will be convenient to distinguish between two fluid-dynamic limits using the terms *hydraulic* and *acoustic*. By *hydraulic*, we refer to the *constant density, incompressible* component of flow generally associated with the inviscid Bernoulli equation, or alternatively, with viscous pipe flows, or with both. These flows may be steady or unsteady.

By *acoustic*, we refer to *transient, compressible* flow effects only. To explain this simply, imagine a row of *infinitely rigid* billiard balls in contact with each other. Any displacement at one end is instantly felt at the opposite end: *incompressibility* is responsible for the *infinite speed* with which information travels. If these balls were made of rubber, with each ball compressing and expanding before passing on the motion, the effect at the opposite end would require a time delay. This acoustic effect is due to compressibility.

Differential equation. As before, let $u(x,t)$ represent the instantaneous axial displacement from equilibrium conditions. The acoustic pressure in a fluid is given by the formula $p(x,t) = - Bu_x$ (this relationship is analogous to Hooke's law $\sigma = E\varepsilon = Eu_x$ from elementary strength of materials).

If the pressure $-Bu_x$ acts on the left side of our one-dimensional control mass, with $-Bu_x + (-Bu_x)_x\, dx$ acting on the right, then following Newton's law $\mathbf{F} = m\mathbf{a}$, the net force differential $Bu_{xx}\, dx$ must equal $u_{tt}\, \rho\, dx$ per on a per unit area basis. Thus, we obtain the classical wave equation for the displacement function $u_{tt} - c^2\, u_{xx} = 0$, where c is the sound speed satisfying $c^2 = B/\rho$.

As we have shown in Chapter 1, the general solution to this equation is a superposition of right and left-going waves. It is tempting to use our earlier results for axial vibrations without modification, on a one-to-one basis, and in many applications for uniform waveguides with unchanging properties in the axial direction, this may be possible and legitimate.

Area and material discontinuities. The analogies, however, break down whenever cross-sectional changes in area or material are found along the one-dimensional system; in drillstrings, this discontinuity always exists, e.g., at the pipe and collar interface. That the correspondence between axial drillstring vibrations and mud column acoustics does not apply to problems with area changes will *not* be apparent from differential equation derivations alone.

To understand why, we must appreciate the fact that one-dimensional wave models really represent approximations to 3D physical phenomena. If we recognize that information is always lost in any averaging process, and that the reduction in the number of spatial dimensions is tantamount to averaging, it is clear that one-dimensional models lead to physical compromises. Lost information, at best, must be re-introduced using auxiliary matching conditions.

Let us consider an example from strength of materials. In stress analysis, a bar with an area discontinuity is associated with a jump in the stress σ. This is so because the force through the transition point does not change; hence, $\sigma^{(p)}A^{(p)} = \sigma^{(c)}A^{(c)}$, assuming that external forces do not act at the point, where the "p" and "c" superscripts denote pipe and collar. In short, we *choose* force continuity in the overall picture, so that stress differs on either side of the jump.

Thus we assumed, following this line of reasoning, that $A^{(p)}E^{(p)}\ \partial u^{(p)}/\partial x = A^{(c)}E^{(c)}\ \partial u^{(c)}/\partial x$ in Equation 5-101. In addition, the displacement function itself was assumed to remain unchanged through the transition point, since *tearing* is disallowed, and we are led to $u^{(p)} = u^{(c)}$ in Equation 5-102.

These two conditions, together, completely defined the pipe-to-collar matching conditions for axial drillstring vibrations, to the extent allowed in one-dimensional wave modeling. These led to the difference approximation $A^{(p)}E^{(p)}\ (u_{i*+1,n} - u_{i*+,n})/\Delta x = A^{(c)}E^{(c)}\ (u_{i*-,n} - u_{i*-1,n})/\Delta x$ derived earlier in Equation 5-103, which supplements the use of separate differential equations for the pipe and collar.

In mud column acoustics, we recognize that fluids physically do *not* support internal stress discontinuities. Thus, acoustic pressure is *assumed* to be continuous through the area discontinuity. Since $p = -B\ \partial u/\partial x$, and B is constant throughout the entire system, we obtain $\partial u^{(p)}/\partial x = \partial u^{(c)}/\partial x$. Continuity of *volume velocity* $A^{(p)}\partial u^{(p)}/\partial t = A^{(c)}\partial u^{(c)}/\partial t$ furnishes another matching condition. These conditions are mathematically different from those for axial vibrations.

It is not obvious that acoustic pressure must be constant through an area change. In *constant density, incompressible flow*, pressures *do* change. The exact level is determined by mass continuity, Bernoulli's equation, and the area ratio; differences between "pressure x area" from one side to the other are assumed to be balanced by stresses acting on the metal structure.

That (acoustic) pressure continuity applies turns out to be realized by long waves only, following rigorous three-dimensional arguments that are beyond the scope of this book. Here, we will summarize the relevant results only. Consider a duct with a sudden areal change, that is, a short transition zone separating two regions, each of which support long plane waves.

In general, the acoustic pressures in both regions differ due to *evanescent modes* (see Chapter 3) excited at the transition, but this difference is small. For circular ducts with radii $r_1 >> r_2$, this difference is not significant unless the frequency $f = \omega/2\pi$ is comparable to or larger than the critical value $f_{crit} = (cr_2{}^2/r_1{}^3)/2\pi$.

Consider $r_2 = 0.25$ ft, $r_1 = 0.5$ ft, and $c_{water} = 4500$ ft/sec as representative *order-of-magnitude* numbers; for this example, we have a high $f_{crit} = 358$ Hz. Thus, for slow pipe movements in swab-surge applications, or for typical MWD mud pulse transmission data rates, the approximation $p_1 \approx p_2$ suffices at area discontinuities. Detailed mathematical discussions are offered in Miles (1944), Miles (1946), Morse and Ingard (1968), and Pierce (1981).

Nor is continuity of *volume* velocity (that is, $v = \partial u/\partial t$), assumed in acoustic textbooks, obvious. Certainly a case can be made for continuity of *mass* flow velocity, and in compressible aerodynamic applications, this does lead to improved agreement with measured results (Chin, 1977). At any rate, we will abide by the conventional acoustic matching conditions in this book.

Unlike our formulation for axial vibrations, where two wave equations are generally needed to characterize different material properties in the drillpipe and drill collar, we emphasize that a single partial differential equation suffices because the same mud is present everywhere. Also, all of our mud acoustic observations and derivations apply to swab-surge problems in the annulus and MWD applications in the drillpipe.

Mud acoustic formulation. In summary, we state the following long wave results for acoustic wave motions in one-dimensional swab-surge and MWD applications,

$$u_{tt} - c^2 u_{xx} = 0 \tag{9.1}$$

$$c^2 = B/\rho \tag{9.2}$$

$$p = -Bu_x \tag{9.3}$$

$$u_x{}^{(p)} = u_x{}^{(c)} \tag{9.4}$$

$$A^{(p)} u_t{}^{(p)} = A^{(c)} u_t{}^{(c)} \tag{9.5}$$

where damping (for now) is ignored. If, in swab-surge problems, the drillbit is off-bottom, a third matching condition $A^{(c)} u_t^{(c)} = A^{(h)} u_t^{(h)}$ would apply at the collar and borehole interface, where "h" refers to the hole radius.

A complete boundary value problem formulation would consist of Equations 9-1 to 9-5, plus boundary conditions at the lower drillbit or mud motor end $x = 0$, boundary conditions at the surface $x = L$, initial conditions at time $t = 0$, *plus* a suitable model for the excitation source. Before dealing with modern applications, we turn to some well known classical results.

Example 9-1: Idealized reflections and transmissions. The complete wave field solving the general formulation can be complicated analytically, and voluminous, if tackled numerically. For these reasons, elementary books often study the problem in the simpler limit where multiple reflections do not appear. We examine these simpler problems because they help us formulate matching conditions at impedance junctions for use in more complicated problems.

Consider two semi-infinite pipe Sections "1" and "2" joined at $x = 0$, having areas A_1 and A_2, containing a fluid with bulk modulus B, density ρ, and sound speed c. No restrictions on the relative sizes of A_1 and A_2 are required below, and the physical problem is sketched in Figure 9-1.

```
---------------+
   Section 1   |
               +--------------
       i --->  x=0  t --->
       <--- r       Section 2
               +--------------
               |
---------------+
```

Figure 9-1. Wave transition at area change.

In Figure 9-1, a plane wave is incident from the left; it approaches the junction $x = 0$, where the incoming wave will partially transmit and partially reflect. Note that $x = 0$ does *not* represent the drill bit in the present analysis (this zero value is taken in order to simplify required algebraic manipulations). Thus, in Section 1, the pressure and displacement functions take the form

$$p_1(x,t) = p_i e^{i(\omega t - kx)} + p_r e^{i(\omega t + kx)} \tag{9.6}$$

$$v_1(x,t) = v_i e^{i(\omega t - kx)} + v_r e^{i(\omega t + kx)} \tag{9.7}$$

whereas in Section 2, the only family of waves that exists is the right-going one given by

$$p_2(x,t) = p_t e^{i(\omega t-kx)} \tag{9.8}$$

$$v_2(x,t) = v_t e^{i(\omega t-kx)} \tag{9.9}$$

In these equations, the subscripts i, r, and t refer to *incident*, *reflected*, and *transmitted*, respectively. At the junction $x = 0$, the acoustic pressure and the volume velocity must be continuous. If we substitute Equations 9-6 to 9-9 in our matching conditions, set $x = 0$ and cancel the $e^{i\omega t}$'s, we obtain

$$p_i + p_r = p_t \tag{9.10}$$

$$A_1(v_i + v_r) = A_2 v_t \tag{9.11}$$

Now, let us divide the first equation by the second, so that

$$(p_i + p_r)/(v_i + v_r) = (A_1/A_2)(p_t/v_t) \tag{9.12}$$

For a right-going wave, the displacement function takes the form $u(x,t) = f(x-ct)$, where f may be arbitrary. Since $p = -B\, u_x = -Bf'$ and $v = u_t = -c\, f'$, where primes denote differentiation with respect to the argument, it follows upon division that $p/v = B/c$, or $p/v = B/c = \rho c$ since $c^2 = B/\rho$. This leads to

$$p = \rho v c \tag{9.13}$$

Now consider the analogous derivation for left-going waves. We replace $-c$ above by $+c$, for which $p = -Bf'$ and $v = cf'$. This leads to $p/v = -B/c$, so that $p = -\rho v c$. If we substitute

$$v_i = p_i/\rho c \tag{9.14a}$$

$$v_r = -p_r/\rho c \tag{9.14b}$$

and

$$v_t = p_t/\rho c \tag{9.15}$$

in Equation 9-12, we have

$$p_r/p_i = (A_1 - A_2)/(A_1 + A_2) \tag{9.16}$$

Thus, in the limit of equal areas when $A_1 = A_2$, the magnitude of the reflected pressure wave vanishes. If the areas are unequal, p_r is completely determined, since the incident pressure p_i and both cross-sectional areas are known. Then, the magnitude of the transmitted pressure wave can be obtained from Equation 9-10. In fact, it is easily seen that $p_t = p_i + p_r = p_i(1 + p_r/p_i)$ or

$$p_t/p_i = 1 + (A_1 - A_2)/(A_1 + A_2) = 2A_1/(A_1 + A_2) \tag{9.17}$$

Example 9-2: Classical water hammer. Equation 9-13 is the well known formula for one-dimensional *water hammer*, and it can be interpreted as follows. If a piston mounted at the end of a semi-infinite pipe containing a stationary fluid with density ρ and speed of sound c is struck with a velocity v, a propagating pressure pulse with magnitude $p = \rho vc$ is created.

The fluid, being compressible, does not move as a rigid column, which would require infinite energy. Rather, it remains undisturbed except during the passage of the wave, when local fluid particles are momentarily displaced a distance u with the speed $v = \partial u/\partial t$. Pressure, fluid velocity, and particle displacement all propagate together at the speed of sound c (note $c >> v$!).

Example 9-3: Acoustic pipe resonances. Again, recall that the displacement u(x,t) satisfies $\partial^2 u/\partial t^2 - c^2\, \partial^2 u/\partial x^2 = 0$, where $p(x,t) = - B\, \partial u/\partial x$ is the acoustic pressure, and $c^2 = B/\rho$. As in Chapter 1, we take

$$u(x,t) = X(x)T(t) \tag{9.18}$$

This separation of variables leads to

$$T''(t)/c^2T(t) = X''(x)/X(x) = - p^2 < 0 \tag{9.19}$$

and the solution $X(x) = A \sin px + B \cos px$. The function X(x) is proportional to the fluid displacement u(x,t), and it vanishes at closed ends. On the other hand, note that

$$p(x,t) = - BX'(x)T(t) \tag{9.20}$$

Thus, p(x,t) is proportional to the spatial derivative X'(x). Our finite system is defined on $0 < x < L$.

Closed-closed ends. The condition $X(0) = 0$ leads to $B = 0$, while $X(L) = A \sin pL = 0$ requires that $pL = n\pi$, where n = 1, 2, 3, ... Thus, the displacement mode shapes are

$$X(x) = A \sin n\pi x/L \tag{9.21}$$

with $T(t) = C \sin n\pi ct/L + D \cos n\pi ct/L$, or

$$T(t) = C \sin \omega t + D \cos \omega t \tag{9.22}$$

where

$$\omega = n\pi c/L \tag{9.23a}$$

$$f = \omega/2\pi = nc/2L \tag{9.23b}$$

with the frequency f being measured in Hertz.

Open-open ends. At open ends in the long wave limit, the acoustic pressure is approximately zero (Morse and Ingard, 1968; Pierce, 1981). Thus, we consider the derivative $X'(x) = Ap \cos px - Bp \sin px$. The condition X'(0)

= 0 implies A = 0, while X'(L) = -Bp sin pL = 0 requires pL = nπ, n = 0, 1, 2, 3, ... and so on. Hence, the mode shapes are

$$X(x) = B \cos n\pi x/L \tag{9.24}$$

with T(t) = C sin nπct/L + D cos nπct/L, or

$$T(t) = C \sin \omega t + D \cos \omega t \tag{9.25}$$

where

$$\omega = n\pi c/L \tag{9.26a}$$

$$f = \omega/2\pi = nc/2L \tag{9.26b}$$

We observe that the eigenfunctions for closed-closed and open-open problems obviously differ, but that the eigenvalues or natural frequencies are identical.

Closed-open ends. Here, X(0) = 0 requires B = 0, so that the function X(x) = A sin px. Then, X'(L) = Ap cos pL = 0 requires that, pL = (2n-1)π/2, with n = 1, 2, 3, ... or p = (2n-1)π/2L. The mode shapes are

$$X(x) = A \sin\ (2n-1)\pi x/2L \tag{9.27}$$

with T(t) = C sin (2n-1)πct/2L + D cos (2n-1)πct/2L, or

$$T(t) = C \sin \omega t + D \cos \omega t \tag{9.28}$$

where

$$\omega = (2n-1)\pi c/2L \tag{9.29a}$$

$$f = \omega/2\pi = (2n-1)c/4L \tag{9.29b}$$

In these solutions, we gave the eigenfunction and eigenvalues only, which are determined from boundary conditions. The full solution, to include initial conditions, can be obtained using the Fourier integral methods fully illustrated in Chapter 1.

We stress that our solutions apply to highly deviated drillstrings and boreholes, despite the appearance of a single "x" in $\partial^2u/\partial t^2 - c^2\ \partial^2u/\partial x^2 = 0$. In swab-surge and MWD applications, the wavelength $\lambda = c/f \approx$ (5,000 ft/sec)/(10 Hz) ≈ 500 ft far exceeds most bend radii (at worst, ≈ 1 ft). Thus, reflections are not incurred at bends, and one-dimensional theory applies. For that matter, our results will apply to coiled tubing wound about a spool of small radius.

Example 9-4: Passage through area obstructions. As indicated in the above paragraph, one-dimensional theory applies to highly deviated geometries, and this is true for problems with multiple area discontinuities as well, where partial transmission and reflection at transition points exist.

```
                    +---------------
                    |
          +---------+
          |
 ---------+
    A1          A2              A3
 ---------+
          |<----L---->
          +----------+
                     |
                     +---------------
```

Figure 9-2. Transmission through three-pipe system.

Consider two semi-infinite Sections "1" and "3" bounding a finite duct of length L (no restrictions are placed on relative areas). An incident wave from the left will partially reflect and transmit at the 1-2 interface; the transmitted wave acts similarly at the 2-3 interface, but a single wave emerges at the right.

The problem of wave transmission through Section "2" is relevant to MWD mud pulse telemetry. An acoustic source is located in this section, which generates waves that propagate uphole and downhole. The latter reflect at the bit and head up towards Section "2" once more. How much of this wave passes and adds to new waves that are progressing uphole?

Transmission efficiency is measured by the *power transmission coefficient* P_t, formed by the ratio of the transmitted power to the incident power. Some algebra shows that, in general, P_t satisfies

$$P_t = 4/\{(A_3/A_1 + 1)^2 \cos^2 kL + (A_2/A_1 + A_3/A_2)^2 \sin^2 kL\} \tag{9.30a}$$

Let us consider an MWD application, and set $A_1 = A_3 = A_c$ and $A_2 = A_{mwd}$, where A_c is the cross-sectional area of the drill collar containing a conventional area-blocking MWD valve, and A_{mwd} is the restricted area at that point. In this limit, we have

$$P_t = 4/\{4 \cos^2 kL + (A_{mwd}/A_c + A_c/A_{mwd})^2 \sin^2 kL\} \tag{9.30b}$$

For any *fixed* value A_{mwd}/A_c, it is clear that in the limit as $kL \rightarrow 0$, complete wave passage is achieved with $P_t \rightarrow 1$. In other words, MWD generated signals and all reverberant signals reflecting at the drill bit and pipe-collar interfaces pass straight through the valve unimpeded. In practice, we might that $k = 2\pi/\lambda \approx 6.28/500$ ft, $L \approx 1$ ft, so that $kL \approx 0.01$.

Example 9-5: Transmission through contrasting media. Let us now refer to Figure 9-1 again, this time allowing *different* materials in Sections 1 and 2. Here, the matching of pressures and volume velocities leads to

$$P_t = 4A_1\rho_1c_1A_2\rho_2c_2/\{A_1\rho_2c_2 + A_2\rho_1c_1\}^2 \tag{9.30c}$$

We emphasize that our matching conditions for continuous pressure and volume velocity are not as general as they appear, e.g., they do not allow nonequilibrium elastic delays that may occur at unusual interfaces (e.g., refer to our discussion in Chapter 1 on elastic boundary conditions). Such problems may arise in transducer design when the pressure elements are isolated from the propagation medium by rubber diaphragm partitions that separate other fluids and gases having different properties.

Bearing in mind such possible limitations, the foregoing formula indicates the possibility of 100% power transmission *if* $A_1\rho_2c_2 = A_2\rho_1c_1$ *even if* $A_1 \neq A_2$. However, if $A_1 = A_2$, then $P_t = 4\rho_1c_1\rho_2c_2/\{\rho_2c_2 + \rho_1c_1\}^2$.

The so-called *acoustic impedance* is simply the product "ρc" between fluid density and sound speed in a pipe of uniform cross-section. It turns out, in modeling plane wave transmission at normal incidence from one media to another, that partial reflections and transmissions are absent whenever the acoustic impedances so defined are continuous.

That is, when the value ρc remains unchanged in the direction of wave motion, complete energy transmission is assured. In Example 9-3, we implicitly assumed that the impedance contrast at a closed "fixed" end was infinite, so that perfect reflection occurs.

We have already identified two means of characterizing acoustic waves, namely, *acoustic pressure* and *acoustic impedance*. Other properties are also used in the literature. The *acoustic intensity* I of a sound wave is defined as the average power transmitted per unit area in the direction of the wave propagation, whereas the *sound energy density* is defined as the energy per unit volume. This energy is partly kinetic, due to the motion of the fluid elements, and partly potential, due to the displacement of the medium.

GOVERNING EULERIAN EQUATIONS

In our derivation above, we ignored any mean background velocities, and dealt with acoustic motions directly. This is justified for two reasons. If the sound speed c far exceeds any local flow velocity V, so that the dimensionless *Mach number* $M = V/c \ll 1$, the effects of acoustic and hydraulic flow physically decouple, and the two fields can be treated independently.

For example, the sound speed in water is approximately 4,500 ft/sec, and downhole fluid speeds are unlikely to exceed 1,000 ft/sec, anywhere. The corresponding Mach number is less than 0.25. In classical applications, e.g., high speed aerodynamics, the dynamic coupling is ignored in all applications where $M < 0.3$ (actual corrections to the *potential equation* derived later are on the order of $O(M^2)$).

This is analogous to ignoring the effects of flow on human speech (and vice-versa) on windy street corners: flow effects are insignificant, except to the extent that a simple kinematic Doppler correction may be required. Calculated pressures are linearly additive, following Equation 9-36 below.

From a different perspective, the effect of a *uniform* background flow is dynamically insignificant, since a Galilean transformation with speed V mathematically transforms $\rho\,(\partial u/\partial t + V\,\partial u/\partial x) = -\,\partial p/\partial x$ into $\rho\,\partial u/\partial t = -\,\partial p/\partial x$, which applies in the no-flow case. In other words, constant background speeds are not consequential to the dynamics.

Flow effects are negligible in sound propagation, but they will have some effect on attenuation, because the effects of background viscous stresses due to flow cannot be completely ignored. But flow effects can be very important in *sound generation*, especially in modeling the *signal creation process* at MWD pulsers. In general, the mechanism is three-dimensional: it occurs over regions very small compared to a wavelength, and it naturally involves significant *acoustic-hydraulic* flow interaction. These issues are discussed in Chapter 10.

Whereas the Lagrangian description above follows a fixed control mass as it moves in space, the Eulerian model focuses its attention at given locations in space. In MWD acoustic modeling, and in studying dynamic swab and surge, both models turn out to be important. Each model provides a convenient vehicle to model different kinds of acoustic sources.

To appreciate the differences between Lagrangian and Eulerian points of view, we must understand that Newton's "**F** = ma" law describes individual moving entities, e.g., a ping-pong ball, a *Frisbee*, or an arrow, tracked as they move. It is, therefore, a Lagrangian description or model, for which the upper-case derivative D/Dt is often used for emphasis; this is also known as the convective, material, or substantive derivative *following* the particle.

Newton's law following a fixed differential amount of control mass can be written, as we have above, in the form $(\rho A dx)\, Du/Dt = \{p(x,t)\, A\} - \{p(x,t) + \partial p/\partial x\, dx\}A$, where p is the acoustic pressure, A is the cross-sectional area, and our *italicized u* is the *fluid speed at a fixed point in space*. The momentum balance simplifies to $\rho\, Du/Dt = -\,\partial p/\partial x$.

Since $u = u(x,t)$ is a function of the two independent variables x and t, the convective derivative can be rewritten in terms of the partial time derivative

"$\partial u/\partial t$" in which space is fixed, and the partial space derivative "$\partial u/\partial x$" in which time is fixed.

Now, the total differential du, from calculus, satisfies $du = (\partial u/\partial t)\,dt + (\partial u/\partial x)\,dx$, so that $du/dt = (\partial u/\partial t) + (\partial u/\partial x)\,(dx/dt)$. Since $dx/dt = u$ following the mass, we obtain $Du/Dt = du/dt = \partial u/\partial t + u\,\partial u/\partial x$, and hence, the required *Euler* momentum equation

$$\rho\,(\partial u/\partial t + u\,\partial u/\partial x) = -\,\partial p/\partial x \qquad (9.31)$$

applicable at fixed points in space.

Note that the only external force considered in deriving Equation 9-31 is the pressure force. Thus, Euler's equation provides an *inviscid* flow model only, and does *not* consider attenuative effects due to *viscosity* or *rheology*. These may be added after-the-fact using imaginary frequency concepts as outlined in Chapters 1 and 2. On the other hand, if application lengths are short, an inviscid model may suffice. The density ρ in Equation 9-31, of course, need not be constant, and it may remain quite general.

The steady hydraulic limit. In a steady, constant density, incompressible flow, the time derivative term $\partial u/\partial t$ disappears and ρ is constant. Then, Equation 9-31 simplifies to $\rho u\,\partial u/\partial x = -\,\partial p/\partial x$, that is, $\partial\{p + 1/2\,\rho u^2\}/\partial x = 0$, so that

$$p(x) + 1/2\,\rho u^2(x) = \text{constant} = p_\infty + 1/2\,\rho u_\infty^{\,2} \qquad (9.32)$$

where we have evaluated the constant of integration at known ambient conditions far upstream (represented by infinity subscripts). Equation 9-32 is the well known *Bernoulli equation*; for further discussion, the reader is referred to elementary fluid mechanics textbooks.

The unsteady hydraulic limit. In unsteady incompressible flow, the transient term $\partial u/\partial t$ remains. When the flow is *irrotational* we can write the velocity $u(x,t)$ in terms of a potential function $\phi(x,t)$ using $u(x,t) = \partial\phi(x,t)/\partial x$ (see Equation 9-47). Since $\partial u/\partial t = \partial^2\phi/\partial t\partial x = \partial(\partial\phi/\partial t)/\partial x$, Equation 9-31 can be written as $\rho\,(\partial(\partial\phi/\partial t)/\partial x + u\,\partial u/\partial x) = -\,\partial p/\partial x$, or $\partial\{\rho\,\partial\phi/\partial t + p + 1/2\,\rho u^2\}/\partial x = 0$, so that $\rho\,\partial\phi/\partial t + p(x) + 1/2\,\rho u^2(x) = f(t)$, where $f(t)$ is an unsteady function of integration. This *unsteady Bernoulli equation* was used earlier in Chapter 3 in our study of water waves (see Equations 3-69 to 3-71), and we will again encounter it in Chapter 12.

Separating hydraulic and acoustic effects. Earlier we introduced the term *hydraulic*, which is used synonymously with *constant density* and *incompressible*. *Acoustic*, again, always refers to transient compressible fluid motions satisfying wave-like equations. These distinctions must be made, especially in the study of mud pulse telemetry, to emphasize the coexistence of two contrasting physical phenomena.

Equation 9-32 is Bernoulli's equation. It does *not* describe wave motion; because the fluid is assumed to be incompressible; disturbances in this limit travel with infinite speed. Also, it does *not* describe viscous flows; Bernoulli's equation assumes *inviscid* (that is, frictionless) flow, since viscous shear stresses are not included in the momentum balance underlying Equation 9-31.

However, Equation 9-32 can be used to relate pressures between two points "1" and "2" in a variable area duct containing viscous flow if the separation distance is small. For such applications, we have the governing equations $p_1(x) + 1/2\ \rho u_1^2(x) = p_2(x) + 1/2\ \rho u_2^2(x)$ and $A_1 u_1 = A_2 u_2$.

We all know from experience that acoustic wave motions can "ride" on hydraulic ones. For example, we can speak and be heard, whether wind is blowing or not. Very often, it is desirable to separate hydraulic from acoustic effects, since these satisfy very different sets of physical laws. This separation helps us understand and better appreciate the actions of different types of forces that can act on the surfaces of or through boreholes and drillpipes.

Now, the one-dimensional equation describing *mass conservation* takes the form

$$\partial \rho/\partial t + \partial(\rho u)/\partial x = 0 \tag{9.33}$$

Equation 9-33 indicates that time-wise increases or decreases in the mass density ρ within a control volume must be balanced by the flux of mass. This equation applies to combined hydraulic and acoustic effects.

In order to separate the two effects, we write the density, velocity, and the pressure as the sum of mean (*m*) and acoustic disturbance (*primed*) parts, namely,

$$\rho(x,t) = \rho_m(x,t) + \rho'(x,t) \tag{9.34}$$

$$u(x,t) = u_m(x,t) + u'(x,t) \tag{9.35}$$

$$p(x,t) = p_m(x,t) + p'(x,t) \tag{9.36}$$

We now assume, in an absolute value sense, that the (primed) disturbances represent small deviations from mean conditions, that is,

$$\rho_m(x,t) >> \rho'(x,t) \tag{9.37}$$

$$u_m(x,t) >> u'(x,t) \tag{9.38}$$

$$p_m(x,t) >> p'(x,t) \tag{9.39}$$

Substitution of Equations 9-37 to 9-39 in Equations 9-31 and 9-33, and subsequent algebraic expansion, lead to a system of equations amenable to simple limit processes.

Limit No. 1. If the primed disturbances vanish identically, and ρ_m is constant, the mean pressure p_m and velocity u_m satisfy Equation 9-32, that is, Bernoulli's equation.

Limit No. 2. If the mean density ρ_m is constant, and the mean speed u_m vanishes, our acoustic disturbances satisfy

$$\rho_m u'_t = - p'_x \tag{9.40}$$

and

$$\rho'_t + \rho_m u'_x = 0 \tag{9.41}$$

to leading order. To obtain Equations 9-40 and 9-41, formally small quadratic quantities in the disturbances have been dropped. Let us now introduce the *compressibility* κ defined by

$$\rho' = \kappa \rho_m p' \tag{9.42}$$

We differentiate Equation 9-40 with respect to x, and Equation 9-41 with respect to t, to find that $\rho_m u'_{tx} = -p'_{xx}$ and $\rho'_{tt} + \rho_m u'_{xt} = 0$. Then, using Equation 9-42, we have the pressure equation

$$\partial^2 p'/\partial t^2 - (B/\rho_m)\, \partial^2 p'/\partial x^2 = 0 \tag{9.43}$$

where the *bulk modulus* B and the *fluid compressibility* κ are related by

$$B = 1/\kappa \tag{9.44}$$

As in the Lagrangian description, the speed of sound c satisfies $c^2 = B/\rho_m$. Thus, Equation 9-43 takes the form $\partial^2 p'/\partial t^2 - c^2\, \partial^2 p'/\partial x^2 = 0$. We could have eliminated pressure as the working variable, and obtained the wave equation for velocity

$$\partial^2 u'/\partial t^2 - (B/\rho_m)\, \partial^2 u'/\partial x^2 = 0 \tag{9.45}$$

Some further analysis shows that the acoustic density also satisfies the wave equation, that is,

$$\partial^2 \rho'/\partial t^2 - (B/\rho_m)\, \partial^2 \rho'/\partial x^2 = 0 \tag{9.46}$$

Velocity potential. Sometimes it is convenient to introduce a *velocity potential* $\phi(x,t)$ defined by

$$u(x,t) = \partial \phi(x,t)/\partial x \tag{9.47}$$

The simplest (and perhaps most illuminating) way to derive a governing equation for $\phi(x,t)$ starts with Equation 9-1, namely $u_{tt} - c^2 u_{xx} = 0$, for the Lagrangian displacement u(x,t). Let us rewrite this equation in the form

$$\partial u_t(x,t)/\partial t = \partial\{c^2 u_x(x,t)\}/\partial x \tag{9.48}$$

The quantity $u_t(x,t)$, representing the time derivative of the Lagrangian displacement $u(x,t)$, is just the (italicized) velocity $u(x,t)$, which from Equation 9-47 is equal to $\partial\phi(x,t)/\partial x$. Thus, the left side of Equation 9-48 can be expressed as $\partial\{\partial\phi(x,t)/\partial x\}/\partial t$.

Now from calculus, orders of differentiation can be interchanged, so that this can be written as $\partial\{\partial\phi(x,t)/\partial t\}/\partial x$. Comparison with the right side of Equation 9-48 shows that $\partial\phi(x,t)/\partial t = c^2\, u_x(x,t)$. In summary, we have the equation pair

$$u_t(x,t) = \partial\phi(x,t)/\partial x \tag{9.49a}$$

$$u_x(x,t) = c^{-2}\, \partial\phi(x,t)/\partial t \tag{9.49b}$$

where the $u(x,t)$ at the left sides of Equations 9-49a and 9-49b represents Lagrangian displacement. If we now take the x derivative of Equation 9-49a, and the t derivative of Equation 9-49b, and eliminate $u_{tx}(x,t)$ (that is, $u_{xt}(x,t)$) between the two results, we obtain the governing *potential equation*

$$\partial^2\phi(x,t)/\partial t^2 - c^2\, \partial^2\phi(x,t)/\partial x^2 = 0 \tag{9.50}$$

The more traditional proof begins with Equation 9-40 rewritten in the form $\rho_m\partial^2\phi(x,t)/\partial x\partial t = -\, p'_x$. Since $u' = \partial\phi(x,t)/\partial x$, or $\partial\{p'+\rho_m\partial\phi/\partial t\}/\partial x = 0$, we find that $p' + \rho_m\partial\phi/\partial t$ = constant, where the constant arises from steady reference conditions. Now, Equations 9-41 and 9-42 (or, $\rho'_t + \rho_m u'_x = 0$ and $\rho' = \kappa\rho_m p'$) imply that $\kappa\rho_m p'_t + \rho_m\partial^2\phi(x,t)/\partial x^2 = 0$ where we have used $u' = \partial\phi(x,t)/\partial x$. From the above paragraph, $p'_t = -\rho_m\partial^2\phi/\partial t^2$, and hence, elimination of p'_t yields $\partial^2\phi/\partial t^2 - c^2\, \partial^2\phi(x,t)/\partial x^2 = 0$, noting $c^2 = B/\rho = 1/\kappa\rho_m$.

Limit No. 3. If ρ_m and u_m are both constant, then $\rho_m(\partial/\partial t + u_m\partial/\partial x)\, u' = -\, p'_x$. Let us introduce the coordinate $\xi = x - u_m t$ and the function $u'(x,t) = \boldsymbol{u}'(\xi,t)$. Then, $\partial u'/\partial t = \partial\boldsymbol{u}'/\partial t + \partial\boldsymbol{u}'/\partial\xi\ \partial\xi/\partial t = \partial\boldsymbol{u}'/\partial t - u_m\ \partial\boldsymbol{u}'/\partial\xi$ and $\partial u'/\partial x = \partial\boldsymbol{u}'/\partial\xi$ allow us to write $\rho_m\partial\boldsymbol{u}'/\partial t = -\, p'_\xi$ which is identical in form to $\rho_m u'_t = -\, p'_x$ in Equation 9-40. This proves our earlier assertion that a constant background speed is dynamically insignificant, since it can be removed by a simple Galilean transformation.

Limit No. 4. When the background mean flow varies weakly with x and t, momentum and energy transfer between wave and mean states in general occurs that cannot be ignored over large scales. This transfer is related to *Reynolds stress* effects in fluid mechanics, responsible for laminar flow instabilities; analogous terms in wave propagation are termed *radiation stresses.* These weak variations do not exist in uniform area pipes and boreholes. They do, however, in the variable mean current ocean applications treated in Chapter 12.

TRANSIENT FINITE DIFFERENCE MODELING

We now discuss finite difference modeling, using our Lagrangian formulation as the host discussion model. Since this involves the displacement function u(x,t), the derivations proceed along the lines of Chapter 5 for axial vibrations. We will replace Equation 5-25 by

$$\rho\, \partial^2 u/\partial t^2 + \zeta\, \partial u/\partial t - B\, \partial^2 u/\partial x^2 = 0 \qquad (9.51)$$

so that Equation 9-51 extends Equations 9-1 and 9-2 to include a simple dissipation model.

Basic difference model. Following steps similar to those leading to Equations 5-88 and 5-89, we derive the finite difference equation

$$U_{i-1,n} - \{2 + (\rho/B)(\Delta x/\Delta t)^2 + (\zeta/2B)((\Delta x)^2/\Delta t)\}\, U_{i,n} + U_{i+1,n}$$
$$= -(\rho/B)(\Delta x/\Delta t)^2\,(2U_{i,n-1} - U_{i,n-2}) - (\zeta/2B)((\Delta x)^2/\Delta t)\, U_{i,n-2} \qquad (9.52)$$

which applies to each of the *internal* nodes $i = 2$ to $i = i_{max} - 1$.

Boundary conditions relating $U_{1,n}$ to $U_{2,n}$ at $x = 0$, and $U_{imax-1,n}$ to $U_{imax,n}$ at $x = L$, complete the system of *tridiagonal* equations. For example, if we are modeling acoustic wave propagation in a borehole, an infinitely rigid bottom can be specified by setting $U_{1,n} = 0$. If far uphole, the annular fluid discharges in a continuous stream into a large tank, the acoustic pressure approximation $p = -B\, \partial u/\partial x \approx 0$ implies that $U_{imax-1,n} - U_{imax,n} = 0$.

More complicated models are easily created. If the pressure uphole near the surface is known with certainty (say, $p_s(t)$), then $p = -B\, \partial u/\partial x$ implies that $p_s(t) = -B(U_{imax,n} - U_{imax-1,n})/\Delta x$ or $U_{imax-1,n} - U_{imax,n} = (p_s(t)\Delta x)/B$. On the other hand, if the uphole conduit is extremely long, and only an outgoing wave *radiation condition* is desired, the fact that u(x,t) must be a function of "x-ct" only faraway means that $\partial u/\partial t + c\, \partial u/\partial x = 0$ at some i = imax, so that we have $(U_{imax,n} - U_{imax,n-1})/\Delta t + c\,(U_{imax,n} - U_{imax-1,n})/\Delta x = 0$.

The finite difference recipes given in Chapter 5, therefore, apply with little modification and are just as easily implemented. The greatest subtleties, as indicated early on in this chapter, are posed by the existence of area discontinuities in the annular wave propagation problems dealing with swab and surge, and in drillpipe acoustics models for MWD mud pulse telemetry. We will discuss the required matching conditions, which we emphasize, are different from those of Chapter 5.

Modeling area discontinuities. Equation 9-52 applies to muds within uniform fluid columns in uniform pipes; also, it applies *within* each section of a drillpipe containing multiple area changes, with the *same* constants ρ, B and ζ because the *same* mud is present throughout. This contrasts with axial vibrations, where we required different partial differential equations for the pipe and drill collar, since the elastic properties in each system can be different.

At an area discontinuity, differential equations do break down and fail to apply, since sudden changes imply rapid wave variations for which the requisite derivatives may not exist. There, they are replaced by matching conditions that analytically continue one solution into the second, while conserving the physically correct quantities.

Axial vibrations. For such problems, *force must be continuous* from one side to the next since external loads do not exist. Since the areas change through the interface, *stress must be discontinuous.* Force continuity can be expressed via Hooke's law "$\sigma = E\varepsilon$" as $A^{(p)}E^{(p)}\, \partial u^{(p)}/\partial x = A^{(c)}E^{(c)}\, \partial u^{(c)}/\partial x$, as shown in Equation 5-101. If we designate the index of the interface location by i*, we can write $A^{(p)}E^{(p)}\,(u_{i^*+1,n} - u_{i^*+,n})/\Delta x = A^{(c)}E^{(c)}\,(u_{i^*-,n} - u_{i^*-1,n})/\Delta x$, where i*+ and i*- refer to positions infinitesimally close to i*.

But since continuity of displacement requires $u^{(p)} = u^{(c)}$, it follows that $A^{(c)}E^{(c)}\, u_{i^*-1,n} - (A^{(p)}E^{(p)} + A^{(c)}E^{(c)})\, u_{i^*,n} + A^{(p)}E^{(p)}\, u_{i^*+1,n} = 0$, as we had obtained in Equation 5-104. In the final recipe, the tridiagonal equations corresponding to all internal nodes are written, but that associated with the pipe-collar location is replaced by the matching condition just given.

Mud acoustics. Here, the matching condition is different, and alternate finite difference results are obtained. Assumed *pressure and volume velocity continuity* imply that $\partial u^{(p)}/\partial x = \partial u^{(c)}/\partial x$ and $A^{(p)}\, \partial u^{(p)}/\partial t = A^{(c)}\, \partial u^{(c)}/\partial t$ (our arguments apply within drillstrings for MWD applications, as well as in borehole annuli for swab-surge modeling). In difference equation form,

$$(u_{i^*+1,n} - u_{i^*+,n})/\Delta x = (u_{i^*-,n} - u_{i^*-1,n})/\Delta x \tag{9.53}$$

$$A^{(p)}\,(u_{i^*+,n+1} - u_{i^*+,n})/\Delta t = A^{(c)}\,(u_{i^*-,n+1} - u_{i^*-,n})/\Delta t \tag{9.54}$$

The second equation indicates that

$$A^{(p)}\, u_{i^*+,n} = A^{(c)}\, u_{i^*-,n} \tag{9.55}$$

in general. Therefore, since $u_{i^*+,n}$ and $u_{i^*-,n}$ are *not* equal as a result of Equation 9-55, the dependent variable $u_{i^*,n}$ must be *double-valued* at the interface. Since we prefer to deal with a single unknown at i = i* in our matrix of algebraic equations, we *define* (without any loss of generality, as we will show) that unknown to be the arithmetic mean value

$$u^m{}_n = (u_{i^*-,n} + u_{i^*+,n})/2 \tag{9.56}$$

so that the pressure matching condition given in Equation 9-53 can be rewritten in the form

$$u_{i^*-1,n} - 2u^m{}_n + u_{i^*+1,n} = 0 \quad (9.57)$$

Equation 9-57 is *completely different* from the matching condition

$$A^{(c)}E^{(c)}\, u_{i^*-1,n} - (A^{(p)}E^{(p)} + A^{(c)}E^{(c)})\, u_{i^*,n} + A^{(p)}E^{(p)}\, u_{i^*+1,n} = 0 \quad (9.58)$$

obtained for axial vibrations. In either case, the matching condition, like the difference approximation to the wave equation, was designed to be *diagonally dominant* and hence numerically stable.

A consequence of our definition and use of a mean displacement is this: the two difference equations at the neighboring sides of the interface, corresponding to the indexes i = i* + 1 and i = i*-1, must be rewritten in terms of u^m.

For example, the u difference equation at i = i*+1 involves the value of u at i = i*+. Since $u_{i^*+,n} = \{A^{(c)}/A^{(p)}\}u_{i^*-,n}$, we find that $u^m{}_n = (u_{i^*-,n} + u_{i^*+,n})/2 = 1/2\ \{1 + A^{(c)}/A^{(p)}\}u_{i^*-,n}$, or

$$u_{i^*-,n} = 2u^m{}_n/\{1 + A^{(c)}/A^{(p)}\} \quad (9.59)$$

Similarly, the u difference equation at i = i*-1 involves the value of u at i = i*-. Since $\{A^{(p)}/A^{(c)}\}u_{i^*+,n} = u_{i^*-,n}$, we determine that $u^m{}_n = (u_{i^*-,n} + u_{i^*+,n})/2 = 1/2\ \{A^{(p)}/A^{(c)} + 1\}u_{i^*+,n}$. Thus,

$$u_{i^*+,n} = 2u^m{}_n/\{A^{(p)}/A^{(c)} + 1\} \quad (9.60)$$

Therefore, the complete set of numerical unknowns, namely, u_1, u_2, u_3, ..., u_{i^*-1}, u_m, u_{i^*+1}, ... , u_{imax-1}, u_{imax-1}, u_{imax}, can be determined from the solution of computationally stable tridiagonal equations.

The procedure described above is taken, incidentally, from a standard aerospace algorithm for *multivalued velocity potentials* modeling *lifting* wing flows. There, the *elliptic* governing equation is not wave-like, but similar tridiagonal equation algorithms are used to iteratively converge the problem in a *time-like* manner. For more details about solutions to elliptic and (time-dependent) parabolic boundary value problems, refer to the author's textbook *Modern Reservoir Flow and Well Transient Analysis* (Chin, 1993).

SWAB-SURGE MODELING

In rotary drilling, *blow-outs* often occur when drillpipe is being *withdrawn* from the hole, despite the fact that while drilling, the muds used contain sufficient hydrostatic head for well control. Reductions in pressure evidently can occur while making a *trip*, which may be greater than the amount by which the hydrostatic pressure exceeds formation pressure.

In the literature, low tripping speeds are recommended. Burkhardt (1961), Fontenot and Clark (1973), and Clark and Fontenot (1974) discuss operational issues and solutions using *steady flow* methods. Modeling parameters include borehole geometry, mud properties, and return area about drill bits, the objective being the prediction of *safe tripping speeds*.

Recent work shows that traditional steady flow models alone may not suffice. Lal (1983) reviews earlier swab-surge models and proposes a *dynamic* acoustical model for the transient compressible flow, which is solved using the *method of characteristics*. Mitchell (1988) also uses a characteristics model, accounting for pipe elasticity, improved turbulent flow friction factors, and variable fluid and formation properties, and gives comparisons with field data.

Wave physics of swab-surge. In order to understand the wave nature of *dynamic swab-surge*, recall from Example 1-1 that stress (or pressure) waves reflect with the same sign at solid boundaries such as wellbore bottoms.

Let us refer to Figure 9-3, and imagine that the drillstring is suddenly lifted upwards, assuming that the mud moves from rest. Immediately uphole of the bit, the fluid is compressed because the bit moves *into* the mud, and an over-pressure (relative to ambient conditions) is transmitted up the annulus.

Just beneath the bit, a low pressure zone is created, because the bit pulls *away* from the drilling mud. This under-pressure propagates downward to the bottom of the rigid hole, where it must reflect with the same negative sign according to Example 1-1.

This low pressure wave travels up the annulus *through* the small bit bypass area (see Example 9-4), where it adds to up-going high pressure wave created after the initial instant. Thus, there is some neutralization of the upgoing signal, at least after the initial instant.

The physics at the bottom of the hole is much more interesting. From Chapter 1, we found that the magnitude of any pressure wave doubles at a rigid termination in a 1D waveguide. Thus, the under-pressure traveling toward the hole bottom from the bit is twice as low there as otherwise. This underbalance means that influx of higher pressure formation fluids is possible, and thus increases the probability of operational hazards.

In general, pulling and running pipes in boreholes induce pressure surges that can cause lost circulation, fractures, and formation fluid influx, resulting in gas kicks and blowouts. Accurate prediction methods are required, especially in wells where pressures must be maintained within narrow limits. The stable and robust difference models given in this book permit the development of field operable software algorithms requiring minimal computing resources.

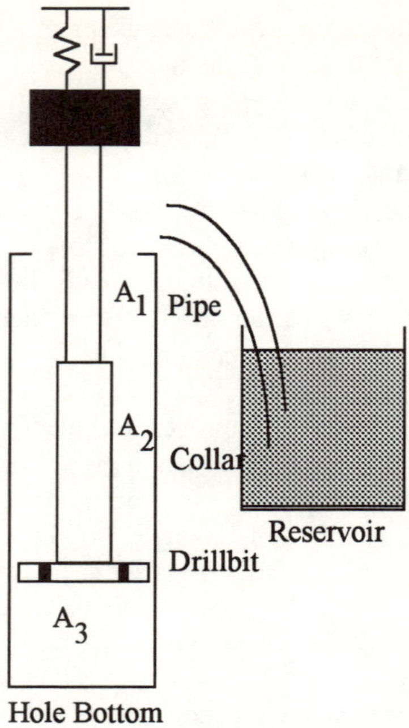

Figure 9-3. Drillstring in borehole.

Designing a swab-surge simulator. The finite difference equations, and their implementation and inversion using TRIDI, were discussed in Chapter 5. Earlier in this chapter, we reviewed several applicable uphole and downhole boundary conditions for swab-surge applications. Here, we summarize the steps necessary in designing a dynamic simulator.

o At the hole bottom $x = 0$, the fluid displacement $u(0,t) = 0$, if the formation is assumed to be rigid, or if the impedance contrast is infinite.

o At the A_1-A_2 and A_2-A_3 area change interfaces, area matching boundary conditions apply.

o Although the bit is moving relative to the borehole, its speed v is small compared to the sound speed c. Computationally, it may be "frozen" during a simulation; its position is updated by moving to the next closest grid block only as needed.

o At the surface, various models are possible. A *capped model* takes $u = 0$, while an *open reservoir model* (see Example 9-3) requires $u_x = 0$. If the drillpipe is extremely long, or if damping is sufficiently high that surface reflection is negligible, or both, the radiation condition $u \sim f(x-ct)$ can be used. Since $u_t+cu_x = -cf' + cf' = 0$, we enforce an outgoing wave using the statement $(u_{imax,n} - u_{imax,n-1})/\Delta t + c\,(u_{imax,n} - u_{imax-1,n})/\Delta x = 0$.

o *At the source*, or *bit*, a "delta-p" (i.e., a jump or discontinuity in pressure) exists. Since $p = - Bu_x$ in general, specifying a jump in p, following Equation 5-95, amounts to prescribing the quantity $(u_x^+ - u_x^-)$, that is, $(U_{ibit+1,n} - U_{ibit+,n})/\Delta x - (U_{ibit-,n} - U_{ibit-1,n})/\Delta x$. Of course, $U_{ibit-,n}$ and $U_{ibit+,n}$ are not equal by virtue of analogous consequences to our Equation 9-55; a mean dependent variable must be used along the lines indicated in Equations 9-56 to 9-60. This logic ensures that the bit, with its bypass area, is transparent to external propagating waves. The strength of the source can be modeled simply. We showed in Example 9-2 that $p = \rho vc$ for a one-dimensional piston acting on a column of fluid. This result applies to source modeling, where in the present context, v represents the pipe speed. The theoretical maximum for delta-p is just the difference at upper and lower sides of the bit, that it, $\rho vc - (-\rho vc)$ or $2\rho vc$. Deviations from this value are due to bypass flow around the bit and through its nozzles. Finally, observe that the delta-p logic and area change logic are implemented near the same spatial index; some experimentation will be required to ensure that all of the relevant physics is captured.

Other research issues. Once the reader familiarizes himself with the wave equation algorithm presented in this book, the details highlighted in various of its applications can be implemented straightforwardly. A swab-surge program constructed in this manner might serve as a test platform for numerical experimentation. Useful computer codes can be devised to solve important field problems, e.g., swab-surge effects induced by bit-bounce, or the effect of more realistic attenuation models on dynamic pressure fields.

STEADY STATE ANNULAR FLOW MODELING

So far, we have ignored the *rheology* of the borehole fluid, concentrating instead on the kinematic modeling of wave propagation; the kinematics is controlled by the geometry of the borehole annulus, and the sound speed of the mud, which depends on bulk modulus and density only.

In reality, two other factors enter: wave attenuation and pressure drops due to steady state flow losses. Attenuation can be introduced after-the-fact, as described in Chapter 1, using imaginary frequency methods that essentially impose weak exponential timewise damping. Attenuation coefficients are conveniently obtained by laboratory acoustic measurement.

Pressure losses due to steady state flow, as various authors have indicated, may be small by comparison to dynamic pressures. Thus, the need to model these accurately is lessened. In modern applications, e.g., eccentric annular flow, slimhole applications, and so on, there is heightened concern about large steady pressure drops, especially because losses for *non-Newtonian* drilling mud (until recently) cannot be accurately predicted.

While this chapter deals principally with *wave propagation* in boreholes, it is in keeping with our field objectives to at least discuss pressure losses due to nonlinear steady-state mechanisms. This is especially so because accurate annular flow modeling capabilities are now available to the industry.

In *Newtonian* fluids such as water and air, stresses are proportional to the rate of strain, and the viscosity is a *constant* independent of the flow rate and the geometry of the holding vessel. This is untrue of *non-Newtonian* fluids, which constitute the great majority of muds; for such fluids, both factors strongly affect viscosity, its variation through the cross-section, and hence pressure losses.

In wave analyses for *Newtonian* fluids, analytical and experimental attenuation studies demonstrate exponential decays like $e^{-x/L}$ with $L \propto cD/\nu\omega$, where D is a radius, ω is the frequency, and ν is the kinematic viscosity. Attenuation behaviour for *non-Newtonian* drilling fluids is often required, and the temptation to apply this exponential decay formula, using viscosity estimates based on surface viscometer measurements, exists. This turns out to be completely inadequate, since actual viscosity values downhole are typically orders of magnitude higher.

For steady state flows, pressure drops through the eccentric boreholes typical of horizontal wells strongly depend on the details of the eccentricity and the mud type. For the same non-Newtonian fluid, order-of-magnitude differences from concentric borehole (or viscometer) values are not atypical.

The book *Borehole Flow Modeling in Horizontal, Deviated, and Vertical Wells* and its accompanying PC-based software (Chin, 1992a,b) for annular flow modeling and field application gives computed "snapshots" showing that actual downhole viscosities can be substantially different from those obtained as surface viscometer readings. Some of these results are reproduced below.

Background. In several early articles published in *Offshore Magazine* in 1991-1992, this author considered general eccentric annular flows of non-Newtonian drilling fluids and cements. Well known practical problems associated with eccentric annuli arise in drilling and completing horizontal wells, and in many aspects of production operations.

A mathematical model was developed to simulate Newtonian, power law, Bingham plastic and Herschel-Bulkley fluids in highly eccentric annular spaces, which may include cuttings beds, severe washouts, significant swelling, square drill collars, and stabilizers.

Exact results for annular velocity, viscous stress, apparent viscosity, and shear rate cross-sectional distributions were calculated and plotted directly over the annular geometry itself to facilitate visual correlation. High resolution finite difference simulations using boundary-conforming grids, incidentally, required about thirty seconds per computation on 386 class machines.

These papers identified *average viscous stress*, defined as the areal mean taken over the lower half annulus, as the relevant correlation parameter for hole cleaning efficiency. Similarly, low *average apparent viscosity* was identified as the parameter that measures the effectiveness of spotting additives used for freeing stuck pipe.

Agreement with hole cleaning and stuck pipe results obtained in the Gulf of Mexico was excellent. Also, the work showed how over one hundred cuttings concentration data points collected at the University of Tulsa fell on linear plots when the correct correlation parameter was used.

The original work assumed an annular flow that was unidirectional and homogeneous, moving in the direction of pressure decrease, parallel to the hole axis. The pressure field was taken to be uniform across the annular cross-section, since numerous flows do behave in this manner.

This last assumption was later removed, to model the *recirculating* annular flows observed in deviated wells with significant density segregation, say due to barite sag. The nondimensional parameter controlling *vortex bubble formation* was identified and the appropriate equations were formulated and solved.

Bubble formation impedes cuttings transport during drilling, and induces flow channeling during cementing due to improper mud displacement. The computations performed agreed well with movies taken at M-I Drilling Fluids' flow loop facility in Houston. This work, which includes recipes for damage avoidance, is described in Chin (1992a,b), and is briefly summarized later following a detailed synopsis on unidirectional flows.

Unidirectional flows. The early model for general, eccentric borehole annular flow supposes that pressure, constant across the annulus, varies only in the streamwise direction. The analysis enforces "no-slip" conditions at all solid boundaries; the finite difference method, which does *not* invoke limiting "slotted flow" or "nearly concentric" approximations, uses "boundary conforming grids" to provide detailed flow resolution in tight spaces.

Again, each simulation requires thirty seconds on a 386 machine. A visual ASCII text output format was developed to overlay computed quantities like velocity, apparent viscosity, stress and shear rate directly on the annular geometry itself, thereby facilitating interpretation and visual correlation. The plotter and numerical algorithm were both coded in standard Fortran for complete portability.

Our model was designed for use in cuttings transport, wellbore instability, cementing, coiled tubing, and sand removal applications. It was first used to correlate the results of numerous cuttings transport experiments conducted at the University of Tulsa.

Fifteen bentonite-polymer, water-base, unweighted muds had been tested, at three different inclinations from vertical (30, 45 and 70 deg), with three average flow speeds (1.91, 2.86 and 3.82 ft/sec). With each possible angle and speed combination fixed, cuttings concentration had been plotted against mud properties like apparent viscosity, gel strength, yield point, plastic viscosity, Fann dial reading, and so on, by the original investigators.

These cross-plots, however, yielded few meaningful physical correlations. This was not surprising, of course, because *the surface rotating viscometer measurements made by the experimenters bore little significance to the actual eccentric annular flow downhole.*

Recognizing this, one early paper determined what correlations, if any, could be established between the empirical data and actual annular flow properties as obtained by exact calculation. Each of the 135 experimental points was simulated using the computer model. Statistical properties for axial velocity, apparent viscosity, shear rate, viscous shear stress, and Stokes product were compared with laboratory data.

The investigation showed that a well-defined, straight line correlation existed between cuttings concentration and "the mean stress". This mean is obtained by averaging calculated viscous stresses over the lower half of the annular area, where cuttings beds are known to form in directional wells. All of the experimental data was successfully correlated in this manner.

This demonstrated that erosive effects due to velocity gradients, acting through the bottom-averaged viscous shear stress, are largely responsible for hole cleaning and cuttings bed removal; this effect, incidentally, is well known in sewage pipeline design. Our correlations, to include qualitative descriptions of the physical processes involved, plus field recipes for improved hole cleaning, appear in Chin (1992a).

Similarly successful correlations were later obtained for a Conoco well in Green Canyon Block 184, also described in the book. The annular flow model correctly predicted that a special "all oil" mud would clean much more efficiently than an "invert emulsion". The "stress rule" was again validated; computed stress levels were three times as high in the all oil mud, while computed velocities increased by an insignificant 20%. Here, the muds evaluated by the computer simulation contrasted sharply in texture, composition and weight with the fluids studied in the previous analysis.

Stuck pipe due to differential pressure between the mud column and the formation often results in costly time delays. In the past, diesel oil, mineral oil, and mixtures of these with surfactants, clays and asphalts were usually spotted to facilitate the release of the drill string. However, the use of these conventional spotting fluids is now stringently controlled by government regulation; environmentally safe alternatives must be found.

Recently, the author examined the mechanical properties of what appears to be a successful, non-toxic, water-based system using the annular model. The new spotting fluid, identified using simple laboratory screening procedures, was used to free a thousand feet of stuck pipe in a 39^{0} hole, from a sand section in the Gulf of Mexico.

Since water-based spotting fluids, being relatively new, have seldom been studied in the literature, it was natural to ask whether or not they really work, and if so, why. The computational work described in Chin (1992a), which compares both spotted and unspotted muds, identifies low average apparent viscosity as the likely reason the additive proved to be useful More work, however, is required to test this hypothesis.

Calculated examples. In this section, we will display annular *velocity pictures* computed for several wellbore configurations. Fluid models and annular geometries have been varied and interchanged to provide a general synopsis and "feel" for the simulation capabilities. Similar pictures are available for the corresponding shear stresses, apparent viscosities, shear rates, and dissipation functions, but these are not shown due to space limitations. Much more complete examples are given in Chin (1992a,b).

We will present our velocity results with minimal discussion, since their implications are quite apparent and obvious. Bear in mind, though, that viscous stress is more likely to be the dominant parameter in hole cleaning considerations. In Figures 9-3a to 9-3p, the numbers shown represent the oncoming annular *velocities* in units of in/sec at the respective locations; the figure numbers and their descriptive titles are given in the following table. Figure 9-4 is discussed in the next section.

Recirculating vortex flows. Figures 9-3a to 9-3p assume homogeneous, unidirectional flow. When density segregation occurs, as a result of barite sag during interrupted circulation, or drilled cuttings falling out of suspension, "bubble-like" vortex recirculation zones can form. Density segregation, of course, also affects the cement slurries used during the completion process.

These flows, which are inertia and gravity dominated, depend little on viscosity and rheology. Figure 9-4 displays calculated *streamfunction* values for one such flow, noting that lines having constant streamfunction define the *streamlines* of the flow. Here we have replaced values 103-107 by "XXX" to highlight the large recirculating zone obtained for this problem.

In Chin (1992a), we show that bubble formation depends on a single nondimensional variable constructed from two simpler ones. The first is a Froude number $U^2/gL \cos \alpha$. Here, U is the oncoming speed, L is the distance between pipe and borehole walls, g is the acceleration due to gravity, and α is the hole inclination angle, zero for horizontal wells.

The second is a relative measure of local density stratification, say $d\rho/\rho_{ref}$. For example, $d\rho$ might represent the density difference between the bottom and top of the annulus, and ρ_{ref} may be taken as a reference arithmetic average. The combined channel parameter of practical significance turns out to be **Ch** = $U^2\rho_{ref}/gLd\rho \cos \alpha$.

For large values of **Ch**, recirculation bubbles will *not* form; the streamlines of motion are essentially straight and the annular model developed in the earlier work applies. For small **Ch**'s of order unity, recirculation zones *do* form, and elongate in the streamwise dimension as **Ch** decreases. For still smaller values, solutions with wavy upstream flows are found, which may or may not be physically realistic.

To suppress recirculating flows, the requirement is simple: *avoid small values of the nondimensional* **Ch** *parameter*. Small values can result from different isolated effects. For example, **Ch** decreases as the hole becomes more horizontal, as density differences become more pronounced, or as pumping rates decrease. But none of these factors alone control the physics; it is the combination shown *taken together* that controls bubble formation and perhaps the fate of a drilling or completion program.

```
                   0       0     0
                         20
             0     20    31    20     0
             20    31    37  3731    20
         0   3137  39    39  39    31        0
         20    39    38  38  38  39    3120
    0      37    38  35  30  353538    37        0
     2031    39  302222  2222  30  3839    3120
       3739    35  12 0   0 0 0122235  3937
0          3830  12             012  3538          0
   203139    2212                 022    39373120
         383522 0                  12303538

020313739383022 0                  12303538373120    0
             22 0                  1222
     37393835   0                 0  30353937
0  2031    353012 0             0  2235      3120 0
         3938  2212 0         0  22    39
       31    3835  2212  1212  22  38    3731
    020      39  35  30  30223035    39      20 0
         3137  3938  35  35  35  39    31
         20    37  3938  38  39    37    20
        0    31    37    37  37    31        0
             20    31    31    31    20
            0      20    20    20     0
                  0             0
                          0
```

Figure 9-3a. Power law fluid, fully concentric annulus.

```
                   0       0     0
                         20
             0     20    31    20     0
             20    31    37  3731    20
         0   3137  39    39  39    31        0
         20    39    38  38  38  39    3120
    0      37    38  35  30  353538    37        0
     2031    39  302222  2222  30  3839    3120
       3738    35  12 0   0 0 0122235  3837
0          3830  12             012  3538          0
   203138    2212                 022    38373120
         383522 0                  12293538

020313638372921 0                  11293437363120    0
           2820 0                  1128
     31363633   0                 0  36373631
0  20        3119 0             0  2631        20 0
         3534  2215 6         6  22    3535
     2030    2926  15 8   8 8  15  29      3020
    0      3031  21  15  1514152121  30          0
           27    20  14  14  14  20    27
         19    10   611   6   6  1710    19
        0       0   0     0   0     0        0
```

Figure 9-3b. Power law model, concentric flow with cuttings bed.

```
                  0       0     0

            0             7           0
                    7           7
        0     6          10         6       0
                   10        10
    0     6     9  11    11  11     9     6     0
            8 10                 10     8
      5              11  11  11               5
0           9    10  10  11  10  10   9           0
        7     9  10   9   9   9  10   9     7
    4     8   9   9   7   8 7   9   9     8     4
      6   8     8   7 5   3 5   7   8   8     6
0       7   7 6 6 4 2 0   0 0 0 2 6   7     7       0
  3       6   5 3 2             0 2 5   6 6     3
      4 5   5 4 2                 0 2 4 5   5 4
0       5 4   3                     1 3 4 5       0
    2 3     4 2 0                   0 3 4     3 2
        3 3   1 0                   1   3 3 3
    0       2 2 0                   0 2 2       0
      1 2 2 2 1 0                 0 1 2 2 2 2 1
                  0             0
        0 0 1 1 1 0 0         0 0 0 1 1 1 0 0
                0 0 0 0   0 0   0 0
            0 0 0   0 0   0   0     0 0
                  0 0     0   0 0
                          0
```

Figure 9-3c. Power law fluid, highly eccentric circular geometries.

```
                  0       0     0
                   15    15    15
            0      25    26    25     0
             14    30    31  30    2214
        0     27  32    33  32    27    12 0
         20    29  3230  31  32  2929  2420
    0      26    27  26  27  26  27  2526       0
     14      23  17  11  12  11   9171423    2214
       2729     7 0   0   0 0   0   7  292927
0          2717 0                   92327         0
 1525303232                               32302515
         302611 0                  112030

015263133312712 0                  1221313333312615 0
                0                   0
     3032322611                    1120303230
01525         9 0                   9        2515 0
         292717                      2329
     222729  14 7 9  10  10 010 9  14    2922
    014    2523  22  18  24  18  22  25      14 0
         2425  2826  28  28  28  28    24
       1220    28  30    30  30    28    12
        0    2226  29    29  29    22       0
             14    24    24    24    14
            0      15    15    15     0
                  0             0
                          0
```

Figure 9-3d. Power law model, square drill collar in circular hole.

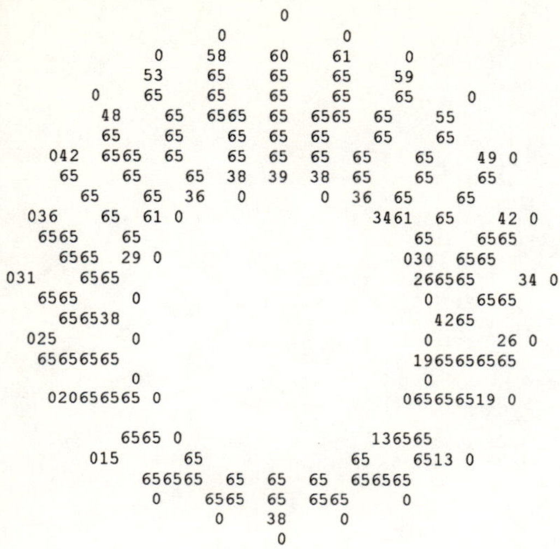

Figure 9-3e. Bingham plastic, slug flow in small, eccentric hole.

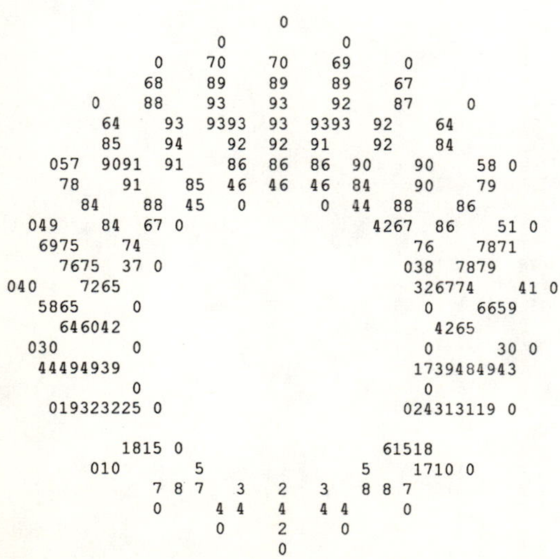

Figure 9-3f. Power law flow in small, eccentric hole.

```
                    0         0      0
              0                            0
                     28      28     28
        0       27           32          27        0
                     32          32
          26     32  32      32  32    32    26
    0       31     32  32    32  32  32    31        0
     24            32  32    32  32  32          24
              32     32  31  3131    32  32
  0    30     32     31  29  2929  31    32     30     0
   22     31     32  28  23  2323  28  32    31    22
     28     3131  27231414    014142327    31    28
        29    292622   0          0  222629    30
  018     3029  2112 0              0  21  2930    19    0
     2527  272418 0                   0  2427  2725
         27     10                      11     27
  015       2421 9                       92124        15 0
     20222221                               21222221
              1712                       717
    0111517171512 5                   0121517171511 0
                    6 0               0 6
            101111 8 4 2          2 4 8111110
        0 7          7   3    3 3   6 7        7 0
                  4 7   4 4   4 4 4   7 4
              0          3    3   3        0
                       0 2    2     0
                              0
```

Figure 9-3g. Power law fluid in large, eccentric hole.

```
                    0         0      0
              0                            0
                     28      28     28
        0       27           32          27        0
                     32          32
          26     32  32      32  32    32    26
    0       31     32  32    32  32  32    31        0
     24            32  32    32  32  32          24
              32     32  31  3131    32  32
  0    30     32     31  29  2929  31    32     30     0
   22     31     32  28  23  2423  28  32    31    22
     28     3131  27231414    014142328    31    28
        30    292622   0          0  222629    30
  019     3029  2112 0              0  21  2930    19    0
     2527  272418 0                   0  2427  2725
         2726   10                     10     27
  015       2420 9                       92124        15 0
     21222221                               21222221
              1512                       715
    0111517171512 4                   0121517171511 0
                    7 2               0 7
          7101110    3 2          1 3  101110 7
        0          5 5 4 1 1    1 1 1 4 5 5        0
                0 4    0 0      0    1 0    4 0
```

Figure 9-3h. Same as Figure 9-3g, with additional cuttings bed.

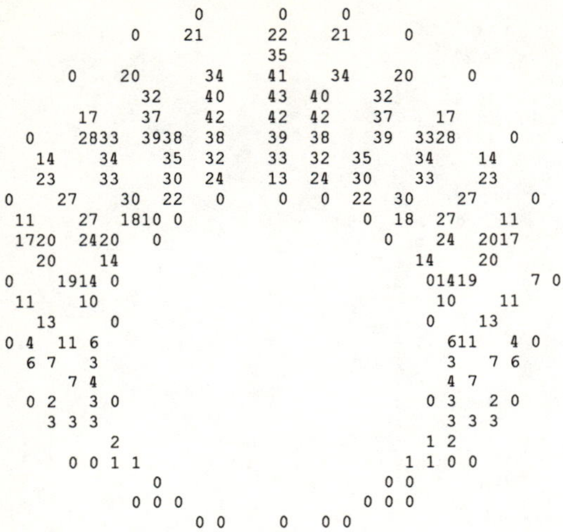

Figure 9-3i. Power law flow, drill pipe resting on bottom.

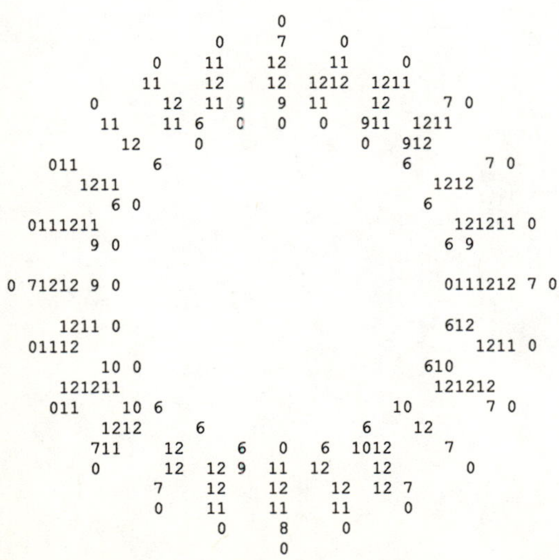

Figure 9-3j. Centralized casing, Class H cement slurry.

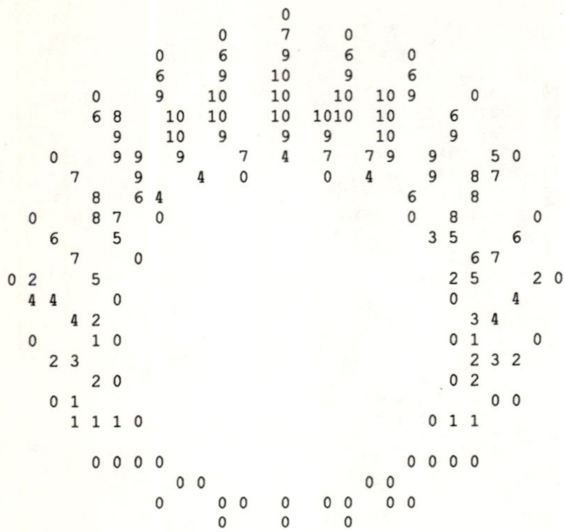

Figure 9-3k. Eccentric casing, Class H cement slurry.

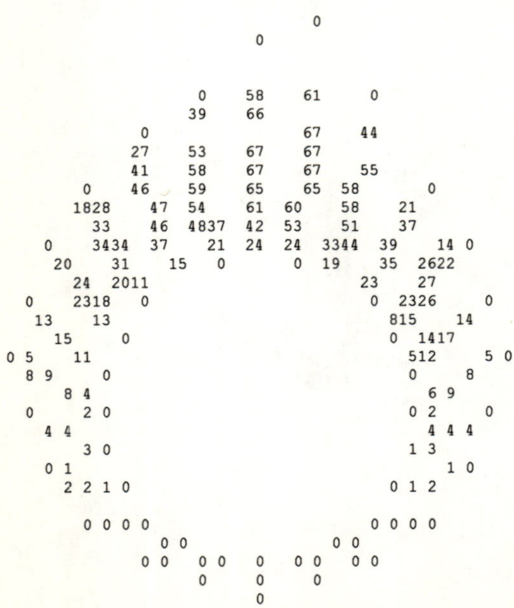

Figure 9-3*l*. Borehole with severe washout.

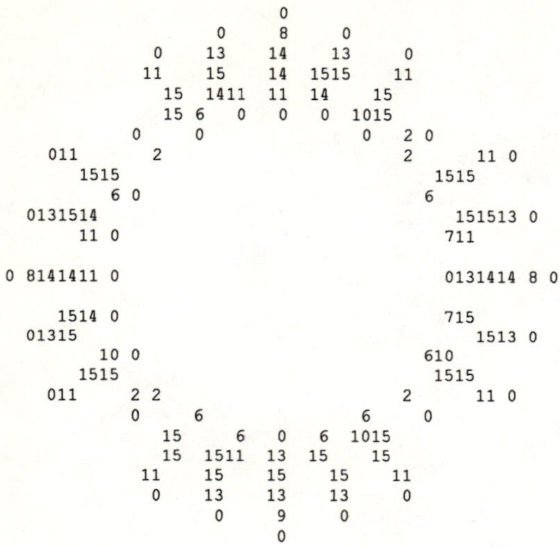

Figure 9-3m. Cement casing with four centralizers.

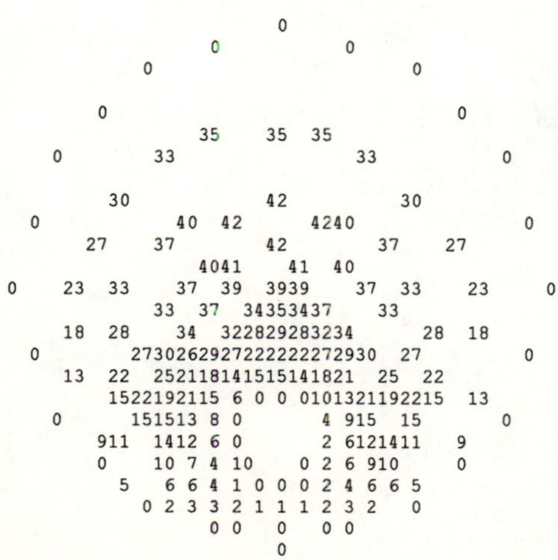

Figure 9-3n. Small diameter coiled tubing in hole.

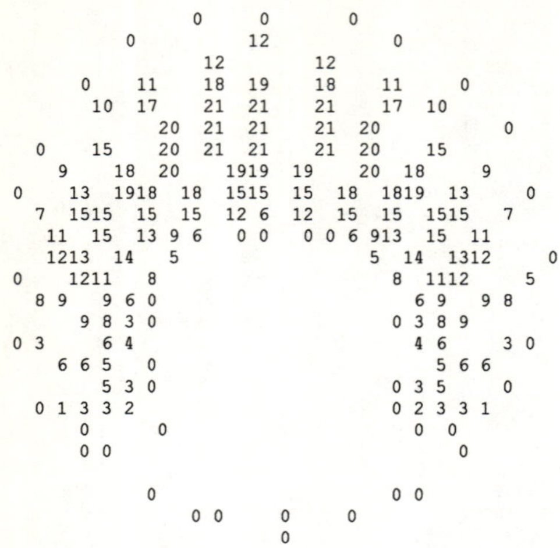

Figure 9-3*o*. Severely stuck pipe in deep cuttings bed.

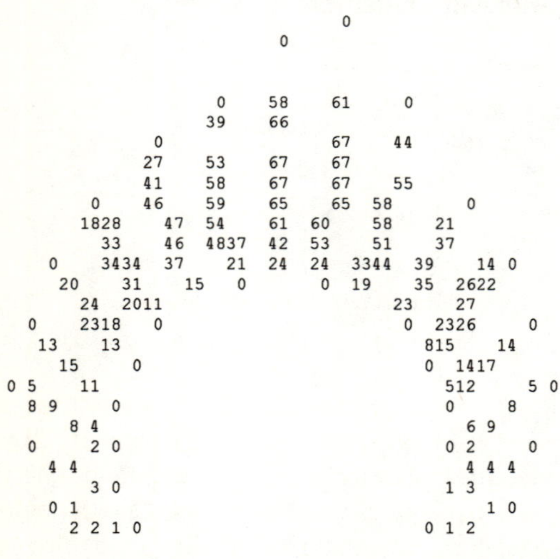

Figure 9-3p. Severe washout with thick cuttings bed.

```
0  75  87  96  98 102  99  97 101 101  98 XXX XXX  99  98 100 100  98 100
0  55  77  87  93  97  99 101 102 XXX XXX XXX XXX XXX 102 102 101 100 100
0  36  61  76  86  93  98 101 XXX XXX XXX XXX XXX XXX XXX XXX 102 101 100
0  27  50  67  80  89  96 101 XXX XXX 108 109 109 108 XXX XXX XXX 102 100
0  23  44  61  75  85  93 100 XXX XXX 109 110 111 110 109 XXX XXX 102 100
0  20  40  57  71  82  91  99 XXX 108 110 112 112 111 110 108 XXX 102 100
0  19  37  53  68  80  90  97 XXX 108 111 112 113 112 111 109 XXX XXX 100
0  18  35  51  65  78  88  96 XXX 108 111 113 114 113 111 109 XXX XXX 100
0  17  34  49  63  76  87  95 102 108 111 113 114 113 112 110 XXX XXX 100
0  16  33  48  62  75  85  94 102 XXX 111 113 114 114 112 110 XXX XXX 100
0  16  32  47  61  73  84  94 101 XXX 111 113 114 114 112 110 XXX XXX 100
0  16  31  46  60  72  83  93 100 XXX 110 113 114 114 112 110 XXX XXX 100
0  15  31  45  59  72  83  92 100 XXX 110 113 114 114 112 110 XXX XXX 100
0  15  30  45  58  71  82  91  99 XXX 110 112 114 113 112 110 XXX XXX 100
0  15  30  44  58  70  81  91  98 XXX 109 112 113 113 112 110 XXX XXX 100
0  15  30  44  57  70  81  90  98 XXX 109 111 113 113 112 110 XXX XXX 100
0  15  29  44  57  69  80  89  97 XXX 108 111 113 113 112 109 XXX XXX 100
0  14  29  43  56  69  79  89  97 XXX 108 111 112 112 111 109 XXX XXX 100
0  14  29  43  56  68  79  88  96 102 XXX 110 112 112 111 109 XXX XXX 100
0  14  29  42  56  68  78  88  96 102 XXX 110 111 112 111 109 XXX XXX 100
0  14  28  42  55  67  78  87  95 101 XXX 109 111 111 110 109 XXX XXX 100
0  14  28  42  55  67  77  87  94 101 XXX 109 110 111 110 108 XXX XXX 100
0  14  28  42  54  66  77  86  94 100 XXX 108 110 111 110 108 XXX XXX 100
0  14  28  41  54  66  76  86  93 100 XXX 108 110 110 109 108 XXX XXX 100
0  14  28  41  54  65  76  85  93  99 XXX XXX 109 110 109 108 XXX XXX 100
0  14  27  41  53  65  75  85  92  99 XXX XXX 109 109 109 XXX XXX 102 100
0  14  27  41  53  65  75  84  92  98 XXX XXX 108 109 109 XXX XXX 102 100
0  13  27  40  53  64  75  84  91  98 102 XXX 108 109 108 XXX XXX 102 100
0  13  27  40  52  64  74  83  91  97 102 XXX XXX 108 108 XXX XXX 102 100
0  13  27  40  52  63  74  83  90  97 102 XXX XXX 108 108 XXX XXX 102 100
0  13  27  40  52  63  73  82  90  96 101 XXX XXX XXX XXX XXX XXX 102 100
0  13  26  39  51  63  73  82  89  96 101 XXX XXX XXX XXX XXX XXX 102 100
0  13  26  39  51  62  72  81  89  95 100 XXX XXX XXX XXX XXX XXX 102 100
0  13  26  39  51  62  72  81  89  95 100 XXX XXX XXX XXX XXX XXX 102 100
0  13  26  39  51  62  72  81  88  94  99 XXX XXX XXX XXX XXX XXX 102 100
0  13  26  38  50  61  71  80  88  94  99 102 XXX XXX XXX XXX XXX 102 100
0  13  26  38  50  61  71  80  87  93  98 102 XXX XXX XXX XXX XXX 102 100
0  13  25  38  50  61  70  79  87  93  98 101 XXX XXX XXX XXX XXX 101 100
0  13  25  38  49  60  70  79  86  93  97 101 XXX XXX XXX XXX XXX 101 100
0  12  25  37  49  60  70  78  86  92  97 101 XXX XXX XXX XXX XXX 101 100
0  12  25  37  49  60  69  78  85  92  97 100 XXX XXX XXX XXX XXX 101 100
```

Figure 9-4. Recirculating vortex flow due to density segregation.

Closing remarks. The two steady annular flow limits considered thus far, that is the viscous, *unidirectional* model, and the inviscid, gravity driven, *recirculating flow* model, provide valuable tools that are useful in selecting muds and cements in drilling and cementing operations. Both are formulated rigorously using the general equations describing continuum fluid flow, and solved accurately using finite difference numerical models.

In both cases, extensive comparisons with field data, laboratory results, and empirical observation show that the physics of both problems are adequately captured. These steady state pressure drops may be significant by comparison with those obtained in dynamic acoustical solutions. If so, they form an integral part of swab-surge analyses.

REFERENCES

Burkhardt, J.A., "Wellbore Pressure Surges Produced by Pipe Movement," Journal of Petroleum Technology, June 1961, pp. 595-605.

Chin, W.C., "Goethert's Rule with an Improved Boundary Condition," AIAA Journal, Oct. 1977.

Chin, W.C., *Borehole Flow Modeling in Horizontal, Deviated, and Vertical Wells*, Gulf Publishing, Houston, 1992.

Chin, W.C., *Petrocalc 14: Horizontal and Vertical Annular Flow Modeling, Petroleum Engineering Software for the IBM PC and Compatibles*, Gulf Publishing, Houston, 1992.

Chin, W.C., *Modern Reservoir Flow and Well Transient Analysis*, Gulf Publishing, Houston, 1993.

Clark, R.K., and Fontenot, J.E., "Field Measurements of the Effects of Drillstring Velocity, Pump Speed, and Lost Circulation Material on Downhole Pressures," SPE Paper No. 4970, 49th Annual Fall Meeting of the Society of Petroleum Engineers of AIME, Houston, Texas, Oct. 6-9, 1974.

Fontenot, J.E., and Clark, R.K., "An Improved Method for Calculating Swab/Surge and Circulating Pressures in a Drilling Well, SPE Paper No. 4521, 48th Annual Fall Meeting of the Society of Petroleum Engineers of AIME, Las Vegas, Nevada, Sept. 30 - Oct. 3, 1973.

Lal, M., "Surge and Swab Modeling for Dynamic Pressures and Safe Trip Velocities," IADC/SPE Paper No. 11412, presented at the IADC/SPE 1983 Drilling Conference, New Orleans, La., Feb. 20-23, 1983.

Miles, J.W., "The Reflection of Sound Due to a Change in Cross Section of a Circular Tube," Journal of the Acoustical Society of America, Vol. 16, 1944, pp. 14-19.

Miles, J.W., "The Analysis of Plane Discontinuities in Cylindrical Tubes, II," Journal of the Acoustical Society of America, Vol. 17, 1946, pp. 272-284.

Mitchell, R.F., "Dynamic Surge/Swab Pressure Predictions," SPE Drilling Engineering Journal, Sept. 1988, pp. 325-333.

Morse, P.M., and Ingard, K.U., *Theoretical Acoustics*, McGraw-Hill, New York, 1968.

Pierce, A.D., *Acoustics: An Introduction to Its Physical Principles and Applications*, McGraw-Hill, New York, 1981.

10
Measurement-While-Drilling

What is *Measurement-While-Drilling*? In conventional rotary drilling, rig operations are periodically interrupted in order to make directional and wireline logging measurements. This can mean removing mile-long, million-pound drillstrings from the hole, section-by-section, an extremely tedious, time-consuming process. By contrast, *MWD* provides surface personnel with *real-time information while drilling* about instantaneous drill bit position and lithological properties of newly penetrated formation.

Real-time directional information is useful because drillers can react more rapidly to unanticipated changes in planned wellbore trajectory; two-way communications, for example, allows steerable bottomhole assemblies. MWD also provides useful safety information related to overpressured zones, lost circulation, gas influx, and blowout prevention.

Formation information obtained from MWD is more reliable than traditional wireline data. The formation is interrogated by downhole sensors before the penetrated rock is altered or damaged due to drilling fluid invasion, mechanical damage during tripping, or wellbore stress relief.

MWD is rapidly gaining in popularity for several reasons. First, modern wells, which tend to be highly deviated or horizontal, cannot be logged by wireline. Second, the fact that modern wells tend to be thousands of feet and even miles long makes real-time MWD logging attractive, since the costs and down times associated with making and unmaking million-pound drillstrings are prohibitive.

Basic MWD system components. Conceptually, a modern MWD tool can be viewed as a stand-alone downhole information system consisting of *sensor*, *telemetry*, and *power* packages that reside within the drill-collar, uncomplicated by wired connections to the surface, situated near the drillbit. We briefly summarize the basic system components.

Position sensors measure depth, direction, and inclination. *Lithology sensors* measure properties associated with electrical resistivity, gamma ray, neutron porosity, acoustic impedance, and so on.

The *telemetry package* encodes sensor information into signals and waveforms that are transmitted to the surface. This process includes downhole processing to compress data, e.g., calculating representative root-mean-square vibrations data so that complete data streams need not be transmitted, plus *data compression* from the communications engineer's point of view.

Different MWD systems will use different telemetry and signal processing schemes, but the basic requirement is a transmission mode that *minimizes bit error* and *maximizes data rate* in the presence of suboptimal *signal-to-noise ratios*. The *modulator* is the hardware that performs the actual transmissions, which are decoded at the surface by complementary *surface receivers* and *signal processors*.

Finally, the *power system* may require no more than batteries for shallow depth and low temperature applications, but more than likely, *downhole turbines* are used which draw energy from the mud stream, converting it into electrical and hydraulic power which drive the MWD modulator and all sensor electronics.

Candidate transmission technologies. So far, we have refrained from discussing specific implementations for Measurement-While-Drilling. Over the years, advances in microelectronics, mechanical and electrical packaging, hydraulics, and materials technology, have continually improved enough such that the meaning of MWD has likewise evolved.

In the late 1960s, the pioneering downhole recorder developed for real-time measurements at the Esso Production Research Company (now, *Exxon*) may be viewed as a forerunner to modern MWD systems (e.g., see Koch (1967), and Deily, Dareing, Paff, Ortloff, and Lynn (1967)).

Denison (1976a,b) described a high-data-rate system where *modified drillpipe* containing special electrical circuitry allows instantaneous transmission from the holes while drilling (the system, capable of high data rates, transmitted at least 600 data samples/sec).

Electric circuits from the surface to the bottom of the borehole allowed continuous transmission of downhole data whether or not drilling was in progress, and did not require the use of downhole power sources. Denison

(1979) later reported a significant increase in data rate to more than 36,000 samples per minute.

As recently as the 1980s, one oil service company offered both *mud pulse telemetry* and *recorded downhole data*, and in research-oriented efforts, developed high-data-rate *wireline transmission* methods for use while drilling (e.g., see Wolf, Zacksenhouse, and Arian (1985)).

Other investigators have considered *drillpipe acoustical telemetry*, that is, the transmission of downhole-to-surface data using longitudinal and torsional drillstring vibrations. Bradbury and Wilhoit (1962) importantly considered the effect of tool joints on the passage of plane longitudinal and torsional waves along a drillpipe. Approximate solutions to the governing partial differential equations for an idealized tool joint showed that tool joints had negligible effect for exciting frequencies of the same order as common rotary speeds, and that the drillpipe can be taken as a uniform pipe with negligible error.

Barnes and Kirkwood (1972) also considered longitudinal and torsional vibrations as a means for MWD communication. For their analysis, only zeroth mode vibrations (i.e., plane wave) were considered, the assumption being that higher modes would be damped out over short distances. The basic ideas on modal eigenfunctions and evanescent waves were discussed in Chapters 1 and 3.

In their idealized model, where the drillstring is taken as an infinitely long cylindrical steel pipe loaded at equal intervals with identical couplings, the authors show that most of the *passbands* for compressional waves are about twice the width of those for torsional waves (passbands are frequencies for which waves propagate without loss). Refer to the paper for further details on the exact geometry used and the numerical results obtained.

In Drumheller (1988), the drillstring is a periodic structure of pipe and threaded tool joints; the transmission characteristics are complex and they exhibit a banded and dispersive structure. This report describes a theoretical model, together with theory, laboratory experiments, and field test results, possibly suggesting the existence of frequency bands which are virtually free of attenuation and hence suitable for high data rate transmission.

Torsional wave propagation along drillstrings, as a means of MWD acoustic telemetry, was reconsidered by Squire and Whitehouse (1979). They importantly studied attenuative effects posed by the drilling mud environment, with and without gelling, and results of field experiments were reported. The conclusions cited in the foregoing papers, which have not been independently verified by this author, are given for completeness only.

Electromagnetic wave telemetry provides still another alternative, although it appears to be attenuative under many practical situations. All of these technologies are manifestations of wave propagation, but unfortunately, even a cursory review of any particular one is beyond the scope of this book.

By far, the most successful commercial MWD systems are based on *mud pulse telemetry*, despite the fact that their data rates have not yet exceeded an extraordinarily low ten bits/sec (bit rate improvements have been largely accomplished through data compression coding techniques rather than hardware advances).

Mud pulse telemetry systems provide the basic focus of this chapter, especially because the mathematics and modeling issues bear strong similarities to our work in axial and torsional vibrations, swab-surge analysis, and dissipation modeling. As we have already developed all of our modeling concepts for vibrations and swab-surge problems, we will discuss only the nuances of MWD acoustic telemetry with brevity, and especially because its readership audience is likely to be limited and highly specialized.

MUD PULSE TELEMETRY: THE ACOUSTIC SOURCE

We begin our discussions on mud pulse telemetry with a description of the acoustic sources in commercial use. Mud pulse signal generation is achieved using three different types of *modulators*, *pulsers*, *valves*, or *signal generators*. Our so-called *positive pressure*, *negative pressure*, and *mud siren* pulsers produce (unfortunately) low data rate signals associated with long acoustic waves that propagate up and down the mud column within the drillpipe.

Wavelengths are typically *hundreds of feet* long, much larger than the physical dimensions of signal modulators, typically measured in *inches*. On this scale, the MWD modulator is an *acoustic source*: like sources in seismics, ultrasonics, and nondestructive testing, they must be accurately characterized if their transmitted signals are to be properly *deconvolved* at the surface receiver.

Positive pressure poppet valves. When the *poppet valve* shown in the schematic of Figure 10-1 closes against a fixed *seat*, a *positive pressure wave* (that is, an *over-pressure* relative to the ambient levels) is created ahead of the valve that travels uphole. This over-pressure results because a significant volume of oncoming nearby drilling fluid is quickly brought to rest.

At the same time, the mud at the downstream side of the valve pulls away, thereby creating an opposite *under-pressure*, or *negative pressure* wave. The net effect is a *differential pressure* acting through or across the valve, or a *delta-p*. This delta-p is created and appears internally within the drill collar waveguide, by virtue of mechanical valve action; it is, in all respects, the *forcing function excitation* introduced in Chapter 1.

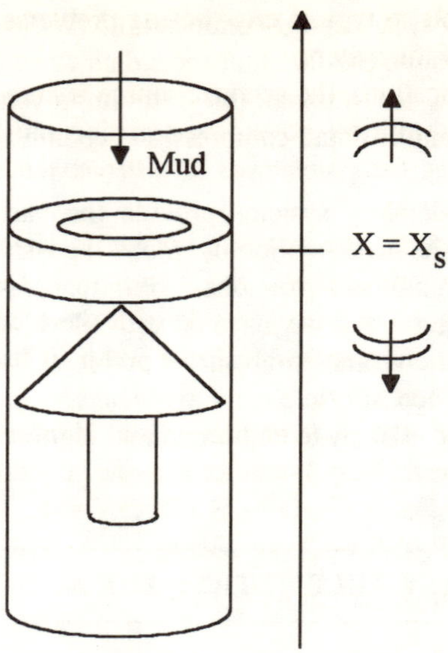

Figure 10-1. Positive pressure poppet pulser.

The delta-p across the valve consists of two parts. Even if the poppet valve is stationary, a pressure drop that is constant with time will exist, due to pure viscous losses. If the poppet valve moves slowly with respect to the valve seat, a quasi-steady hydraulic pressure drop varies slowly with time, and can be detected uphole as a quasi-steady change in simple hydraulic pressure.

If the valve moves quickly, e.g., as the diaphragm in a telephone transmitter does, an additional contribution due to the fluid compressibility of the mud arises. For example, a thin disk or diaphragm that completely blocks a duct will create a water-hammer delta-p equal to $2\rho vc$ when it is suddenly jerked with a speed v (see Example 9-2). Thus, delta-p's comprise of hydraulic (DC) and acoustic (or, AC sound) flow components.

Put differently, not all delta-p's propagate as sound: only acoustic signals propagate rapidly with any efficiency, and the greatest practical potential is reaped through transmission via "AC" rather than hydraulic "DC" means (again, refer to Chapter 1 for an explanation of our AC/DC nomenclature).

Accordingly, it is this acoustic delta-p or discontinuous jump in acoustic pressure that should be created mechanically and modeled analytically. So far, we have developed many wave equation models for propagation in one-dimensional systems. Dependent variables have included pressure, velocity potential, density, velocity, and so on. All of these are viable candidates as mathematical models in typical engineering problems where excitations occur at the *boundaries* of the system.

In MWD applications, the acoustic source is found *within* the transmission channel, and the appropriate governing equation must account for the *physics* of the source, that is, the *symmetries or antisymmetries in created velocity or pressure* (see Chapter 1).

For poppet valves, the acoustic source is characterized by a jump in pressure through the source point. In Chapter 1, we showed that the addition of a delta function $\delta(x-x_s)$ to the right side of a wave equation in the dependent variable, say, h(x,t), is responsible for a jump in the first spatial derivative $\partial h/\partial x$ through the point $x = x_s$.

Now consider the Lagrangian *fluid displacement function*, which measures displacements from (static or flowing) equilibrium. From Chapter 9, we find that it satisfies $\rho\, \partial^2 u/\partial t^2 - B\, \partial^2 u/\partial x^2 = 0$. If we add to it the delta function excitation $P_s(t)\, \delta(x-x_s)$, we obtain,

$$\rho\, \partial^2 u/\partial t^2 - B\, \partial^2 u/\partial x^2 = P_s(t)\, \delta(x-x_s) \qquad (10.1)$$

Equation 10-1 clearly implies a jump in the first derivative $\partial u/\partial x$. If we recall that the acoustic pressure is $p = -B\, \partial u/\partial x$, integration across the source point (using the delta function formalism of Chapter 1) shows that the jump in pressure equals the *pressure source strength* $P_s(t)$.

It is important, invoking results from Chapter 1, to observe that *the overpressure upstream is equal in magnitude to the underpressure downstream.* Thus, the *created signal pressure* is *antisymmetric* with respect to the source point. Also, in an infinite system, half the delta-p will propagate uphole away from the source, while the remaining half will propagate downhole.

By contrast, the created acoustic *velocity* field is symmetric with respect to the source; that is, upstream and downstream velocities always possess the same signs, their directions being physically identical. The propagating pressures, displacements, and velocities in the above paragraph will travel with a sound speed $c = (B/\rho)^{1/2}$. As the wave passes any fixed point within the drillpipe, it momentarily induces a fluid displacement u(x,t), associated with a kinematic strain $\partial u/\partial x$, and a *material velocity* $\partial u/\partial t$ satisfying $|\partial u/\partial t| \ll c$.

Negative pressure valves. Negative pulsers may be conceptually visualized as *corked holes* in the drill collar that periodically open and shut. When this "door" opens suddenly, the rush of mud through the orifice creates a pressure drop or negative pressure in the drillpipe; the pressure drop between pipe and annular flows is responsible for signal generation. When the "door" closes, an over-pressure is created, that propagates up and downhole. A negative pulser is schematically shown in Figure 10-2.

In either case, the physical corking nature of the signal generation implies that *created signal pressures*, by contrast to those for positive pressure valves, *are symmetric with respect to the source point*, the signs in pressures being the same immediately above and below the pulser. Thus, negative pressure valves create MWD signals detectable at the surface, but they are *not* associated with nonzero delta-p's. Instead, they are associated with nonzero *delta-velocities* (whether the cork is plugging or unplugging, the velocities at the sides of the valve are always opposite in direction). The conventional maxim that nonzero delta-p's are always required to transmit information is therefore incorrect.

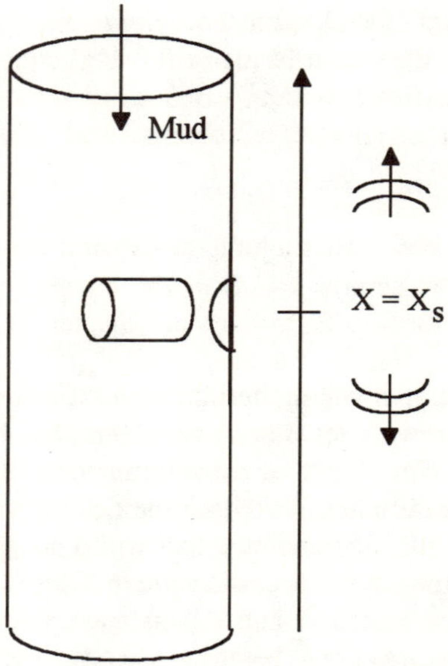

Figure 10-2. Negative pressure pulser.

The physical picture described above applies, as well, to a balloon that is "popped" in a long conduit. Again, because the acoustic pressure distribution is symmetric about the source point, there can be no pressure discontinuity, but the existence of equal and opposite velocities at all times leads to jumps in velocity or nonzero *delta-velocities*. How are these modeled?

It is clear that Equation 10-1 does not apply: as shown in Chapter 1, the *material velocity* $\partial u/\partial t$ *is continuous through the source point, so that it is never antisymmetric*. We do know that some sort of delta function is required, the only problem being, the choice of dependent variable that correctly yields the velocity jumps characteristic of negative pulsers. It turns out the *velocity potential* formulation discussed in Chapter 9 works elegantly (see Morse and Feshbach (1953) for detailed developments and discussion).

The main velocity potential results of Chapter 9 can conveniently summarized by Equations 9-49a, 9-49b, and 9-50, which we re-order as,

$$\partial^2\phi(x,t)/\partial t^2 - c^2\, \partial^2\phi(x,t)/\partial x^2 = 0 \tag{10.2}$$

$$\partial\phi(x,t)/\partial x = u_t(x,t) \tag{10.3}$$

$$\partial\phi(x,t)/\partial t = c^2\, u_x(x,t) \tag{10.4}$$

That is, $\phi(x,t)$ satisfies the classical wave equation; the first spatial derivative physically equals the material velocity, while the first time derivative is proportional to the strain as shown in Equation 10-4. The velocity potential, therefore, provides an alternative but equivalent vehicle for solving mud pulse acoustic problems.

Now, let us introduce the delta function term $c^2\, V_s(t)\, \delta(x-x_s)$ to the right side of Equation 10-2, that is, add an internal source to the potential wave equation in the form

$$\partial^2\phi(x,t)/\partial t^2 - c^2\, \partial^2\phi(x,t)/\partial x^2 = c^2\, V_s(t)\, \delta(x-x_s) \tag{10.5}$$

Following Chapter 1, an integration through the source point $x = x_s$ shows that the right-side term $c^2\, V_s(t)\, \delta(x-x_s)$ is responsible for a discontinuity in the quantity $c^2\, \partial\phi(x,t)/\partial x$. That is, substituting from Equation 10-3, we have a jump in the term $c^2\, u_t(x,t)$ with the strength $c^2\, V_s(t)$; canceling out c^2, we find that the jump in the material velocity $u_t(x,t)$ is specified by assigning the *velocity source strength* $V_s(t)$. *An arbitrary MWD source can, in general, be modeled as the superposition of appropriate positive and negative pulsers.*

Mud siren sources. A mud siren consists of a turbine-like *rotor and stator pair* mounted in a low-clearance housing, through which drilling mud passes, as shown in Figure 10-3. The *rotor* rotates, while a stationary *stator* remains in place; as it periodically interrupts the mud flow, acoustic pressure waves are created that propagate both upwards and downwards.

Sirens are only "turbine-like" to the extent that they rotate. Their blades are chunky as shown, rather than thin and cambered, as in aircraft turbines or turbodrills. Real sirens create high pitch whining sounds; mud sirens, by contrast, periodically block the complete cross-sectional area, thus creating acoustic plane waves in the eigenfunction sense of Chapters 1 and 3 (higher order effects are evanescent with space). These plane waves, like those for poppet valves and negative pulsers, are amenable to wave modeling.

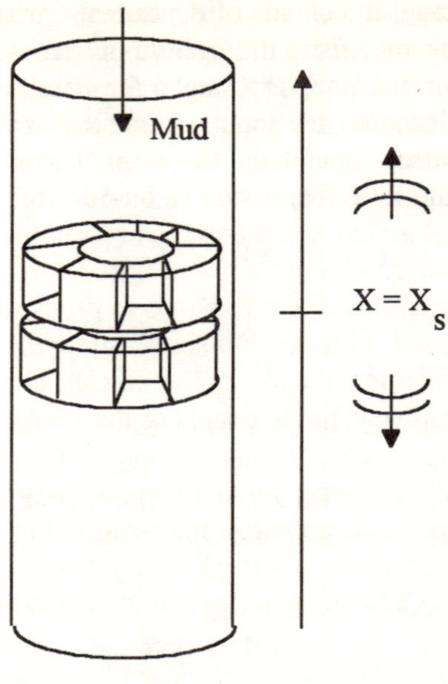

Figure 10-3. Mud siren.

As the siren closes, flow at the uphole side is brought to rest, and an over-pressure relative to ambient conditions is created. At the same time, flow pulls away from the siren at the downstream side, creating an underpressure. In this sense, the physics of the siren acoustic source is identical to that for positive pressure poppet pulsers: the mathematical discussion leading to Equation 10-1 applies without modification.

When the siren opens, the over-pressure at the top is relieved, but the under-pressure at the bottom increases; the above description reverses, and Equation 10-1 again applies. Continuous mud siren rotation creates a continuous periodic pulse, but we note that poppet valves and negative pulsers can likewise function in continuous mode if required. Acoustically speaking, positive pressure valves and mud sirens function identically.

Signal generation at the source. So far, we have described the symmetries and antisymmetries associated with pressure and velocity fields for positive pressure valves, negative pulsers, and mud sirens. We indicated that poppet valves and sirens satisfy Equation 10-1, while negative pulsers satisfy Equation 10-5. The *delta-p* strength $P_s(t)$ appears in the former, while the *delta-velocity* strength $V_s(t)$ appears in the latter.

These wave differential equations can be solved using the analytical and numerical methods devised thus far for axial and torsional vibrations, and swab-surge modeling in the annulus, directly without change. Typical boundary conditions may be solid reflectors, acoustic free ends, radiation conditions, and so on (see Chapters 1 and 5 for details).

We have not discussed the exact values for $P_s(t)$ and $V_s(t)$ yet. The *time history* will, of course, depend on the *telemetry scheme* used to represent downhole information for transmission to the surface. For example, analogies to AM and FM transmission in radio and television broadcasts are easily made, which we will expand upon later.

The magnitude of the respective jumps in pressure and velocity are crucial to a good MWD signal with audible signal strengths. Here, the mechanical design, to include size and shape, can be crucial, since the signal generation process depends sensitively on the details of the unsteady hydraulic flow acting at the valve.

For example, if the poppet valve shown in Figure 10-1 does not approach a fully closed position, or if the siren rotor-stator gap in Figure 10-3 is large, signal strength will be weak. With geometry fixed, signal strength will vary with mud flow rate, and to some extent with data rate, sound speed and rheology.

The key to a good MWD system depends on two factors. The first is a *nearfield* problem: a reliable mechanical valve, shaped or modulated in such a way that $P_s(t)$ or $V_s(t)$ is as strong as possible, so that the greatest achievable source signal is produced. The second challenge is *farfield*: a good telemetry strategy, which sorts the *information* encoded in the transmitted $P_s(t)$ or $V_s(t)$ waveform from its reflections and distortions in the transmission channel and from pump noise.

Mechanical design considerations. Although the schematics in Figures 10-1 to 10-3 depict rather simple valves, actual downhole implementation poses significant engineering challenges. These are briefly described below.

Packaging constraints. In practice, the MWD tool must be packaged within the drill collar, and it must be placed near the bit in order to measure real-time information. The entire space within the collar is not available to the MWD tool, since it must allow mud flow to bypass. In addition, since drill collars do bend while drilling highly deviated and horizontal wells, tool tolerances must be large enough that turbines and acoustic signal generators do not jam at bends and doglegs, and thereby interrupt the transmission process.

Shock and vibration. MWD tools must be ruggedized to survive the downhole shock and vibration environment, which consists of combined axial, torsional, and lateral vibrations, and transient shocks. Shock and vibration considerations are extremely important because of their detrimental effect on downhole microelectronics, seals, sensors, and so on.

Mud erosion. Drilling mud, because it contains abrasive weighting materials like barite, will severely erode MWD tools. Acoustic signal generators are routinely designed using tungsten carbide and other surface hardening techniques to minimize the "sandpapering" prevalent at high fluid speeds encountered during signal generation.

Mud turbines also erode quickly, in part, a consequence of mechanical packaging limitations. In aerospace turbines, blade angles of attack are restricted to 15 degrees or less to ensure aerodynamic efficiency, and multiple stator-rotor stages are allowed. By contrast, length constraints limit the number of allowed stages; in order to create enough power to run the pulser and sensor electronics, mud turbine blades are designed with a high degree of pitch. This in turn creates high surface velocities which shorten the life of the turbine.

Power requirements. Depending on rate-of-penetration, kinds of data required, and data rate, power requirements can be substantial. Thus, batteries do not provide the best alternative for power usage, and downhole turbines are desirable and more flexible. But regulating power production, in an engineering sense, is challenging. Generated power varies as the cube of flow rate, and slight increases or decreases in flow rate relative to design conditions can be detrimental to the power system.

High pressure and temperature. MWD tools must be built to withstand hundreds of degrees Fahrenheit and tens of thousands of psi. This harsh environment is detrimental to sensor survival, electronics, batteries, seals, and materials. Under severe conditions, heightened concern with respect to corrosion exists.

Fluid mechanics problems. Fluid mechanics can be extremely unforgiving to pulser design. Obvious problems arise from increased signal strength requirements: sirens and poppet valves, for example, must operate at tighter gaps, thus creating increased local flow speeds and consequent erosion.

Unusual problems exist as well. Consider, say, the mud siren. Thus far, we have not stated which of the two siren plates shown in Figure 10-3 is the rotor or stator. We have only noted that rotation of the rotor relative to the stator in circulating mud flow produces a cyclic signal that travels up the mud column to be detected at the drill site surface, and that information is encoded by selectively varying rotor rotation.

Actually, mud sirens can act in irrational ways that nature condones: the drilling mud naturally tends to "blow" the rotor to a stable closed position which in turn blocks the mud flow. This jams the MWD tool, as well as the drillstring; high pressure builds up at the mud pumps, and dangerous field situations are created at interrupt drilling and logging operations.

This "stable closed" behavior is well known to MWD designers, and over the years, proposed fixes have be purely "brute force" and mechanical. Chin and Trevino (1988) uncovered the basic principles for "stable open" mud siren design, noting that rotor placement in the downstream position, *plus* special aerodynamically tapered cuts will cause the rotor to bias in the open position.

At any rate, we wish only to emphasize the important role of fluid mechanics in downhole hardware design. The magnitudes of hydraulic torques, for example, will affect power requirements, electric versus hydraulic motor selection, and the quality of the signal spectrum in the generated pressure. Similar anomalous behavior is found for positive pressure poppet valves and negative pressure pulsers.

More practical considerations. Finally, there are the issues of low manufacturing cost, ease of maintenance and repair, long "mean time between failure" or servicing, and so on, that can be crucial to the commercial viability of any MWD system.

MUD PULSE TELEMETRY: THE TRANSMISSION CHANNEL

In the foregoing discussion, we spoke of signal generation at the source point, and assumed that, once created, singals were discernible to receivers at the surface. This is not the case, since reflections within the transmission channel introduce complications that affect reception. The nature of the reflections depends, to a large extent, on the type of pulser used.

Reflection problems near the pulser. Figure 10-4 illustrates the reflection modes associated with poppet valves, negative pulsers, and mud sirens, respectively, from left to right. In the first schematic for poppet valves, a created positive pulse is shown traveling uphole, while its complementary antisymmetric pressure pulse travels downhole.

This latter *pressure* signal reflects at the solid boundary shown, and turns upward with the same negative sign. If the pulse length or duration is long enough, it interferes with the tail end of the upgoing positive pulse being generated, thus causing *destructive interference* of the created signal. A similar process applies during the opening cycle of the poppet pulser operation.

Negative pulsers act differently. Because the pressures created at the source point are symmetric about it, thus having the same sign, reflection at the bottom and subsequent superposition enhances the tail end of the created signal, leading to *constructive interference*.

Mud siren operation is even more complicated. While sirens physically behave like positive pressure pulsers, it is clear that at the higher frequencies that they are usually operated at, both constructive and destructive interference are possible. Such acoustic effects can degrade field operations, but they can, of course, also be applied to advantage.

In addition to these effects, partial reflections and transmissions at area discontinuities (e.g., at the pipe-to-collar interface) will affect transmitted signal quality. It is clear that high-data-rate mud pulse telemetry systems of the future, whether they are poppet, negative pressure, or siren in variety, will require a detailed understanding of and intelligent application of acoustic source properties as elucidated in Chapter 1 and impedance characteristics as described in Chapter 9.

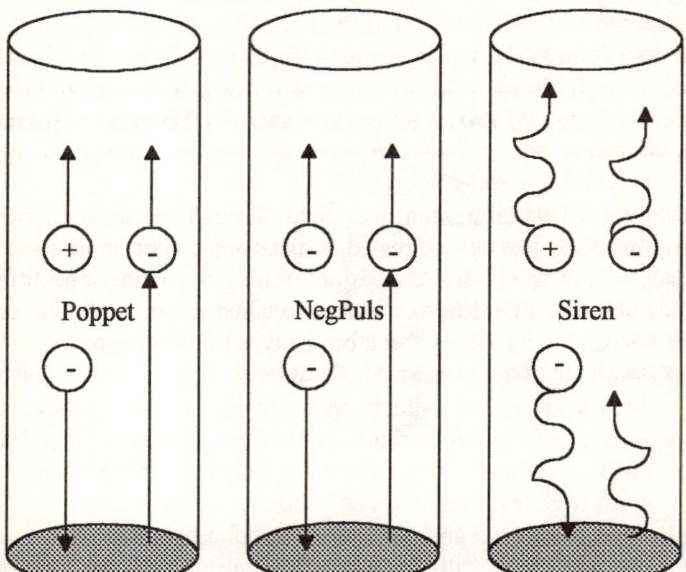

Figure 10-4. Downhole pulser reflection models.

The transmission channel uphole. As if the downhole environment were not challenging enough, we introduce additional complications arising from the remainder of the *transmission channel*. The pressure transducers shown in the schematic of Figure 10-5 will receive signals that consist of the wave signals described above, plus their downward reflections from the mud pump or desurger, plus mud pump noise, drilling noise, and general random noise. The surface receiver must take into account all of these non-ideal conditions, and *deconvolve* (or, "unscramble") the raw data in such a way as to recover the original information-bearing $P_s(t)$ or $V_s(t)$ waveform.

Pump noise and MWD reflections do not constitute true noise in the communications sense, since they are not random: they represent coherent waves with well defined acoustic properties. Thus, they can be filtered using software methods, many of which are described in the open patent literature. These range from techniques that subtract complete "snapshots" of the mud pump signature from the total received signal, to standard filters that eliminate pump noise based on frequency spectrum, and to more novel filters that discriminate on the basis of direction rather than frequency.

Such *directional filters* are described in the now classic paper of Clayton and Engquist (1977), which gave wave-equation-based *differential operators* that were capable of accurately filtering or selecting waves originating from any direction in a completely transient three-dimensional field. These authors then expressed their findings in terms of *finite difference operators* in space and time; the resulting *difference molecule clusters*, of course, translate directly into requirements for *pressure transducer spacing* and *data acquisition sampling time* in the MWD context.

In the late 1970s, this author applied the new ideas to computational unsteady transonic flows at Boeing. In particular, special filtering schemes were used to eliminate inward reflections at the boundaries of finite size solution domains, so that smaller computational boxes requiring less computing resources could be used. This specialized subject and its application to MWD signal processing is beyond the scope of this book.

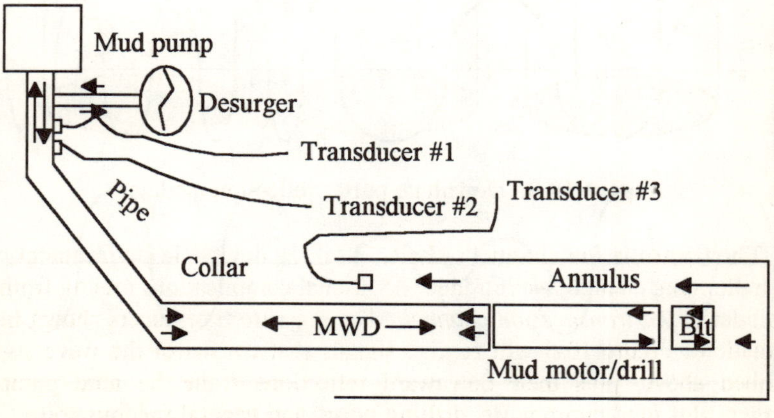

Figure 10-5. Drillpipe system as an acoustical waveguide.

Telemetry design objectives. The downhole and surface acoustic problems described above, plus additional *attenuation issues* associated with viscous drilling muds, limit the ability of existing MWD mud pulse systems to extremely low data rates, at least by comparison to transmission rates achievable by satellite systems or simple systems on computer networks.

At the present time, MWD bit rates at best approach ten bits per second, but in this author's opinion, these commercial claims are questionable and somewhat exaggerated.

The end objective in telemetry design is *high data rate* with low probability of bit error in the presence of noise. One quantitative measure of transmission error is achieved by a *dimensionless plot* relating the *probability of bit error* on the vertical axis to the *signal-to-noise (S/N) ratio* on the horizontal, for different host telemetry schemes.

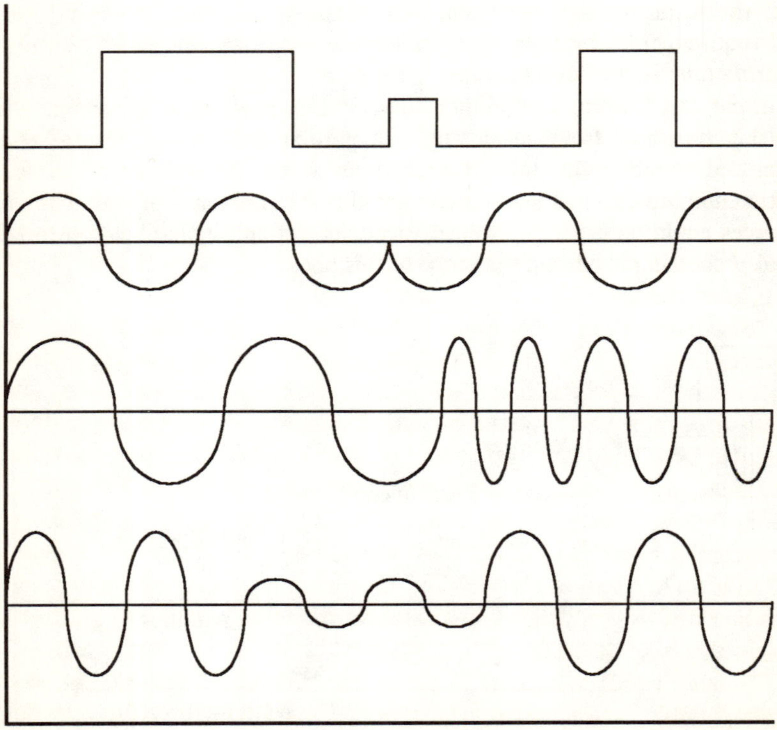

Figure 10-6. Conceptual telemetry schemes for MWD sources.

In a true sense, the "smoke signal method" used by American Indians during the last century represents an early prototype telemetry scheme. Modern methods number in the *thousands*, as simple literature searches will show, on scanning communications journal publications. Figure 10-6 shows, conceptually, several types of telemetry methods available to MWD designers.

The top trace illustrates *pulse width modulation*, while the second demonstrates *phase shift keying* (or, "PSK"). The third introduces FM *frequency modulation*, while the fourth shows PSK modulation due to a low torque MWD motor. Downhole hardware limitations clearly, in this example, inflict profound consequences on surface signal processing.

Additional practical considerations. We have described many of the design variables likely to affect telemetry rate, but additional practical issues remain. These include the effects of mudpump pistons with worn O-rings, unsteady flow rates due to positive displacement effects of duplex and triplex pumps, loss of signal synchronization due to drill bit bounce and momentary axial shock, attenuation and frequency dispersion in the modern viscoelastic muds used in horizontal wells, and real-time signal processing capabilities and limitations in view of allowed costs.

Lear and Dareing (1990) considered the effect of drillstring vibrations on MWD pressure pulse signals. Vibrations of the metal at the bit face or mud motor face will excite the mud column and introduce high frequency background noise. In this sense, drillstring axial vibrations and MWD acoustics are coupled; thus, integration of the algorithms described in Chapters 5 and 9 should provide a first step in understanding the physics of the solids-fluids interaction.

The theoretical maximum. In Chapter 3, we studied wave propagation in three-dimensional rectangular ducts, and derived separable eigenfunction solutions in (x,y,z) coordinates in terms of trigonometric functions. Similar results are possible for circular ducts, but the analogous eigenfunctions (available in the classical literature) are related to Bessel's equation. Since the basic ideas remain unchanged, we only quote results that are relevant to MWD.

For a duct radius R, and a sound speed c, the dispersive modes are all evanescent if $\omega < 1.841\ c/R$. This requires that the frequency be less than 1,000 Hz for a 0.1 m radius duct (or, an 8 inch pipe) containing *air* at 20° C. For water and typical drilling muds, where the sound speed is about five times greater, this is achieved for $\omega < 5{,}000$ Hz.

In other words, it appears that a bit rate of at least 5,000 bits per second is ideally possible, and substantially higher, if multi-level amplitude modulation is allowed. We are, therefore, a long way from what nature permits! This suggests that, perhaps, the obvious mechanical pulser concepts shown in Figures 10-1 to 10-3 represent no more than crude "Stone Age" attempts toward achieving ideal data rates. The key therefore rests with intelligent and clever applications of acoustics principles rather than able implementations of brute force mechanical excitation.

Acoustic signals in the annulus. We note that acoustic signals traveling in the drill pipe, due to MWD pulsers or mudpump noise, *do* propagate through the bit and up the annulus (e.g., see Example 9-4). This physical fact can be used to advantage in several applications.

By and large, propagating drillpipe signals are correlatable with those in the borehole annulus under ideal conditions . But the correlations break down when gas is present in the annulus; severe signal attenuation and phase delays result. This fact can be used to infer gas influx while drilling (e.g., Bryant, Grosso, and Wallace (1991), Hamilton, Swanson, and Wand (1992), Codazzi, Till, Starkey, Lenamond, and Monaghan (1992)).

Even when gas is not present in the annulus, propagation characteristics such as wave speed can differ from drillpipe to annulus, because the non-rigid nature of the borehole wall introduces changes to the mud sound speed in the annular waveguide. Jeffryes (1993), for example, suggests means of obtaining downhole borehole information by interpreting measured differences via an acoustic model that takes wall flexibility into account.

Finally, we suggest two papers for additional reading, by Desbrandes, Bourgoyne, and Carter (1987), and Montaron, Hache, and Voisin (1993). The first describes experimental results for mud attenuation, showing the weakly exponential time behaviour assumed in the modeling of Part I of this book (the data applies to simple water base muds under surface conditions only). The second paper describes other interesting aspects of MWD acoustic telemetry and practical design issues.

REFERENCES

Barnes, T.G., and Kirkwood, B.R., "Passbands for Acoustic Transmission in an Idealized Drill String, Vol. 51, No. 5, Journal of the Acoustical Society of America, 1972, pp. 1606-1608.

Bradbury, R.E., and Wilhoit, J.C., "Effect of Tool Joints on Passage of Plane Longitudinal and Torsional Waves Along a Drill Pipe," Paper No. 62-PET-32, presented at the Petroleum Mechanical Engineering Conference, American Society of Mechanical Engineers, Dallas, Tx., Sept. 23-26, 1962.

Bryant, T.M., Grosso, D.S., and Wallace, S.N., "Gas-Influx Detection With MWD Technology," SPE Drilling Engineering Journal, December 1991, pp. 273-278.

Chin, W.C., and Trevino, J.A., "Pressure Pulse Generator", U.S. Patent No. 4,785,300, awarded Nov. 15, 1988.

Clayton, R., and Engquist, B., "Absorbing Boundary Conditions for Acoustic and Elastic Wave Equations," Bulletin of the Seismological Society of America, 1977, Volume 67, pp. 1529-1540.

Codazzi, D., Till, P.K., Starkey, A.A., Lenamond, C.P., and Monaghan, B.J., "Rapid and Reliable Gas Influx Detection," IADC/SPE Paper No. 23936, presented at the 1992 IADC/SPE Drilling Conference, New Orleans, Louisiana, February 18-21, 1992.

Deily, F.H., Dareing, D.W., Paff, G.H., Ortloff, J.E., and Lynn, R.D., "Downhole Measurements of Drill String Forces and Motions," presented at the Petroleum Mechanical Engineering Conference, American Society of Mechanical Engineers, Philadelphia, Pa., Sept. 17-20, 1967.

Denison, E.B., "Downhole Measurements Through Modified Drill Pipe," presented at the Petroleum Mechanical Engineering and Pressure Vessels and Piping Conference, of the American Society of Mechanical Engineers (ASME), Mexico City, Mexico, Sept. 19-24, 1976.

Denison, E.B., "Making Downhole Measurements Through Modified Drill Pipe," World Oil, Oct. 1976, pp. 86-89.

Denison, E.B., "High Data-Rate Drilling Telemetry System," Journal of Petroleum Technology, Feb. 1979, pp. 155-163.

Desbrandes, R., Bourgoyne, A.T., annd Carter, J.A., "MWD Transmission Data Rates Can Be Optimized," Petroleum Engineer International, June 1987, pp. 46-50.

Drumheller, D.S., "Acoustical Properties of Drill Strings," Sandia National Laboratory Report No. SAND-88-0502, August 1988.

Hamilton, T., Swanson, B., and Wand, P., "Use of New Kick Simulator Will Increase Wellsite Safety, World Oil, September 1992, pp. 71-77.

Jeffryes, B.P., "Method and Apparatus for Obtaining Borehole Information Downhole," U.K. Patent Application GB 2 257 785 A, dated January 20, 1993.

Koch, W.M., "Analysis of the First Field Measurements of Downhole Drill Bit Motions and Forces," presented at the Twenty-Seventh Annual Meeting of the American Association of Oilwell Drilling Contractors, Oklahoma City, Oklahoma, Oct. 11-13, 1967.

Lear, W.E., and Dareing, D.W., "Effect of Drillstring Vibrations on MWD Pressure Pulse Signals," Journal of Energy Resources Technology, Transactions of the ASME, June 1990, pp. 84-89.

Montaron, B.A., Hache, J.M.D., and Voisin, B., "Improvements in MWD Telemetry: The Right Data at the Right Time," SPE Paper No. 25356, 1993 SPE/IADC Drilling Conference, Amsterdam, Feb. 23-25, 1993.

Morse, P.M., and Feshbach, H., *Methods of Theoretical Physics, Parts I and II*, McGraw-Hill, New York, 1953.

Squire, W.D., and Whitehouse, H.J., "A New Approach to Drill-String Acoustical Telemetry," SPE Paper No. 8340, presented at the 54th Annual Fall Technical Conference and Exhibition of the Society of Petroleum Engineers of the AIME, Las Vegas, Nevada, Sept. 23-26, 1979.

Wolf, S.F., Zacksenhouse, M., and Arian, A., "Field Measurements of Downhole Drilling Vibrations," SPE Paper No. 14330, 60th Annual Technical Conference and Exhibition of the Society of Petroleum Engineers, Las Vegas, Nevada, Sept. 22-25, 1985.

11
Geophysical Applications

This chapter introduces fundamental concepts from *kinematic wave theory* (see Chapter 2) as they may apply to geophysics problems at borehole, interlayer, and crosswell scales; that is, we deal with the physics of *slowly varying wavetrains* in modern seismic applications.

Elementary notions in geophysics, e.g., eikonal methods, ray tracing, Fermat's principle, and so on, are also developed from first principles as needed. We emphasize that our treatment is not intended to replace standard geophysics curriculum materials. Instead, our complementary discussions aim at highlighting the underlying formulation aspects of a mathematical subject.

Thus, we will not develop eikonal methods in their fullest detail, leaving that task to many well written books already available. We will not derive the general equations governing P and S waves, introduce specialized techniques used in migration, nor will we delve into the attenuative behaviour of seismic waves in oil saturated rocks.

The subjects presented for discussion were selectively chosen to guide novices and experts in geophysics on key ideas essential to formulating correct boundary value problems. An attempt is made to expand upon traditional equation-based modeling methods, by introducing more powerful kinematic wave formalisms that allow immediate use of empirically measured $\Omega^r(k)$ and $\Omega^i(k)$ functions in seismic modeling. The limitations of conventional modeling methods are examined in light of modern kinematic wave theory.

CLASSICAL WAVE MODELING: EIKONAL METHODS AND RAY TRACING

A suitable starting point for an elementary discussion of eikonal methods and ray tracing is the classical three-dimensional wave equation introduced earlier in Part I of this book,

$$\partial^2 P/\partial t^2 - c^2 (\partial^2 P/\partial x^2 + \partial^2 P/\partial y^2 + \partial^2 P/\partial z^2) = 0 \tag{11.1}$$

describing *P waves* in an isotropic, homogeneous, nondissipative earth. In Equation 11-1, P(x,y,z,t) represents *pressure* or *normal stress*, with the effects of *shear* ignored in the model. The speed of sound c is determined from the formula $c = \sqrt{(E/\rho)}$, where ρ is the mass density, and E is Young's modulus.

Note that Equation 11-1 is identical to the pressure equation used to model acoustic waves in liquids in Chapter 3. There, we considered waveguide applications leading to eigenvalue problems, but in this section, we will specifically consider wave propagation in unbounded domains.

The plane wave. In Chapters 1 and 2, we examined the *uniform plane wave* "sin (kx-ωt)" propagating in a single dimension. Its generalization in three-dimensions is the *monochromatic* wave

$$P(x,y,z,t) = P_0 \sin (k_x x + k_y y + k_z z - \omega t) \tag{11.2}$$

where P_0 is a constant amplitude. Substitution in Equation 11-1 leads to the exact *dispersion relation*

$$\omega^2 = c^2 (k_x{}^2 + k_y{}^2 + k_z{}^2) \tag{11.3}$$

connecting the wave frequency ω to each of the wavenumbers k_x , k_y, and k_z, and the speed of sound c.

High frequency limit. Equation 11-2 is a *plane wave*, since P(x,y,z,t) is constant whenever the argument $k_x x + k_y y + k_z z - \omega t$ is constant, that is, P does not vary on the *plane* $k_x x + k_y y + k_z z = \omega t +$ constant. But Equation 11-2 is not the only substitution formally possible, for it is common to consider expansions of the form

$$P(x,y,z,t) \ = \ \exp[i\omega(t-\tau)] \sum_{k=0}^{\infty} (i\omega)^{-k} P_k \tag{11.4}$$

whenever periodic solutions having a specified frequency ω are desired. Note that the leading k = 0 term in the summation is simply $P_0(x,y,x)$.

Here, τ and the functions P_k are independent of ω and t, depending only on x, y, and z. Also, the exponential factor in Equation 11-4 is analogous to our previous "sin (kx - ωt)", but the constant P_0 in Equation 11-2 has been replaced by a more general summation.

Since the expansion proceeds inversely with ω, it is a *high frequency* assumption. Equation 11-4 is called a *ray series*, $\tau(x,y,z)$ is known as the *phase function*, and the subscripted P_k coefficients are the *amplitude coefficents* of the series.

When t is a given constant, the contours $\tau(x,y,z) = t$ describe lines in space characterized by constant phase (refer to Equation 11-4 above). And because $t = (k_x x + k_y y + k_z z - \text{constant})/\omega$ is the equation of the planar wave front for the Equation 11-2, the moving surfaces of *constant phase* $t = \tau(x,y,z)$ are called the *wave fronts* of Equation 11-4, and the orthogonal trajectories of these surfaces are known as *rays* (a more precise definition is given later). For Equation 11-2, we have $\tau(x,y,z) = (k_x x + k_y y + k_z z)/\omega$ which, in a single dimension, reduces to $\tau(x) = x/c$. For a comprehensive treatment of eikonal methods, the reader is referred to the books by Cerveny and Ravindra (1971) and Bleistein (1984).

Eikonal equation in nonuniform media. We next consider a weakly inhomogeneous medium, retaining the isotropic assumption, and in particular, Equation 11-1 with our constant c replaced by the variable sound speed c(x,y,z). Substitution of Equation 11-4 in Equation 11-1 shows that, to leading order,

$$(\partial\tau/\partial x)^2 + (\partial\tau/\partial y)^2 + (\partial\tau/\partial z)^2 = 1/c(x,y,z)^2 \quad (11.5)$$

which is analogous to Equation 11-3, although not quite the same. The solution to Equation 11-4, therefore, requires that Equation 11-5 in turn be solved. Equation 11-5 is known as the *eikonal equation*, and it is the nonlinear first-order partial differential equation governing $\tau(x,y,z)$.

Continuing the series. To obtain formulas governing τ and P_k in the ray series of Equation 11-4, we substitute Equation 11-4 into Equation 11-1 with our variable c(x,y,z), and carry out all required expansions. Since the result must hold for any value of ω, the coefficient of each power of ω must vanish.

Equation 11-5 for the phase function τ is obtained as the very first equation in this procedure, and it describes wave kinemetics only. To the next order, the *transport equation* governing the amplitude function P_0 is obtained; it contains, in addition to P_0, the effects of propagation speeds, expressed in terms of functionals of τ and its derivatives.

Integrating the eikonal equation. In Chapter 1, a general solution method for first-order partial differential equations was developed (see

Equation 1-181, and Equations 1-210 to 1-218). We will use it to integrate the eikonal equation. It will be convenient to rearrange Equation 11-5 in the form

$$\{c(x,y,z)/2\}\ \{(\partial\tau/\partial x)^2 + (\partial\tau/\partial y)^2 + (\partial\tau/\partial z)^2\} - 1/2\ \{c(x,y,z)\}^{-1} = 0 \quad (11.6)$$

and, consistently with Equation 1-181, define the function H satisfying

$$H = \{c(x,y,z)/2\}\{(\partial\tau/\partial x)^2 + (\partial\tau/\partial y)^2 + (\partial\tau/\partial z)^2\} - 1/2\ c(x,y,z)^{-1} \quad (11.7)$$

Using Equations 1-210, 1-211, and 1-212, with the parameter λ from Chapter 1 set to s, and the dependent variable θ set to τ, the characteristic equations become

$$dx/ds = \partial H/\partial(\partial\tau/\partial x) = c\ (\partial\tau/\partial x) \quad (11.8)$$

$$dy/ds = \partial H/\partial(\partial\tau/\partial y) = c\ (\partial\tau/\partial y) \quad (11.9)$$

$$dz/ds = \partial H/\partial(\partial\tau/\partial z) = c\ (\partial\tau/\partial z) \quad (11.10)$$

Observe that Equations 1-213 to 1-217 are not required because c(x,y,z) does not depend on time.

Equations 1-214, 1-215, and 1-216 are used to derive the next group of ray equations. We will illustrate the steps required in order to implement Equation 1-214. If we take partial derivatives of Equation 11-7 with respect to x, and simplify the result using Equation 11-5, recognizing that $\partial H/\partial\tau = 0$, we obtain successively

$$\begin{aligned} d(\partial\tau/\partial x)/ds &= -\ \partial H/\partial x \\ &= -\ \{1/2\ c_x\{(\partial\tau/\partial x)^2 + (\partial\tau/\partial y)^2 + (\partial\tau/\partial z)^2\} + 1/2\ c^{-2}c_x\} \\ &= -\ \{1/2\ c_x\{1/c(x,y,z)^2\} + 1/2\ c^{-2}c_x\} \end{aligned} \quad (11.11a)$$

and hence

$$d(\partial\tau/\partial x)/ds = -\ c^{-2}c_x \quad (11.11b)$$

Similarly, Equations 1-215 and 1-216 lead to

$$d(\partial\tau/\partial y)/ds = -\ c^{-2}c_y \quad (11.12)$$

$$d(\partial\tau/\partial z)/ds = -\ c^{-2}c_z \quad (11.13)$$

Finally, since $\partial H/\partial(\partial\tau/\partial t) = 0$, Equation 1-218 requires that

$$\begin{aligned} d\tau/ds &= (\partial\tau/\partial x)\ \partial H/\partial(\partial\tau/\partial x) + (\partial\tau/\partial y)\ \partial H/\partial(\partial\tau/\partial y) \\ &\quad + (\partial\tau/\partial z)\ \partial H/\partial(\partial\tau/\partial z) \\ &= (\partial\tau/\partial x)\ \{c(\partial\tau/\partial x)\} + (\partial\tau/\partial y)\ \{c(\partial\tau/\partial y)\} + (\partial\tau/\partial z)\ \{c(\partial\tau/\partial z)\} \\ &= c\ \{(\partial\tau/\partial x)^2 + (\partial\tau/\partial y)^2 + (\partial\tau/\partial z)^2\} \end{aligned} \quad (11.14a)$$

or, using Equation 11-5,

$$d\tau/ds = 1/c \tag{11.14b}$$

Summary of ray tracing results. We have engaged in a substantial amount of algebra, and it is worthwhile to recapitulate the main results and recipes. In summary, when c(x,y,z) is given, the solution to Equation 11-5 requires our integrating seven nonlinearly coupled ordinary differential equations in the unknowns x, y, z, $\partial\tau/\partial x$, $\partial\tau/\partial y$, $\partial\tau/\partial z$, and τ along a ray, in particular,

$$dx/ds = c(x,y,z)\ (\partial\tau/\partial x) \tag{11.15a}$$

$$dy/ds = c(x,y,z)\ (\partial\tau/\partial y) \tag{11.15b}$$

$$dz/ds = c(x,y,z)\ (\partial\tau/\partial z) \tag{11.15c}$$

$$d(\partial\tau/\partial x)/ds = -\ c^{-2}c_x(x,y,z) \tag{11.15d}$$

$$d(\partial\tau/\partial y)/ds = -\ c^{-2}c_y(x,y,z) \tag{11.15e}$$

$$d(\partial\tau/\partial z)/ds = -\ c^{-2}c_z(x,y,z) \tag{11.15f}$$

$$d\tau/ds = 1/c(x,y,z) \tag{11.15g}$$

subject to prescribed initial conditions on each dependent variable. A simple computational scheme was outlined in Chapter 2 for systems of first order ordinary differential equations.

The foregoing recipe is commonly used in seismic ray tracing. It goes without saying that the implementation is numerically intensive. We will not dwell upon these computational details in this book, but instead, concentrate on a number of basic issues which do arise, that are amenable to theoretical analysis and improvement.

FERMAT'S PRINCIPLE OF LEAST TIME (VIA CALCULUS OF VARIATIONS)

Fermat's *Principle of Least Time* formally states that *time is stationary compared with neighboring paths between the two points.* In order to apply standard methods in the *calculus of variations* (Hildebrand, 1952; Gelfand and Fomin, 1963), we conveniently normalize our independent variable so that the integration is taken over the same fixed range for all paths.

Let an arbitrary path between two points (x_a, y_a, z_a) and (x_b, y_b, z_b) be specified parametrically by $x = x(\mu)$, $y = y(\mu)$, and $z = z(\mu)$, where $0 < \mu < 1$. This parametric representation (with fixed limits $\mu = 0$ and 1) means that a single independent variable can be used to formulate the minimization problem to follow, thus simplifying the mathematics.

Travel time along a ray. Equation 11-15g can be formally integrated along a ray. If τ is the time of travel of the wavefront along the ray, we can certainly write

$$\tau = \int_{ray} ds/c(x,y,z) \tag{11.16}$$

where ds represents the infinitesimal arc length given by

$$ds = \sqrt{(dx^2 + dy^2 + dz^2)} \tag{11.17a}$$

$$= \sqrt{\{(dx/d\mu)^2 + (dy/d\mu)^2 + (dz/d\mu)^2\}}d\mu \tag{11.17b}$$

Then, Equation 11-16 becomes

$$\tau = \int_{ray} \sqrt{(dx^2 + dy^2 + dz^2)}\ /c(x,y,z)$$

$$= \int_{ray} c\{x(\mu),y(\mu),z(\mu)\}^{-1}\sqrt{\{(dx/d\mu)^2 + (dy/d\mu)^2 + (dz/d\mu)^2\}}d\mu \tag{11.18}$$

so that the *total time* T on each wave path is

$$T = \int_0^1 c\{x(\mu),y(\mu),z(\mu)\}^{-1}\{(dx/d\mu)^2 + (dy/d\mu)^2 + (dz/d\mu)^2\}^{1/2}d\mu \tag{11.19}$$

Calculus of variations. *Calculus of variations* is the branch of advanced mathematics aimed at solving maximal and minimal problems in continuum mechanics (Hildebrand, 1952; Gelfand and Fomin, 1963). From a standard argument, the $x(\mu)$, $y(\mu)$, and $z(\mu)$ values for a stationary value of *any* integral

$$\int_0^1 F\{x(\mu),y(\mu), z(\mu),dx(\mu)/d\mu,dy(\mu)/d\mu,dz(\mu)/d\mu\}d\mu \tag{11.20}$$

must satisfy the variational *Euler equations* defined by

$$d\{\partial F/\partial(dx/d\mu)\}/d\mu\ - \partial F/\partial x = 0 \tag{11.21}$$

$$d\{\partial F/\partial(dy/d\mu)\}/d\mu\ - \partial F/\partial y = 0 \tag{11.22}$$

$$d\{\partial F/\partial(dz/d\mu)\}/d\mu\ - \partial F/\partial z = 0 \tag{11.23}$$

Let us consider Equation 11-21 in detail, in order to understand its implications more fully. We now wish to explore the requirements for minimal total travel time. If we set F equal to the travel time *definition* found in Equation 11-19, that is,

$$F = c\{x(\mu),y(\mu),z(\mu)\}^{-1}\{(dx/d\mu)^2 + (dy/d\mu)^2 + (dz/d\mu)^2\}^{1/2} \tag{11.24}$$

we straightforwardly find that

$$\partial F/\partial(dx/d\mu) = c^{-1}\ (1/2)\{\ \}^{-1/2}\ 2\ (dx/d\mu) = c^{-1}\{\ \}^{-1/2}\ dx/d\mu \quad (11.25)$$

and

$$\partial F/\partial x = -\ c^{-2}\ c_x\ \{\ \}^{1/2} \quad (11.26)$$

Then, Equation 11-21 becomes, using Equations 11-25 and 11-26,

$$d/d\mu\ [\ c^{-1}\{\ \}^{-1/2}\ dx/d\mu] + c^{-2}\ c_x\ \{\ \}^{1/2} = 0 \quad (11.27)$$

or, upon division by the length $\{\ \}^{1/2}$,

$$\{\ \}^{-1/2}\ d/d\mu\ [\ c^{-1}\{\ \}^{-1/2}\ dx/d\mu] + c^{-2}\ c_x\ = 0 \quad (11.28)$$

Now, since Equation 11-17 implies that

$$\{(dx/d\mu)^2 + (dy/d\mu)^2 + (dz/d\mu)^2\}^{1/2} d\mu = ds \quad (11.29)$$

Equation 11-28 simplifies to the requirement

$$d/ds\ \{c^{-1}dx/ds\} + c^{-2}\ c_x\ = 0 \quad (11.30)$$

for the actual path between two points, taken by the signal which renders the time of travel stationary (or less correctly, the ray representing the path which requires the least time for a signal to travel from the first point to the second).

Eikonal solution satisfies least time condition. So far, we have only asked for the constraints needed to minimize the integral or definition in Equation 11-19, and the result is the requirement defined by Equation 11-30. How can this requirement be achieved in reality?

The dynamics of the physical problem have not yet been invoked in deriving Equation 11-30. *At this point*, we consider Equation 11-15a, which states that $(1/c)\ dx/ds\ = \partial\tau/\partial x$. Together with Equation 11-15d, this implies that $d\{(1/c)dx/ds\}/ds = -\ c^{-2}c_x(x,y,z)$, which is *identical* to variational condition required by Equation 11-30*!*

Similar arguments apply to the y and z components of the wave flow. But the bottom line is this: *solutions of the eikonal equation in Equation 11-5, which arise from Equation 11-1, satisfy Fermat's Principle of Least Time.* Equation 11-1 with a variable c(x,y,z), again, assumes isotropic, nondissipative media, and it is *this* specialized wave model that is consistent with Fermat's principle.

FERMAT'S PRINCIPLE REVISITED VIA KINEMATIC WAVE THEORY

In the previous section, we found using the calculus of variations that the integral in Equation 11-19 can be minimized provided Equation 11-30 is satisfied. Then, we showed that solutions to the eikonal equation based on the classical undamped wave model in Equation 11-1 satisfied this requirement.

Here, we will rederive Fermat's Principle starting with the *kinematic wave theory* formalism of Chapter 2. The three-dimensional KWT ray equations are given in Equations 2-147 to 2-152; for simplicity, we discuss the x-component of wave propagation only. Equations 2-147 and 2-150 are

$$dk_x/dt = -\Omega^r_x \tag{11.31}$$

$$dx/dt = \Omega^r_{kx} \tag{11.32}$$

Consistently with Equation 11-3, which is strictly applicable to Equation 11-1, we assume that

$$\Omega^r = c(x,y,z)\{k_x^2 + k_y^2 + k_z^2\}^{1/2} \tag{11.33}$$

Equations 11-32 and 11-33 yield $dx/dt = c(x,y,z)\, k_x\{k_x^2 + k_y^2 + k_z^2\}^{-1/2}$ or

$$\{k_x^2 + k_y^2 + k_z^2\}^{1/2}/c\; dx/dt = k_x \tag{11.34}$$

Now, on combining Equations 11-31 and 11-33, we obtain the relationship $dk_x/dt = -\, c_x(x,y,z)\{k_x^2 + k_y^2 + k_z^2\}^{1/2}$. This result can be rewritten as

$$1/c\; dk_x/dt = -\, \{c_x(x,y,z)/c\}\{k_x^2 + k_y^2 + k_z^2\}^{1/2} \tag{11.35}$$

Next, we substitute Equation 11-34 for k_x into Equation 11-35, to obtain

$$\begin{aligned} 1/c\; d/dt\, &\{k_x^2 + k_y^2 + k_z^2\}^{1/2}/c\; dx/dt \\ &= -\, \{c_x(x,y,z)/c\}\{k_x^2 + k_y^2 + k_z^2\}^{1/2} \end{aligned} \tag{11.36}$$

For dynamically steady problems, the expression in Equation 11-33 is equal to the constant excitation frequency ω_0, that is, $c(x,y,z)\{k_x^2 + k_y^2 + k_z^2\}^{1/2} = \omega_0$, so that

$$\{k_x^2 + k_y^2 + k_z^2\}^{1/2} = \omega_0/c(x,y,z) \tag{11.37}$$

Along a ray trajectory, we have

$$c(x,y,z)\; dt = ds \tag{11.38}$$

where ds is a differential length. Substitution of Equations 11-37 and 11-38 into Equation 11-36 leads to

$$d/ds \{1/c(x,y,z)\} dx/ds = -c_x(x,y,z)/c^2 \qquad (11.39)$$

which is exactly the same as the condition $d/ds \{c^{-1}dx/ds\} + c^{-2} c_x = 0$ given by Equation 11-30 (the frequency ω_0 importantly drops out entirely). We have shown, this time using our kinematic wave theory formalism, that wave motions satisfying Equation 11-37 also follow Fermat's Principle.

In general, wave motions need not follow minimization rules, since these apply to restrictive classes of problems such as that governed by Equation 11-1. However, kinematic wave theory will always embody the correct physics relevant to a particular formulation, and consistently reproduce these rules, if warranted. That is, *there is nothing in Fermat's Principle that is not already contained in the wave conservation formalism developed in Chapter 2.*

Again, we emphasize that, like Equation 11-1, the assumptions underlying Equations 11-31 and 11-32 preclude wave dissipation; hence, Fermat's Principle does *not* apply over large space and time scales when the cumulative effects of small dissipation due to oil and gas bearing formations become important. We will develop an alternative model later.

MODELING WAVE DISSIPATION

To motivate our discussion of dissipation modeling, and to accentuate the limitations of conventional analysis, it is useful to review some elementary examples where exact results can be obtained.

Example 11-1: A simple model. In Chapter 1, we considered two simple dissipative wave examples. In the first, we studied the one-dimensional damped wave equation

$$\rho\, \partial^2 u/\partial t^2 + \gamma\, \partial u/\partial t - T\, \partial^2 u/\partial x^2 = 0 \qquad (11.40)$$

The substitution $u(x,t) = A \exp i(kx-\omega t)$ led to the complex dispersion relation $\rho\omega^2 = Tk^2 - i\,\gamma\omega$, which motivated us to assume a complex frequency $\omega(k) = \omega_r(k) + i\,\omega_i(k)$. Direct substitution gave the real dispersion relation

$$\rho\, (\omega_r^2 - \omega_i^2) = Tk^2 + \gamma\, \omega_i \qquad (11.41)$$

and the imaginary frequency

$$\omega_i = -\gamma /2\rho \qquad (11.42)$$

For weakly dissipative problems, Equation 11-41 can be approximated by the nondissipative solution $\rho\omega_r^2 = Tk^2$. But the more complete relation indicates that propagation characteristics *are* affected by dissipation over large space and time scales. The result $\omega_i = -\gamma/2\rho$ is interesting from a different point of view. The dissipation rate is a *constant* independent of k, and therefore, independent of excitation frequency (this may not apply to specific applications).

Example 11-2: Another case history. In this second review example, recall that we previously sought *dynamically steady* solutions to the damped equation

$$\partial^2u/\partial t^2 + \zeta\, \partial u/\partial t - c^2\, \partial^2u/\partial x^2 = 0 \tag{11.43}$$

satisfying $u(0,t) = A \cos \omega t$. The result took the form

$$u(x,t) = A \exp \{ (\omega^4/c^4 + \omega^2\zeta^2/c^4)^{1/4}\cos\{\tfrac{1}{2} \arctan(-\zeta/\omega) + \pi\}\}x$$
$$\times \cos \{\omega t + (\omega^4/c^4 + \omega^2\zeta^2/c^4)^{1/4}\sin\{\tfrac{1}{2} \arctan(-\zeta/\omega) + \pi\}x\} \tag{11.44}$$

Equation 11-44 is reminiscent of the $u(x,t) \cong A \cos \omega(t - x/c)$ for undamped waves. But the exact argument shows that dissipation *does* affect the wave phase "$\omega(t - x/c)$", and in addition, yields the expected exponential decrease in amplitude.

Example 11-3: Motivating damped wave study. Let us now consider the three-dimensional wave equation with classical damping, that is,

$$\partial^2u/\partial t^2 + \zeta\, \partial u/\partial t - c^2 (\partial^2u/\partial x^2 + \partial^2u/\partial y^2 + \partial^2u/\partial z^2) = 0 \tag{11.45}$$

Following Example 11-1, the substitutions

$$u(x,y,z,t) = A \exp i(k_x x - k_y y - k_z z - \omega t) \tag{11.46}$$

$$\omega(k) = \omega_r(k) + i\, \omega_i(k) \tag{11.47}$$

lead to

$$\omega_r^2 - \omega_i^2 = c^2(k_x^2 + k_y^2 + k_z^2) + \zeta\, \omega_i \tag{11.48}$$

and the imaginary frequency

$$\omega_i = -\zeta/2 \tag{11.49}$$

The damping rate, as in Equation 11-42, is again a *constant* independent of k, and therefore, the wave frequency.

The quality factor Q. In geophysical problems, the *dimensionless measure of attenuation Q* is commonly used (Aki and Richards, 1980; Toksoz and Johnston, 1981; Kennett, 1983; Bourbie, Coussy, and Zinszner, 1987). If a volume of material is cycled in stress at a fixed frequency, Q can be defined as

$$1/Q = -\Delta E/2\pi E \tag{11.50}$$

where E is the peak strain energy stored in the volume, and $-\Delta E$ is the energy lost *per cycle* due to material imperfections.

A detailed discussion on attenuation modeling is given in Chapter 2 dealing with kinematic wave formalisms. There, this laboratory Q is recast into a form useful for standing and propagating wave analyses. The propagation result for wave *amplitude* (as opposed to *energy*) was obtained as

$$A(x) = A_0 \exp(-\Omega_r x/2cQ) \tag{11.51}$$

Standard geophysical conventions (Clay, 1990) often take propagating amplitude in the form

$$A(x) = A_0 \exp(-\alpha x) \tag{11.52}$$

with the *definition*

$$\alpha = \pi f/cQ \tag{11.53}$$

where f is the frequency in Hertz and Q may additionally depend on frequency (since $\omega = 2\pi f$, an alternative to Equation 11-53 is $\alpha = \omega/2cQ$, which is consistent with the conclusion drawn by directly comparing Equations 11-51 and 11-52).

The kinematic wave model developed in Chapter 2 renders Q's determined from standard experimental procedures for uniform media (Toksoz and Johnston, 1981) usable in general wave propagation modeling studies over heterogeneous formations covering large space and time scales.

A simple example. Wave dissipation occurs because of viscous fluid losses in rock and frictional losses in solids. The *damping* or *quality factor* Q depends on the type of rock; it is nearly constant in the seismic frequency range for most rocks (its value varies from ten to several hundred for common rocks). Geophysicists are interested in the dependence of Q on on rock structure, pore fluid, and external forces such as overburden pressure.

Fluids, by contrast, are characterized by viscous losses, and α is approximately proportional to f^2 (Clay, 1990). Toksoz and Johnston (1981), for example, give an approximate law for *seismic* damping at low frequencies,

$$A = A_0 e^{-\alpha x} \cos 2\pi f(t - x/c) \tag{11.54}$$

$$\alpha = \pi f^2/cf_0 \tag{11.55}$$

where α is an absorption coefficient, f represents frequency (in Hertz), c is the sound speed, and the positive *transition frequency* f_0 is empirically determined constant of the formation.

We stress that more detailed models taking into account fluid saturation, pore pressure, rock properties, and so on, are available in the literature, e.g.,

Yew and Jogi (1976), White (1984), De la Cruz and Spanos (1985), McCann and McCann (1985), and Hassanzadeh (1991).

The use of Equations 11-54 and 11-55, for illustrative purposes only, does *not* imply any agreement or disagreement with present research views on dissipation modeling. Our objective is a tutorial one, emphasizing the development of high order corrections to the eikonal equation, which will be different for different dissipation models.

Because our kinematic wave formalism for damped oscillations was developed entirely in terms of the imaginary frequency function $\Omega^i(k)$, we need to determine the $\Omega^i(k)$ consistently with Equations 11-54 and 11-55. Now, since $f = \omega/2\pi$ and $\omega = ck$, we can write rewrite Equation 11-55 in the wavenumber-explicit form $\alpha = \omega^2/(4cf_0\pi) = c^2k^2/(4cf_0\pi) = ck^2/(4f_0\pi)$. Then, because $x = ct$, the exponential term in Equation 11-54 can be manipulated to give an expression for the imaginary frequency $\Omega^i(k)$, that is,

$$\begin{aligned} \exp(-\alpha x) &= \exp\{-\{ck^2/(4f_0\pi)\}\, x\} \\ &= \exp\{-\{c^2k^2/(4f_0\pi)\}t\} \\ &= \exp\{\Omega^i(k)t\} \end{aligned} \tag{11.56}$$

where we have followed the sign convention of Chapter 1 (refer to the discussion immediately following Equation 2-30). Thus,

$$\Omega^i(k) = -\, c^2k^2/(4f_0\pi) < 0 \tag{11.57}$$

as required. For this model,

$$\Omega^i{}_k(k) = -\, c^2k/(2f_0\pi) < 0 \tag{11.58}$$

$$\Omega^i{}_{kk}(k) = -\, c^2/(2f_0\pi) < 0 \tag{11.59}$$

$$\Omega^i{}_{kkk}(k) = \Omega^i{}_{kkkk}(k) = \Omega^i{}_{kkkkk}(k) = \ldots = 0 \tag{11.60}$$

so that simplifications to the general high-order kinematic wave modulation equations of Chapter 2 are definitely possible. In the next section, we will show how formulas such as these will lead to improved modulation laws that apply, even when Fermat's Principle does not.

RAY TRACING OVER LARGE SPACE-TIME SCALES

The low-order kinematic wave theory described by Equations 2-41 to 2-70 is consistent with the fact that wave dissipation does *not* affect phase at early times. Over large space and time scales, the examples of Chapter 1 show that ray trajectories are affected, as are the corresponding amplitudes calculated along wave fronts.

In *surface seismics*, typical propagation distances are on the order of *tens* of feet, and phase changes are not important. In *crosswell seismics*, however, ray paths extend hundreds and thousands of feet, so that nondissipative interpretation models, e.g., $(\partial\tau/\partial x)^2 + (\partial\tau/\partial y)^2 + (\partial\tau/\partial z)^2 = 1/c(x,y,z)^2$, can be inherently incorrect whenever attenuation arising from oil and gas-bearing formations exists. Higher order accuracy is required.

In tomographic analysis, care must be taken to account for the cumulative effects of locally small dissipative terms, a consideration emphasized in many areas of continuum mechanics (Whitham, 1974). For simplicity, we will restrict ourselves to one-dimensional applications; extensions to three dimensions, implicit in Equations 2-130 to 2-158, will be left as exercises for reader.

High-order modulation equations. In simple one-dimensional problems, the uniform plane wave is in general characterized by a complex dispersion relation of the form

$$\Omega(k) = \Omega^r(k) + i\Omega^i(k) \tag{11.61}$$

We showed in Chapter 2 that the slowly varying wavetrain (for example, the large-time asymptotic solution for an exploding point source) satisfies the coupled phase and amplitude laws given by Equations 2-94 and 2-95. Again, the phase relation and amplitude law shown

$$\begin{aligned}\omega = {} & \Omega^r(k) + a_x\Omega^i{}_k(k)/a + 1/2\, k_x\Omega^i{}_{kk} - 1/2\, a_{xx}\Omega^r{}_{kk}/a \\ & - 1/2\, a_x k_x \Omega^r{}_{kkk}/a - 1/8\, k_x{}^2\Omega^r{}_{kkkk} - 1/6\, k_{xx}\Omega^r{}_{kkk}\end{aligned} \tag{11.62}$$

$$\begin{aligned}\partial a^2/\partial t + \partial(\Omega^r{}_k a^2)/\partial x = {} & 2\Omega^i(k)a^2 \\ & - (aa_{xx}\Omega^i{}_{kk} + aa_x k_x \Omega^i{}_{kkk} \\ & + 1/4\, a^2 k_x{}^2\Omega^i{}_{kkkk} + 1/3\, a^2 k_{xx}\Omega^i{}_{kkk}) \\ & + (1/3\, aa_{xxx}\Omega^r{}_{kkk} + 1/2\, aa_{xx}k_x\Omega^r{}_{kkkk} \\ & + 1/3\, aa_x k_{xx}\Omega^r{}_{kkkk} + 1/12\, a^2 k_{xxx}\Omega^r{}_{kkkk} \\ & + 1/6\, a^2 k_x k_{xx}\Omega^r{}_{kkkkk} + 1/4\, aa_x k_x{}^2\Omega^r{}_{kkkkk} \\ & + 1/24\, a^2 k_x{}^3\Omega^r{}_{kkkkkk})\end{aligned} \tag{11.63}$$

are coupled through the wave crest conservation law

$$\partial k/\partial t + \partial \omega/\partial x = 0 \tag{11.64}$$

The low-order limit. Let us explore the consequences of Equations 11-62 to 11-64 in the low-order limit, and understand exactly how Fermat's Principle arises as a consequence. Following Chapter 2, Equation 11-64 is expanded in the form $\partial k/\partial t + \Omega^r{}_k \partial k/\partial x = -\Omega^r{}_x$ using the chain rule. Then, since the total differential $dk = \partial k/\partial t\ dt + \partial k/\partial x\ dx$, we have $dk/dt = \partial k/\partial t + dx/dt\ \partial k/\partial x$. Direct comparison leads to $dk_x/dt = -\Omega^r{}_x$ and $dx/dt = \Omega^r{}_{kx}$ ... that is, Equations 11-31 and 11-32, which led to Fermat's Principle of Least Time.

Extended eikonal equations. Now that we understand how the principle of least time arises as a low-order consequence of kinematic wave theory, we are in a position to extend the overall model so that problems with both dissipation and dispersion can be simulated.

To do this, we understand that the frequency ω in Equation 11-64 must allow an amplitude dependence which is important over large scales, in addition to the usual dependence on k and possibly x, as suggested in Equation 11-62. Thus, we need to write $\partial k/\partial t + \partial \omega/\partial x = 0$ as

$$\partial k/\partial t + \partial \omega(k,a,x,t)/\partial x = 0 \tag{11.65}$$

$$\partial k/\partial t + \omega_k \partial k/\partial x + \omega_a \partial a/\partial x + \omega_x = 0 \tag{11.66}$$

When Equation 11-62, modified to include any heterogeneities that may exist in a particular application, is available explicitly, Equation 11-66 and Equation 11-63 (similarly altered to include the low order radiation stress terms implied in $\partial E/\partial t + \partial(\Omega^r{}_k(k,x,t)E)/\partial x = E\{2\Omega^i + \Omega^r{}_t(k,x,t)/\Omega^r\}$) provide the required *extended eikonal equations.* At this order, an additional coupling with wave amplitude exists, so that the motion in not entirely kinematic or amplitude-independent.

Extended eikonal equation in homogeneous medium. To illustrate the basic ideas with a concrete example, we consider large space-time wave propagation in a uniform homogeneous medium. In this limit, the explicit dependence on x disappears, that is, $\omega(k,a,x,t)$ becomes $\omega(k,a)$, and ω_x vanishes. The governing equations for k(x,t) and a(x,t) reduce to

$$\partial k/\partial t + \omega_k \partial k/\partial x + \omega_a \partial a/\partial x = 0 \tag{11.67}$$

where

$$\omega = \Omega^r(k) + a_x \Omega^i{}_k(k)/a + 1/2\ k_x \Omega^i{}_{kk} - 1/2\ a_{xx}\Omega^r{}_{kk}/a \tag{11.68}$$
$$- 1/2\ a_x k_x \Omega^r{}_{kkk}/a - 1/8\ k_x{}^2 \Omega^r{}_{kkkk} - 1/6\ k_{xx}\Omega^r{}_{kkk}$$

and

$$\begin{aligned}\partial a^2/\partial t + \partial(\Omega^r{}_k a^2)/\partial x = \; & 2\Omega^i(k)a^2 \\ & - (aa_{xx}\Omega^i{}_{kk} + aa_x k_x \Omega^i{}_{kkk} \\ & + 1/4\, a^2 k_x{}^2 \Omega^i{}_{kkkk} + 1/3\, a^2 k_{xx} \Omega^i{}_{kkk}) \\ & + (1/3\, aa_{xxx}\Omega^r{}_{kkk} + 1/2\, aa_{xx} k_x \Omega^r{}_{kkkk} \\ & + 1/3\, aa_x k_{xx} \Omega^r{}_{kkkk} + 1/12\, a^2 k_{xxx} \Omega^r{}_{kkkk} \\ & + 1/6\, a^2 k_x k_{xx} \Omega^r{}_{kkkkk} + 1/4\, aa_x k_x{}^2 \Omega^r{}_{kkkkk} \\ & + 1/24\, a^2 k_x{}^3 \Omega^r{}_{kkkkkk})\end{aligned} \tag{11.69}$$

Equations 11-67 to 11-69 must be solved subject to the usual initial and boundary conditions, which vary from problem to problem.

The seismic limit. Despite the simplifications, Equations 11-67 to 11-69 are nonetheless complicated, and therefore form the basis for ongoing research. However, they can be simplified, using facts obtained from empirical observation.

Example 11-4: Simple rock formations. The above transient equations apply to *dissipative and dispersive* waves in one-dimensional *homogeneous* media. Again, in kinematic wave modeling, the exact differential equations for the particular seismic motion need not be available: the functions $\Omega^r(k)$ and $\Omega^i(k)$ may be determined empirically in the laboratory if desired.

Examples 11-1 to 11-3 show how damping affects phase and amplitude, but they apply specifically to "$\partial u/\partial t$" dissipation, and then only to the classical wave equation. This damping may or may not be relevant to all facets of seismics. In general, different fluid-bearing rocks may damp differently, depending on pore pressure, fluid type, rock matrix, and so on, and $\Omega^i(k)$ must be left free in a general formulation. The real frequency $\Omega^r(k)$, likewise, need not be simple.

We will explore one specific limit of the foregoing modulation equations. We will assume that the waves in question are *nondispersive*, their phase and group velocities being identical and independent of k. This assumption is noncontroversial. Thus, when "c" is a constant sound speed, we may write

$$\Omega^r(k) = ck \tag{11.70}$$

so that

$$\Omega^r{}_k(k) = c \tag{11.71}$$

$$\Omega^r{}_{kk}(k) = \Omega^r{}_{kkk}(k) = \Omega^r{}_{kkkk}(k) = \ldots = 0 \tag{11.72}$$

We also invoke the attenuation model used previously, that is, take the imaginary frequency in the form

$$\Omega^i(k) = -\, c^2k^2/(4f_0\pi) < 0 \tag{11.73}$$

For this model, the following derivatives are found,

$$\Omega i_k(k) = - c^2k/(2f_0\pi) < 0 \tag{11.74}$$

$$\Omega i_{kk}(k) = - c^2/(2f_0\pi) < 0 \tag{11.75}$$

$$\Omega i_{kkk}(k) = \Omega i_{kkkk}(k) = \Omega i_{kkkkk}(k) = \ldots = 0 \tag{11.76}$$

All of these simplifications serve to reduce our general high-order kinematic wave modulation to mathematically tractable form. In particular, Equations 11-62 and 11-63 simplify to

$$\omega = ck - c^2ka_x/(2f_0\pi a) - 1/2\ c^2k_x/(2f_0\pi) \tag{11.77}$$

$$\partial a^2/\partial t + c\ \partial a^2/\partial x = - c^2k^2a^2/(2f_0\pi) + c^2aa_{xx}/(2f_0\pi) \tag{11.78}$$

For *dynamically steady problems* excited by a constant frequency ω_0, the propagation equations reduce to

$$ck - c^2ka_x/(2f_0\pi a) - 1/2\ c^2k_x/(2f_0\pi) = \omega_0 \tag{11.79}$$

$$c\ da^2/dx = - c^2k^2a^2/(2f_0\pi) + c^2aa_{xx}/(2f_0\pi) \tag{11.80}$$

which is a set of nonlinearly coupled ordinary differential equations in a(x) and k(x) easily solved by conventional numerical methods. We will not pursue computational solutions in this chapter, but rather, discuss the consequences associated with different high-order terms in a general manner, and establish their special relevance to ray tracing, cross-well tomography, and *first arrival time* analysis. Before delving into these seismic issues, we set the stage for our discussion by referring to some fundamental ideas.

SUBTLE HIGH-ORDER EFFECTS

The role of high-order derivatives in the theory of nonlinear equations forms a relatively new area of study in applied mathematics just decades old. Many of the already established ideas, understandably, have not yet found their way into engineering practice, by virtue of their highly specialized nature.

The following discussion expands upon ideas elucidated in Chin (1980), developed in the contexts of water waves and general continuum mechanics, and Chin (1993a,b), developed in the context of low-order saturation discontinuities and high-order capillary pressure effects in reservoir simulation.

A low-order nonlinear wave equation. Let us consider the first-order, convective, nonlinear wave equation for a function u(x,t) satisfying

$$\partial u/\partial t + u\ \partial u/\partial x = 0 \tag{11.81}$$

Equation 11-81 is not unlike

$$\partial k/\partial t + \Omega'_k(k)\ \partial k/\partial x = 0 \tag{11.82}$$

describing low-order wave conservation in a homogeneous medium.

Equation 11-81 possesses a simple general solution assuming very arbitrary initial conditions. To construct it, we recall that the total differential for any function $u(x,t)$ satisfies $du = \partial u/\partial t\ dt + \partial u/\partial x\ dx$. Thus,

$$du/dt = \partial u/\partial t + dx/dt\ \partial u/\partial x \tag{11.83}$$

If we compare Equations 11-81 and 11-83, we find that $du/dt = 0$ provided we identify $dx/dt = u$; in other words, $u(x,t)$ does not change if we follow the ray path $dx/dt = u(x,t)$. This can be succinctly expressed by the equivalent functional statement

$$u(x,t) = G(x-ut) \tag{11.84}$$

which emphasizes that $u(x,t)$ must be a function of "x-ut". G can be any function, of course. If, in the problem at hand, we are to enforce the initial condition

$$u(x,0) = F(x) \tag{11.85}$$

where $F(x)$ is given, then it is clear that the choice $G = F$ solves the problem. Hence, we have the general solution

$$u(x,t) = F\{x-u(x,t),t\} \tag{11.86}$$

Singularities in the low-order model. As in the case of Equation 11-82, Equation 11-86 may or may not lead to shock formation, infinities, or singularities in the first derivatives. To see how these may arise, let us differentiate Equation 11-86 with respect to x; using the chain rule, we obtain $\partial u/\partial x = F'\{x-u(x,t)t\}\{1 - t\partial u/\partial x\}$. If we solve for $\partial u/\partial x$, we find that

$$\partial u/\partial x = F'\{x-u(x,t)t\}/\{1 + tF'\} \tag{11.87}$$

If the initial condition $u(x,0) = F(x)$ is such that $F' > 0$, it follows that the denominator $1 + tF' > 0$ is positive, and $\partial u/\partial x$ is well behaved. On the other hand, if F' is negative, shockwave solutions with infinite values of $\partial u/\partial x$ form in a finite amount of time. These shocks are mathematically analogous to the saturation discontinuities encountered in reservoir flow and the pressure jumps obtained in the theory of gasdynamic shocks.

Existence of the singularity. We have shown that Equation 11-86 embodies a class of solutions containing shockwaves. *But do these exist in reality?* Although shocks can form as a mathematical consequence of Equation

11-86, it is often the case that Equation 11-81 arises only as a rough model to a more accurate problem formulation containing more detailed information.

Fortunately, for us, *exact* solutions to two higher-order equations whose low-order terms are identical to Equation 11-81 are available; these are, namely, Burger's equation and the Korteweg de Vries equation. *Burger's equation*, which arises in the description of gasdynamic shocks in high speed aerodynamics, is given by

$$\partial u/\partial t + u\ \partial u/\partial x = \varepsilon\ \partial^2 u/\partial x^2 \qquad (11.88)$$

where $\varepsilon > 0$, a "small number", is related to the viscosity of the fluid, and the term $\varepsilon\ \partial^2 u/\partial x^2$ as a whole represents *viscous diffusion*.

The exact *Cole-Hopf solution* is discussed extensively in Whitham (1974). It is possible to show that physical models that satisfy Equation 11-88 can be modeled by Equation 11-81, which is considerably simpler. In other words, shocked solutions of Equation 11-81 are also obtained from the more detailed description offered by Equation 11-88.

Why is this *not* intuitively obvious? This issue is subtle because we have not stated *what* ε is small with respect to; when spatial gradients, as obtained from Equation 11-81 become large, the $\varepsilon\ \partial^2 u/\partial x^2$ term in the more complete model of Equation 11-88 may no longer be small by comparison to the left hand side. Thus, Equation 11-81 may or may not apply near the shock, and direct recourse to the detailed physical model must be made.

Now consider the *Korteweg deVries equation*, obtained in the study of long, inviscid, water waves. Instead of Equation 11-88, we have

$$\partial u/\partial t + u\ \partial u/\partial x = \delta\ \partial^3 u/\partial x^3 \qquad (11.89)$$

where $\delta > 0$ is also a "small number". While Equations 11-88 and 11-89 differ only in the order of the "small" right-side term, the solution structure governing Equation 11-89 is completely different.

An exact solution to the general initial value problem, using recent methods in *inverse scattering* theory, is again available and described in Whitham (1974). It turns out, interestingly, that Equation 11-89 does not admit any solutions with shocks. Thus, even if δ *is* small, Equation 11-81 is *never* relevant as a simplified model.

Our point is this: the high-order terms in Equations 11-88 and 11-89 control the dynamics of the solution at all scales. If shocks exist, their structure and thickness are completely determined by the diffusive term $\varepsilon\ \partial^2 u/\partial x^2$. In gasdynamics, this term is related to the viscous stress tensor; in reservoir engineering, the analogous high-order term is dictated by the capillary pressure function.

If high-order *dispersive* terms such as that in Equation 11-89 exist, e.g., the terms related to the high order wavenumber derivatives of the real frequency in Equations 11-62 and 11-63, shocks may never form. At any rate, the interplay between high-order diffusion (as in Equation 11-88) and high-order dispersion is very subtle.

Entropy conditions. We have emphasized that once the high-order model is agreed upon, say Equation 11-88 or 11-89, the complete physical description of the problem is self-contained. That is, the *entropy conditions* one normally "pulls from hats" in thermodynamics courses can be straightforwardly obtained through simple integration by parts.

To see that this is so, let us consider Equation 11-88. For simplicity, we will take the position of an observer moving with the shock speed, so that the model

$$u \, \partial u/\partial x = \varepsilon \, \partial^2 u/\partial x^2 \tag{11.90}$$

applies locally in that coordinate system. We can certainly rewrite Equation 11-90 in the conservation form

$$\partial\{1/2 \, u^2 - \varepsilon \, \partial u/\partial x\}/\partial x = 0 \tag{11.91}$$

If we integrate from one side of the shock to the other, where each side is represented by uniform thermodynamic conditions with vanishing $\partial u/\partial x$'s, it is clear that $\{1/2 \, u^2 - \varepsilon \, \partial u/\partial x\}_{upstream} = \{1/2 \, u^2 - \varepsilon \, \partial u/\partial x\}_{downstream}$, or more conveniently,

$$u^2_- = u^2_+ \tag{11.92}$$

This "jump condition" is analogous to the global mass conservation constraint enforced through the Buckley-Leverett constraint in reservoir engineering, or the enthalpy laws used in modeling gasdynamic discontinuities.

Now let us, for example, multiply Equation 11-90 by u(x) throughout, so that $u^2 \, \partial u/\partial x = \varepsilon \, u \, \partial^2 u/\partial x^2$. This can be recast in the form $\partial(1/3 \, u^3)/\partial x = \varepsilon \, u \, \partial^2 u/\partial x^2$. If we now integrate by parts, we have

$$(1/3 \, u^3)_+ - (1/3 \, u^3)_- = \varepsilon \, [\{u \, \partial u/\partial x - \textstyle\int(\partial u/\partial x)^2 \, dx \}_+ - \{u \, \partial u/\partial x - \textstyle\int(\partial u/\partial x)^2 \, dx \}_-] \tag{11.93}$$

The $\partial u/\partial x$ terms on either side of the shock vanish identically, but the positive definite integral does not. This leaves

$$(1/3 \, u^3)_+ - (1/3 \, u^3)_- = - \varepsilon \textstyle\int (\partial u/\partial x)^2 \, dx < 0 \tag{11.94}$$

and hence the *entropy condition* $(1/3 \, u^3)_- > (1/3 \, u^3)_+$ or

$$(u^3)_- > (u^3)_+ \tag{11.95}$$

Thus, we have shown that entropy conditions need not be derived at "independently" via thermodynamic considerations; they, and indeed all of the physics, can be obtained naturally once the structure of the high-order derivative is known with confidence.

Additional entropy conditions can, obviously, be generated by multiplying Equation 11-90 by other powers of u(x), or even other functionals of u(x), and then integrating; this is left as an exercise to the reader. We might note that the more complete model in Equation 11-88 furnishes the complete physical structure of the narrow zone where the characteristics of Equation 11-81 would have intersected. Its solution is expected to be unique, whereas a unique solution to Equation 11-81 will involve additional contraints ("jump conditions" or "shock-fitting" formulas) related to anticipated global conserved quantities and entropy conditions.

TRAVEL TIME MODELING

While the subject of high-order derivative terms is in itself of fundamental interest, and perhaps insofar as accurate ray tracing in the presence of dissipation and dispersion is concerned, it is its role in explaining some controversial results encountered by recent authors that is by far the most fascinating.

Applications to crosswell tomography. In crosswell tomography, an array of acosutic transmitters (or, a single transmitter moved among several different physical locations) located within a wellbore, transmits signals that are received at nearby wells instrumented along their lengths with receiver arrays.

First arrival time data is used to provide *velocity estimates* for the formation *velocity model* between the transmitters and the receivers. In most applications, the velocity function c(x,y,z) is assumed by trial and error, and the ray equations

$$dx/dt = c(x,y,z) \tag{11.96}$$

$$dy/dt = c(x,y,z) \tag{11.97}$$

$$dz/dt = c(x,y,z) \tag{11.98}$$

are numerically integrated in time. If the computed first arrival times corresponding to the guess for c(x,y,z) are correct for all transmitter and receiver points, the conjecture for c(x,y,z) is assumed to be representative of the actual velocity distribution in the underground formation.

Of course, the solution may be nonunique, but that represents a separate issue. Additional constraints, beyond those called for in first arrival time matching, can be invoked. One possibility involves the matching of *relative received energies* detected at the receivers. These can be calculated using the energy and momentum equations obtained in Chapter 2, in particular,

$$\partial E/\partial t + \partial(\Omega^r_k(k,x,t)E)/\partial x = E\{2\Omega^i + \Omega^r_t(k,x,t)/\Omega^r\} \quad (2.104)$$

$$\partial M/\partial t + \partial(\Omega^r_k(k,x,t)M)/\partial x = M\{2\Omega^i - \Omega^r_x(k,x,t)/k\} \quad (2.105)$$

There are two methods for accomplishing the required ray integration discussed earlier. In *shooting methods* for ray tracing, a *fan* of rays is shot from one point in the general direction of the other. The denser the fan, the greater the accuracy, and the higher the probability of successful integration (there is no guarantee that any particular ray will reach the desired receiver).

In *bending methods*, an (incorrect) guess for the ray path provides an initial guess. This conjectured path is bent by perturbation methods until it satisfies a minimum travel-time criterion; thus, the method, following our earlier arguments, implicitly assumes a nondissipative system. Additional information about the formation can be obtained from the more complete time trace containing reflections, but this subject is beyond the scope of this chapter.

These methods have, in addition to computation intensive requirements, significant disadvantages. Shooting methods often experience difficulty finding correct rays in shadow zones, while bending methods do give answers there. However, in both methods, it is possible that the solution is only a local minimum, with the global minimum travel time and corresponding ray path remaining unknown.

Other methods have also appeared in the literature. Like the bending method, a recent study by Moser (1991) also takes advantage of Fermat's Principle of Least Time. However, the *network theory based* implementation is quite novel.

The author views the seismic ray trace problem as analogous to the well known *traveling salesman problem*, where a hypothetical salesman (hopefully) selects the shortest possible routes from city to city in order to minimize total travel time. With this analogy in mind, the author also chooses methods traditionally used to solve the traveling salesman problem, namely, methods from *network theory*. We refer the reader to the original paper for details.

Applications to surface seismics. Traveltime calculations play important roles in seismic processing. For example, Kirchhoff methods of migrating and modeling seismic data require the calculation of Green's functions, which in turn depend on traveltimes between surface survey points and depth points in the assumed velocity model. Ray tracing is often used, but the extensive interpolations required to extrapolate data to points on regular grids are impractical. For complicated velocity models, rays may cross each other, or they may not penetrate shadow zones.

Finite difference calculation of travel times. Rather than directly integrating the ray equations, using shooting or bending methods, or working with network theory approaches, several authors recently suggested a direct *finite difference* attack on the *eikonal equation* itself, solving the nonlinear equation

$$(\partial\tau/\partial x)^2 + (\partial\tau/\partial y)^2 + (\partial\tau/\partial z)^2 = 1/c(x,y,z)^2 \qquad (11.99)$$

which we may recognize as Equation 11-5 (see Chapters 5-8 for a review of finite difference methods).

In this approach, *wavefronts, rather than traditional rays, are tracked*; head waves are properly treated, according to early authors, and shadow zones are filled by appropriate diffractions. We will refer the details of the implementation to the original papers, e.g., Vidale (1988), Vidale and Houston (1990), Qin *et al* (1992), and Coultrip (1993), since they do not fall within the objectives of the present chapter.

Difficulties with simple difference formulation. There are subtleties and problems in the basic approach. To understand these them, let us consider the details that appear in the formulation, and take as our starting point, Equation 11-99. For simplicity, we define the *slowness* function s(x,y,z) by

$$s(x,y,z) = 1/c(x,y,z) \qquad (11.100)$$

and consider two space dimensions first. In the coordinate space (x,z), Equation 11-99 becomes

$$(\partial\tau/\partial x)^2 + (\partial\tau/\partial z)^2 = s^2(x,z) \qquad (11.101)$$

Now, we introduce the notation

$$u(x,z) = \tau_x(x,z) \qquad (11.102)$$

so that Equation 11-101 becomes

$$u^2 + (\partial\tau/\partial z)^2 = s^2(x,z) \qquad (11.103)$$

Equation 11-103 can therefore be solved to yield

$$\tau_z(x,z) = \sqrt{\{s^2-u^2\}} \equiv F(u) \qquad (11.104)$$

Next, we differentiate Equation 11-102 with respect to z, and combine the result with the x derivative of Equation 11-104. This successively gives

$\partial u(x,z)/\partial z = \tau_{zx}(x,z) = \tau_{xz}(x,z) = \partial F(u)/\partial x$, and hence, the first-order, nonlinear, wave-like governing differential equation

$$\partial u/\partial z - \partial F(u)/\partial x = 0 \tag{11.105}$$

Why is Equation 11-105 physically significant? Let us consider the total differential for the phase function $\tau(x,z)$,

$$d\tau = \tau_x\, dx + \tau_z\, dz \tag{11.106}$$

If we use Equations 11-102 and 11-104, we can rewrite Equation 11-106 as

$$d\tau = u\, dx + F(u)\, dz \tag{11.107}$$

Thus, if the function u(x,z) is known, numerically say, so that F(u) is likewise known, then the travel-time between any two points A and B can be performed by the extremely simple integration

$$\tau\Big|_A^B = \int_A^B u\, dx + \int_A^B F(u)\, dz \tag{11.108}$$

Three space dimensions. Now, with a clearer motivation in mind, we reconsider the three-dimensional eikonal equation

$$(\partial\tau/\partial x)^2 + (\partial\tau/\partial y)^2 + (\partial\tau/\partial z)^2 = s^2(x,y,z) \tag{11.109}$$

and introduce

$$u = \partial\tau/\partial x \tag{11.110}$$

$$v = \partial\tau/\partial y \tag{11.111}$$

so that Equation 11-109 can be rewritten as

$$\partial\tau/\partial z = \surd\{s^2 - u^2 - v^2\} \equiv G(u,v) \tag{11.112}$$

Then, taking x derivatives of Equation 11-112 leads to

$$\partial(\tau_z)/\partial x = \partial G(u,v)/\partial x = G_u\, \partial u/\partial x + G_v\, \partial v/\partial x \tag{11.113}$$

Now, in Equation 11-113, $\partial(\tau_z)/\partial x = \partial(\tau_x)/\partial z = \partial u/\partial z$. Also, we find from Equations 11-110 and 11-111 that $\partial v/\partial x = \partial(\tau_y)/\partial x = \partial(\tau_x)/\partial y = \partial u/\partial y$. Thus, Equation 11-113 becomes

$$\partial u/\partial z - G_u\, \partial u/\partial x - G_v\, \partial u/\partial y = 0 \tag{11.114}$$

which is analogous to Equation 11-105. Again, we consider the total differential for the phase function $\tau(x,y,z)$,

$$d\tau = \tau_x\, dx + \tau_y\, dy + \tau_z\, dz \tag{11.115}$$

If we next use Equations 11-110 to 11-112, we obtain

$$d\tau = u\,dx + v\,dy + G(u,v)\,dz \tag{11.116}$$

Thus, the total travel-time between any two points A and B is

$$\tau\Big|_A^B = \int_A^B u\,dx + \int_A^B v\,dy + \int_A^B G(u,v)\,dz \tag{11.117}$$

Analysis of the problem. Consider the two-dimensional problem. From Equation 11-104, $F(u) = \{s^2-u^2\}^{1/2}$, and we obtain $F'(u) = 1/2\ \{s^2-u^2\}^{-1/2}(-2u) = -u\ \{s^2-u^2\}^{-1/2}$. Since Equation 11-105 can be rewritten as $\partial u/\partial z - F'(u)\ \partial u/\partial x = 0$, we actually have

$$\partial u/\partial z + u\ \{s^2-u^2\}^{-1/2}\ \partial u/\partial x = 0 \tag{11.118}$$

Similar considerations apply to the three-dimensional problem. From Equation 11-112, $G(u,v) = \{s^2 - u^2 - v^2\}^{1/2}$, and $G_u = 1/2\ \{\ \}^{-1/2}\ (-2u) = -u\{\ \}^{-1/2}$. Similarly, $G_v = -v\{\ \}^{-1/2}$, so that $\partial u/\partial z - G_u\ \partial u/\partial x - G_v\ \partial u/\partial y = 0$ in Equation 11-114 becomes

$$\partial u/\partial z + u\{s^2 - u^2 - v^2\}^{-1/2}\ \partial u/\partial x + v\{s^2 - u^2 - v^2\}^{-1/2}\ \partial u/\partial y = 0 \tag{11.119}$$

The derivation leading up to Equations 11-119 was motivated by the approach of Van Trier and Symes (1991). Equations 11-118 and 11-119 are first-order, nonlinear, partial differential equations with variable coefficients in $s(x,z)$ and $s(x,y,z)$.

But as we noted in our analysis of Equation 11-81 (or, $\partial u/\partial t + u\ \partial u/\partial x = 0$), shock formation and ray coalescence are in general possible, and some additional information on the structure of the "small" right sides of Equations 11-118 and 11-119 is required. This is so if we recall that $\partial u/\partial t + u\ \partial u/\partial x = \varepsilon\ \partial^2 u/\partial x^2$ and $\partial u/\partial t + u\ \partial u/\partial x = \delta\ \partial^3 u/\partial x^3$ have completely different solution properties.

But where do we obtain information about these right-hand-sides? The answer is clear if we recall that the classical eikonal equation shown in Equation 11-99 is not the most general description for our seismic waves. In fact, it turns out that the neglected high-order derivatives take a mathematical form analogous to the $\varepsilon\ \partial^2 u/\partial x^2$ in Burger's equation, and that the exact structure of the terms can be ascertained by extending one higher order our assumed high frequency expansion.

It is not necessary, though, to take this brute force approach. As we had shown earlier, all of the information contained in the eikonal equation is likewise embedded in the kinematic wave description, including Fermat's theorem, and more. Thus, the straightforward extension of Equations 11-77 and 11-78 to two and three spatial dimensions provides high-order partial differential equations that completely define the physics everywhere. These equations, again, couple the effects of phase and amplitude.

Van Trier and Symes (1991) also noted that their analogues to Equations 11-118 and 11-119 may contain discontinuous gradients. Thus, they add *artificial viscosity* to the governing equation, to render its solution unique, by using upwind differencing in their numerical solution of the eikonal equation.

This introduces a "$\varepsilon\ \partial^2 u/\partial x^2$" type correction, and, as does Burger's equation, smooths out all shocks. The ε though, depends on details of the mesh, the truncation error of the numerical approximation scheme used, and the details of the programming and coding (see Dablain (1984) for a discussion of truncation error forms used in seismics). It does *not* represent a true physical effect, based on real attenuation and viscous diffusion, as Equations 11-77 and 11-78 would.

Thus, the upwinding yields results that may *appear* correct, but which in fact are not. This same point is expressed in Chin (1993a), in the context of numerical reservoir simulation, commenting on models where upwind differencing is used in conjunction with true capillary pressure based diffusion to stabilize steep saturation gradients.

Similar concerns were also recently expressed in a paper by Geoltrain and Brac (1993). These authors experienced difficulties with traveltimes computed by finite differencing, noting that many of the observed irregularities have no physical basis for the velocity models used. Their problems, of course, arose because they solved an incorrect form of the eikonal equation; the proper high-order amplitude model, or Equations 11-77 and 11-78 (appropriately modified to contain the heterogeneities of the velocity model), should yield correct physical results.

REFERENCES

Aki, K., and Richards, P.G., *Quantitative Seismology, Volumes I and II: Theory and Methods*, W.H. Freeman and Company, San Francisco, 1980.

Bleistein, N., *Mathematical Methods for Wave Phenomena*, Academic Press, Orlando, 1984.

Bourbie, T., Coussy, O., and Zinszner, B., *Acoustics of Porous Media*, Gulf Publishing Company, Houston, 1987.

Cerveny, V., and Ravindra, R., *Theory of Seismic Head Waves*, University of Toronto Press, Toronto, 1971.

Chin, W.C., "Effect of Dissipation and Dispersion on Slowly Varying Wavetrains," AIAA Journal, Feb. 1980, pp. 149-158.

Chin, W.C., *Modern Reservoir Flow and Well Transient Analysis*, Gulf Publishing, Houston, 1993.

Chin, W.C., *3D/SIM: 3D Petroleum Reservoir Simulation for Vertical, Horizontal, and Deviated Wells, Petroleum Engineering Software for the IBM PC and Compatibles*, Gulf Publishing, Houston, 1993.

Clay, C.S., *Elementary Exploration Seismology*, Prentice-Hall, Englewood Cliffs, New Jersey, 1990.

Coultrip, R.L., "High Accuracy Wavefront Tracing Traveltime Calculation,", Geophysics, Vol. 58, No. 2, Feb. 1993, pp. 284-292.

Dablain, M.A., "The Role and Application of Entropy Terms for Acoustic Wave Modeling by the Finite Difference Method," Geophysics, Vol. 49, No. 9, Sept. 1984, pp. 1457-1465.

De la Cruz, V., and Spanos, T.J.T., "Seismic Wave Propagation in a Porous Medium," Geophysics, Vol. 50, No. 10, Oct. 1985, pp. 1556-1565.

Gelfand, I.M., and Fomin, S.V., *Calculus of Variations*, Prentice-Hall, Englewood Cliffs, New Jersey, 1963.

Geoltrain, S., and Brac, J., "Can We Image Complex Structures With First Arrival Traveltime?," Geophysics, Vol. 58, No. 4, April 1993, pp. 564-575.

Hassanzadeh, S., "Acoustic Modeling in Fluid-Saturated Porous Media," Geophysics, Vol. 56, No. 4, April 1991, pp. 424-435.

Hildebrand, F., *Methods of Applied Mathematics*, Prentice-Hall, Englewood Cliffs, New Jersey, 1952.

Kennet, B.L.N., *Seismic Wave Propagation in Stratified Media*, Cambridge University Press, Cambridge, 1983.

McCann, C., and McCann, D.M., "A Theory of Compressional Wave Attenuation in Noncohesive Sediments," Geophysics, Vol. 50, No. 8, Aug. 1985, pp. 1311-1317.

Moser, T.J., "Shortest Path Calculation of Seismic Rays," Geophysics, Vol 56, No. 1, Feb. 1991, pp. 59-67.

Qin, F., Luo, Y., Olsen, K.B., Cai, W., and Schuster, G.T., "Finite Difference Solution of the Eikonal Equation Along Expanding Wavefronts," Geophysics, Vol. 57, No. 3, March 1992, pp. 478-487.

Toksoz, M.N., and Johnston, D.H., *Seismic Wave Attenuation*, Geophysics Reprint Series No. 2, Society of Exploration Geophysicists, Tulsa, Oklahoma, 1981.

Van Trier, J., and Symes, W.W., "Upwind Finite Difference Calculation of Traveltimes," Geophysics, Vol. 56, No. 6, June 1991, pp. 812-821.

Vidale, J., "Finite Difference Calculation of Travel Times," Bulletin of the Seismological Society of America, Vol. 78, No. 6, Dec. 1988, pp. 2062-2076.

Vidale, J.E., and Houston, H., "Rapid Calculation of Seismic Amplitudes," Geophysics, Vol. 55, No. 11, Nov. 1990, pp. 1504-1507.

White, J.E., "Seismic Waves in Fluid-Saturated Rocks: An Examination of the Biot Theory," Colorado School of Mines Quarterly, Vol. 1, 1984, pp. 1-32.

Whitham, G.B., *Linear and Nonlinear Waves*, John Wiley and Sons, New York, 1974.

Yew, C.H., and Jogi, P.N., "Study of Wave Motions in Fluid-Saturated Porous Rocks," Journal of the Acoustical Society of America, Vol. 60, No. 1, July 1976, pp. 2-8.

12
Wave and Current Interaction in the Ocean

In this chapter, we will describe ocean applications for *kinematic wave theory* using simple examples, assuming an elementary background in hydrodynamics (see Chapter 3). The solution methodologies offered are powerful, but the specific results are no less important. They are, for example, particularly useful in estimating wave forces for offshore applications.

Some of these may include loadings on drilling platforms, their effects on dynamic positioning, marine riser cables, and pipelines. Also, we will show how results which are in themselves limited can be integrated to form usable global models for other applications.

Recapitulation. Again, when the complex dispersion relation corresponding to an elementary plane wave

$$\Omega(k,x,t) = \Omega^r(k,x,t) + i\Omega^i(k,x,t) \tag{12.1}$$

is known, either analytically or experimentally, the wave energy and momentum densities E and M evolve in space-time according to

$$\partial E/\partial t + \partial(\Omega^r_k(k,x,t)E)/\partial x = E\{2\Omega^i + \Omega^r_t(k,x,t)/\Omega^r\} \tag{12.2}$$

$$\partial M/\partial t + \partial(\Omega^r_k(k,x,t)M)/\partial x = M\{2\Omega^i - \Omega^r_x(k,x,t)/k\} \tag{12.3}$$

These results, given in Chapter 2, apply to *all* varieties of elastic, fluid-dynamic, electromagnetic wave propagation, and so on.

Damped waves in deep water. Let us apply these equations to water waves. In Chapter 3, we replicated a well known hydrodynamic solution, and showed that gravity-capillary waves in deep *quiescent* water propagate according to

$$\Omega^r(k) = \sqrt{(gk + Tk^3/\rho)} \tag{12.4}$$

where g, ρ, and T represent gravitational acceleration, mass density, and surface tension.

The velocity potential solution, which is nontrivial, was contingent upon the existence of *nondissipative* inviscid flow. Now suppose that the frequency-dependent effects of viscosity must be estimated. In *separate* studies, it is known that water waves damp like $e^{\Omega i t}$, with

$$\Omega^i(k) = -2\nu k^2 \tag{12.5}$$

where ν is the kinematic viscosity (Lamb, 1945).

We now wish to determine how a wave will propagate *and* damp in space and time, using the separate results in Equations 12-4 and 12-5, and next, the additional effect of a variable background mean flow.

Following the ideas introduced in Chapter 2, the first objective is straightforwardly accomplished. Since the wave kinematics, to leading order, is unaffected by dissipation, the wavenumber field k(x,t) and position x(t) *must* satisfy dk/dt = 0 and dx/dt = $\Omega^r_k(k)$. But energy *is* affected by viscosity: the governing differential equation is obtained by "plugging into" the general result in Equation 12-2; thus, $\partial E/\partial t + \partial(\Omega^r_k(k)E)/\partial x = -4\nu k^2 E$ since $\Omega^r_t(k,x,t)$ vanishes identically. The solution to the general initial value problem can be determined along the lines described in Chapter 2. Similar comments apply to wave momentum.

Effect of variable background flow. Now consider waves propagating parallel to a background current U(x,t), where variations with respect to x and t are weak. Using variational arguments similar to those in Whitham (1974), Chin (1976, 1980a) showed that E(x,t) and M(x,t) now satisfy

$$\partial E/\partial t + \partial((U(x,t) + \Omega^r_k(k,x,t))E)/\partial x = E\{2\Omega^i + \{\Omega^r_t(k,x,t) + U\Omega^r_x(k,x,t) - k\Omega^r_k(k,x,t)\, U_x\}/\Omega^r\} \tag{12.6}$$

$$\partial M/\partial t + \partial((U(x,t) + \Omega^r_k(k,x,t))M)/\partial x = M\{2\Omega^i - \Omega^r_x(k,x,t)/k - U_x\} \tag{12.7}$$

The wave and current superposition is nontrivial, because dynamical interactions occur through a so-called *radiation* or *Reynolds stress tensor*. (Longuet-Higgins and Stewart, 1964). Equations 12-6 and 12-7, of course, apply to applications other than water waves; for example, they can be used to model acoustic wave motions in nonuniform winds, or MWD signal propagation in ducts with slowly varying cross-sections.

Note that the kinematic ray equations corresponding to Equations 12-6 and 12-7 are

$$dk/dt = \Omega^r_x(k,x,t) + U_x k \tag{12.8}$$

and

$$dx/dt = U(x,t) + \Omega^r_k(k,x,t) \tag{12.9}$$

where an obvious Doppler shift has been invoked. That is, to Equations 2-102 and 2-103, namely, $dk/dt = -\Omega^r_x(k,x,t)$ and $dx/dt = \Omega^r_k(k,x,t)$, we simply replace $\Omega^r(k,x,t)$ by $U(x,t)k + \Omega^r(k,x,t)$.

Equations 12-8 and 12-9, and then Equations 12-6 and 12-7, may be integrated numerically as discussed in Chapter 2. The function U(x,t) may be prescribed analytically, but it may also be specified from known climactic conditions, say from meteorological data.

Continuing with our deep water example, where $\Omega^r_t(k,x,t)$ and $\Omega^r_x(k,x,t)$ both vanish, we clearly have

$$\begin{aligned} \partial E/\partial t + \partial((U(x,t) + \Omega^r_k(k))E)/\partial x &= \\ &= -4\nu k^2 E - \{k\Omega^r_k(k)/\Omega^r\} U_x E \end{aligned} \tag{12.10}$$

where $k\Omega^r_k(k)/\Omega^r = 1/2$ for pure gravity waves (T = 0), and 3/2 for pure capillary waves (g = 0).

But in general, $k\Omega^r_k(k)/\Omega^r$ takes the value $(1+3\beta)/(2+2\beta)$, where $\beta = Tk^2/\rho g$ is a dimensionless number characterizing relative capillary-to-gravity forces. Equation 12-8 shows how these factors modify U_x, and through the sign of U_x, how local flow acceleration or deceleration affects wave energy growth. The right side of Equation 12-10 shows that wave growth or decay depends on the relative combination of dissipation and heterogeneity.

Waves in finite depth water. Consider next the more difficult problem for *gravity* waves in water of finite depth h(x,t). It can be shown that

$$\Omega^r(k,x,t) = (gk \tanh kh)^{1/2} \tag{12.11}$$

following Lamb (1945). The terms $\Omega^r_t(k,x,t)$ and $\Omega^r_x(k,x,t)$ in Equation 12-6 do not vanish, since Ω^r depends explicitly upon x and t through h(x,t). Expanding Equation 12-6, and simplifying with the mass conservation law

$$\partial h/\partial t + \partial\{U(x,t)h\}/\partial x = 0 \tag{12.12}$$

leads to

$$\begin{aligned} \partial E/\partial t + \partial((U(x,t) + \Omega^r_k(k,x,t))E)/\partial x &= \\ &= -4\nu k^2 E - \{1/2 + 2kh/\sinh(2kh)\} U_x E \end{aligned} \tag{12.13}$$

In deep water, kh → ∞, and "2kh/sinh(2kh)" vanishes, thus leaving the "1/2" obtained in the deep water example above. Equation 12-13 agrees with Longuet-Higgins and Stewart (1964), who apply detailed perturbation methods.

Sources of hydrodynamic loading. In offshore applications, three types of loading are important: hydraulic pressures imposed by mean currents (that is, U(x,t)), induced loads due to vortex oscillations arising from unsteady flow separation, and dynamic forces that arise from wave interactions.

The first two do not fall within the scope of this book. The last may be estimated in various ways. For example, direct tabulations for E(x,t) and M(x,t) will provide useful statistical information on local wave elevation.

Numerical solution of the 3D KWT initial value problem formulated in Chapter 2, for example, may determine how storm effects are likely to propagate in oceans with variable currents, and how dissipation will affect different types of initial disturbances.

On the other hand, the results of wave impact are more accurately assessed by following the net energy

$$\Sigma(t) = \int_{x_1}^{x_2} E(x,t)\, dx \qquad (12.14)$$

between two rays. If the ray integration discussed in Chapter 2 is applied to Equations 12-6 and 12-14, we have

$$d\Sigma(t)/dt = \int_{x_1}^{x_2} E\{2\Omega^i + \{\Omega^r_t(k,x,t) + U\Omega^r_x(k,x,t) - k\Omega^r_k(k,x,t)U_x\}/\Omega^r\}dx \qquad (12.15)$$

and similarly for total wave momentum. Once k(x,t) and E(x,t) are determined, Equation 12-15 can be integrated to provide potential impact information.

Instabilities due to heterogeneity. In Chapter 3, we showed that gravity waves in deep water satisfy the dispersion relation

$$\Omega^r(k) = \pm (gk)^{1/2} \qquad (12.16)$$

where we have retained the "±" to describe two possible wave families propagating in opposite directions. When a background flow with mean speed U(x,t) exists, our dispersion relation is modified by a Doppler shift, so that

$$\omega = U(x,t)k \pm (gk)^{1/2} \qquad (12.17)$$

In the limit when U(x,t) depends on x only, and dynamically steady waves having a constant frequency ω_0 are considered, Equation 12-17 leads directly to

$$U(x)k + s(gk)^{1/2} = \omega_0 > 0 \qquad (12.18)$$

where s = ± 1 (see Equation 2-116). That is, $k^{1/2}(x)$ can be explicitly solved, using the quadratic formula for second-order algebraic equations.

The consequences of Equation 12-18, the result of an innocuous "Uk" background term, may be surprising, but they describe the *kinematic barrier* well known to oceanographers (Chin, 1976, 1979, 1980b, 1981). The discussion proceeds generally if we introduce the nondimensional variables U^* and k^* defined by

$$U^* = 4U\omega_0/g \tag{12.19}$$

and

$$(k^*)^{1/2} = (gk)^{1/2}/2\omega_0 \tag{12.20}$$

Thus, Equation 12-18 can be written in the concise form

$$Uk^* + 2sk^{*1/2} = 1 \tag{12.21}$$

Now, we introduce the corresponding dimensionless group velocity $C_g{}^*$

$$C_g{}^* = U^* + sk^{*-1/2} \tag{12.22}$$

Equations 12-21 and 12-22 can be rearranged to form $U^* = (1- 2sk^{*1/2})/k$ and $C_g{}^* = U^* + sk^{*-1/2} = 1/k^* - s/k^{*1/2}$. A complete family of solutions is produced by varying the parameter k^* throughout a range of positive values.

The solutions so obtained are shown in Figure 12-1, where $C_g{}^*$ is plotted against U^*. Zeros in the group velocity $C_g{}^*$, as discussed earlier in this book, identify areas where energy propagation ceases. The resulting trapping of wave energy implies high amplitude wave motions.

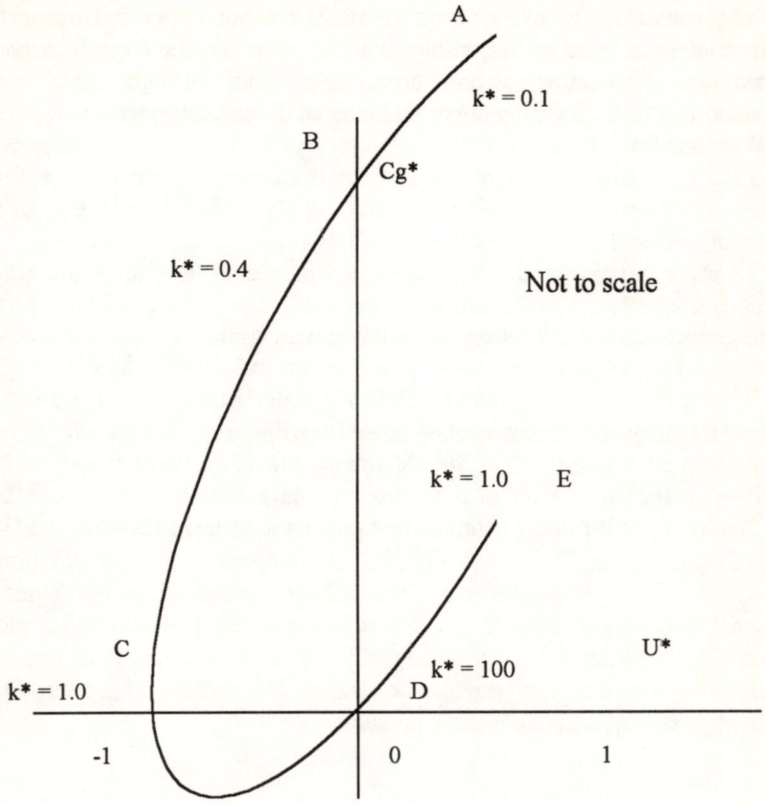

Figure 12-1. Linear gravity waves on variable currents.

Let us discuss the implications behind Figure 12-1. First, we note that along ABCDE, the segments ABCD and DE are derived using s = +1 and -1, respectively; the corresponding (dimensionless) wavenumber solutions along each of these curves are directly indicated.

Now, consider a wave approaching C from A. As the wave nears C, the group velocity decreases to zero, thus producing singular values in the wave energy density E, while the dimensionless wavenumber increases to unity.

In the velocity range $-1 < U^* < 0$, the medium supports two sets of waves, one with positive and the other with negative group velocity. Thus, the wave approaching C from A may reflect from the focus along the second branch, but it then encounters a second focus at D where it must remain trapped.

For positive U^*, two sets of waves exist, but both have positive group velocity relative to a ground-fixed observer. A wave approaching D from E, therefore, cannot reflect and is once more trapped; but the high energy waves found at D have infinite wavenumber and may be damped by viscosity.

Wave solutions do not exist for U^* less than -1 within the framework of linear theory. An extension of the present discussion to account for finite amplitude nonlinear effects is offered in Chin (1979, 1980b, 1981), but for the purposes of this book, linear results will suffice.

The wave trapping and instability ideas uncovered here for water waves are directly analogous to the lateral wave instabilities described in Chapter 6 for bending disturbances propagating on variable axial strain fields. That analysis was motivated by the earlier work on gravity waves just described.

The *kinematic barrier* predicted here for water waves, and in Chapter 6 for bending instabilities at the neutral point in drillstrings, are both quite real, and represent manifestations of simple linear kinematic wave theory. Used properly with the correct physical picture, we have shown in this book how simple models describe many complex phenomena observed in nature.

REFERENCES

Chin, W.C., *Physics of Slowly Varying Wavetrains in Continuum Systems, Ph.D. Thesis*, Massachusetts Institute of Technology, Cambridge, Massachusetts, 1976.

Chin, W.C., "Kinematic Wave Approach to Hydraulic Jumps with Waves," Journal of Hydronautics, Oct. 1979.

Chin, W.C., "Effect of Dissipation and Dispersion on Slowly Varying Wavetrains," AIAA Journal, Feb. 1980.

Chin, W.C., "Wave Focusing and Hydraulic Jump Formation," Journal of Hydronautics, July 1980.

Chin, W.C., "Kinematic Barrier for Gravity Waves on Variable Currents, Journal of Hydronautics," Jan. 1981.

Lamb, H., *Hydrodynamics*, Dover Publications, New York, 1945.

Longuet-Higgins, M.S., and Stewart, R.W., "Radiation Stresses in Water Waves: A Physical Discussion, with Applications," Deep Sea Research, Vol. 11, 1964.

Whitham, G.B., *Linear and Nonlinear Waves*, John Wiley and Sons, New York, 1974.

Cumulative References

Abramson, H.N., Plass, H.J., and Ripperger, E.A., "Stress Wave Propagation in Rods and Beams," in *Advances in Applied Mechanics, Vol. V*, edited by Hugh Dryden and Theodore von Karman, Academic Press, New York, 1958.

Aki, K., and Richards, P.G., *Quantitative Seismology, Volumes I and II: Theory and Methods*, W.H. Freeman and Company, San Francisco, 1980.

Aldred, W.D., and Sheppard, M.C., "Drillstring Vibrations: A New Generation Mechanism and Control Strategies," SPE Paper No. 24582, presented at the 67th Annual Technical Conference and Exhibition of the Society of Petroleum Engineers, Washington, D.C., Oct. 4-7, 1992.

Ames, W.F., *Numerical Methods for Partial Differential Equations*, Academic Press, New York, 1977.

Angona, F.A., "Drill String Vibration Attenuation and Its Effect on a Surface Oscillator Drilling System," Journal of Engineering for Industry, Transactions of the ASME, May 1965, pp. 110-114.

Bailey, J.J., and Finnie, I., "An Analytical Study of Drill String Vibration," Journal of Engineering for Industry, Transactions of the ASME, May 1960, pp. 122-128.

Baird, J.A., Caskey, B.C., Tinianow, M.A., and Stone, C.M., "GEODYN: A Geological Formation/Drillstring Dynamics Computer Program," SPE Paper No. 13023, presented at the 59th Annual Technical Conference and Exhibition, Houston, Texas, Sept. 16-19, 1984.

The cumulative references given in this final section contain all previous chapter references, and also, additional papers and monographs deemed significant to the material presented in this book. No attempt has been made to provide a comprehensive listing covering the subject materials in their entirety, but hopefully, the representative cross-section of the open literature given here, will prove to be a good first start for the reader interested in further study.

Baird, J.A., Caskey, B.C., Wormley, D.N., and Stone, C.M., "GEODYN2: A Bottomhole Assembly/Geological Formation Dynamic Interaction Computer Program," SPE Paper No. 14328, presented at the 60th Annual Technical Conference and Exhibition of the Society of Petroleum Engineers, Las Vegas, Nevada, Sept. 22-25, 1985.

Barnes, T.G., and Kirkwood, B.R., "Passbands for Acoustic Transmission in an Idealized Drill String, Vol. 51, No. 5, Journal of the Acoustical Society of America, 1972, pp. 1606-1608.

Bickley, W.G., and Talbot, A., *An Introduction to the Theory of Vibrating Systems*, Oxford University Press, London, 1961.

Bishop, R.E.D., "The Vibration of Rotating Shafts," Journal of Mechanical Engineering Science, Vol.1, No. 1, 1959, pp. 50-65.

Bishop, R.E.D., and Gladwell, G.M.L., "The Vibration and Balancing of an Unbalanced Flexible Rotor," Journal of Mechanical Engineering Science, Vol.1, No. 1, 1959, pp. 66-77.

Bleistein, N., *Mathematical Methods for Wave Phenomena*, Academic Press, Orlando, 1984.

Bobrosky, D., and Osmak, G., "Antiwhirl PDC Bits Increased Penetration Rates in Alberta Drilling," Oil & Gas Journal, July 5, 1993, pp. 41-46.

Bolotin, V.V., *The Dynamic Stability of Elastic Systems*, Holden-Day, Inc., San Francisco, 1964.

Booer, A.K., and Meehan, R.J., "Drillstring Imaging: An Interpretation of Surface Drilling Vibrations," IADC/SPE Paper No. 23889, presented at the 1992 IADC/SPE Drilling Conference, New Orleans, Louisiana, Feb. 18-21, 1992.

Booer, A.K., and Meehan, R.J., "Drillstring Imaging - An Interpretation of Surface Drilling Vibrations, SPE Drilling and Completion Journal, June 1993, pp. 93-98.

Bourbie, T., Coussy, O., and Zinszner, B., *Acoustics of Porous Media*, Gulf Publishing Company, Houston, 1987.

Bradbury, R.E., and Wilhoit, J.C., "Effect of Tool Joints on Passage of Plane Longitudinal and Torsional Waves Along a Drill Pipe," Paper No. 62-PET-32, presented at the Petroleum Mechanical Engineering Conference, American Society of Mechanical Engineers, Dallas, Tx., Sept. 23-26, 1962.

Brett, J.F., "The Genesis of Bit-Induced Torsional Drillstring Vibrations," SPE/IADC Paper No. 21943, presented at the 1991 SPE/IADC Drilling Conference, Amsterdam, Mar. 11-14, 1991.

Brett, J.F., "The Genesis of Torsional Drillstring Vibrations," SPE Drilling Engineering Journal, Sept. 1992, pp. 168-174.

Bryant, T.M., Grosso, D.S., and Wallace, S.N., "Gas-Influx Detection With MWD Technology," SPE Drilling Engineering Journal, December 1991, pp. 273-278.

Burgess, T.M., McDaniel, G.L., and Das, P.K., "Improving BHA Tool Reliability with Drillstring Vibration Models: Field Experience and Limitations," SPE/IADC Paper No. 16109, 1987 SPE/IADC Drilling Conference, New Orleans, March 15-18, 1987.

Burkhardt, J.A., "Wellbore Pressure Surges Produced by Pipe Movement," Journal of Petroleum Technology, June 1961, pp. 595-605.

Carrier, G.F., Krook, M., and Pearson, C.E., *Functions of a Complex Variable*, McGraw-Hill, New York, 1966.

Cerveny, V., and Ravindra, R., *Theory of Seismic Head Waves*, University of Toronto Press, Toronto, 1971.

Cheatham, C.A., and Comeaux, B.C., "General Guidelines for Predicting Fatigue Life of MWD Tools," IADC/SPE Paper No. 23906, presented at the 1992 IADC/SPE Drilling Conference, New Orleans, Louisiana, Feb. 18-21, 1992.

Chen, B.C.M., and Gair, R., "Flowing Fluid Dynamics Can No Longer Be Ignored in Deepwater Operations," Offshore Magazine, April 1992, pp. 54-59.

Chin, W.C., *Physics of Slowly Varying Wavetrains in Continuum Systems, Ph.D. Thesis*, Massachusetts Institute of Technology, Cambridge, Massachusetts, 1976.

Chin, W.C., "Supersonic Wave Drag for Nonplanar Singularity Distributions," AIAA Journal, June 1977.

Chin, W.C., "Numerical Solution for Viscous Transonic Flow," AIAA Journal, Sept. 1977.

Chin, W.C., "Goethert's Rule with an Improved Boundary Condition," AIAA Journal, Oct. 1977.

Chin, W.C., "Pseudo-Transonic Equation with a Diffusion Term," AIAA Journal, Jan. 1978.

Chin, W.C., "Some Singular Aspects of Three-Dimensional Transonic Flow," AIAA Journal, Mar. 1978.

Chin, W.C., "Supersonic Wave Drag of Planar Singularity Distributions," AIAA Journal, May 1978.

Chin, W.C., "Nonlinear Formulation for Low-Frequency Transonic Flow," AIAA Journal, June 1978.

Chin, W.C., "Similar Solutions for Unsteady Transonic Flow," AIAA Journal, June 1978.

Chin, W.C., "Type-Independent Solutions for Mixed Compressible Flows," AIAA Journal, Aug. 1978.

Chin, W.C., "Algorithm for Inviscid Flow Using the Viscous Transonic Equation," AIAA Journal, Aug. 1978.

Chin, W.C., "On the Design of Thin Subsonic Airfoils," ASME Journal of Applied Mechanics, Mar. 1979.

Chin, W.C., "Some Exact Solutions to Guderley's Equations," AIAA Journal, April 1979.

Chin, W.C., "Stability of Inviscid Shear Flow Over Flexible Membranes," AIAA Journal, June 1979.

Chin, W.C., "Effect of Frequency in Unsteady Transonic Flow," with D. Rizzetta, AIAA Journal, July 1979.

Chin, W.C., "Kinematic Wave Approach to Hydraulic Jumps with Waves," Journal of Hydronautics, Oct. 1979.

Chin, W.C., "Airfoil Design in Subcritical and Supercritical Flow," with D. Rizzetta, ASME Journal of Applied Mechanics, Dec. 1979.

Chin, W.C., "Effect of Dissipation and Dispersion on Slowly Varying Wavetrains," AIAA Journal, Feb. 1980, pp. 149-158.

Chin, W.C., "Class of Shockfree Airfoils Producing the Same Surface Pressure," Journal of Aircraft, April 1980.

Chin, W.C., "Inviscid Parallel Flow Stability with Mean Profile Distortion," Journal of Hydronautics, July 1980.

Chin, W.C., "Wave Focusing and Hydraulic Jump Formation," Journal of Hydronautics, July 1980.

Chin, W.C., "Transonic Nacelle Inlet Analyses," with W. Presz, D. Ives, D. Paris and D. Golden, in NASA Lewis Workshop on Application of Advanced Computational Methods, Nov. 1980.

Chin, W.C., "Optimal Coordinates for Squire's Jet," AIAA Journal, Jan. 1981

Chin, W.C., "Kinematic Barrier for Gravity Waves on Variable Currents, Journal of Hydronautics," Jan. 1981.

Chin, W.C., "Harmonic Analysis of Unsteady Transonic Flow," AIAA Journal, Feb. 1981.

Chin, W.C., "Direct Approach to Aerodynamic Design Problems," ASME Journal of Applied Mechanics, Dec. 1981.

Chin, W.C., "An Axisymmetric Nacelle and Turboprop Inlet Analysis with Flow-Through and Power Simulation Capabilities," with D. Golden and T. Barber, AIAA Paper No. 82-0256, AIAA 20th Aerospace Sciences Meeting, Orlando, Florida, Jan. 1982.

Chin, W.C., "Superpotential Solution for Jet Engine External Potential and Internal Rotational Flow Interaction," ASME Journal of Applied Mechanics, June 1983

Chin, W.C., "An Axisymmetric Nacelle and Turboprop Inlet Analysis Including Power Simulation," with D. Golden and T. Barber, Journal of Aircraft, June 1983.

Chin, W.C., "Engine Power Simulation for Transonic Flow-Through Nacelles," AIAA Journal, Oct. 1983.

Chin, W.C., "Thin Airfoil Theory for Planar Inviscid Shear Flow," ASME Journal of Applied Mechanics, Mar. 1984.

Chin, W.C., "Comment on "PAN AIR Applications to Aero-Propulsion Integration," Journal of Aircraft, Nov. 1984.

Chin, W.C., "Inviscid Steady Flow Past Turbofan Mixer Nozzles," ASME Journal of Applied Mechanics, Dec. 1984.

Chin, W.C., "Formation Evaluation Using Repeated MWD Logging Measurements," with A. Suresh, P. Holbrook, L. Affleck and H. Robertson, SPWLA 27th Annual Logging Symposium, Houston, Texas, June 9-13, 1986.

Chin, W.C., "Fatal Turbular Bending Motions Difficult to Detect Uphole," Offshore Magazine, April 1988, pp. 32-36.

Chin, W.C., "Why Drill Strings Fail at the Neutral Point," Petroleum Engineer International, May 1988, pp. 62-67.

Chin, W.C., "Simulating Horizontal Well Fracture Flow," Offshore Magazine, Aug. 1988.

Chin, W.C., "Automating the Acquisition Planning Process at the Johnson Space Center (via Artificial Intelligence Methods)," with J. Golej, MITRE Corporation Report MTR-88D00065, NASA/JSC Contract No. NAS9-18057, Sept. 1988.

Chin, W.C., "Comparative Studies in Dual Porosity Reservoir Simulation," BP Exploration Reservoir Technology Report No. H090.0005, Houston, TX, Jan. 1990.

Chin, W.C., "Advances in Annular Borehole Flow Modeling," Offshore Magazine, February 1990, pp. 31-37.

Chin, W.C., "Exact Cuttings Transport Correlations Developed for High Angle Wells," Offshore Magazine, May 1990, pp. 67-70.

Chin, W.C., "Annular Flow Model Explains Conoco's Borehole Cleaning Success," Offshore Magazine, October 1990, pp. 41-42.

Chin, W.C., "Model Offers Insight into Spotting Fluid Performance," Offshore Magazine, February 1991, pp. 32-33.

Chin, W.C., *Borehole Flow Modeling in Horizontal, Deviated, and Vertical Wells*, Gulf Publishing Company, Houston, 1992.

Chin, W.C., *Petrocalc 14: Horizontal and Vertical Annular Flow Modeling, Petroleum Engineering Software for the IBM PC and Compatibles*, Gulf Publishing Company, Houston, 1992.

Chin, W.C., *Modern Reservoir Flow and Well Transient Analysis*, Gulf Publishing Company, Houston, 1993.

Chin, W.C., *3D/SIM: 3D Petroleum Reservoir Simulation for Vertical, Horizontal, and Deviated Wells, Petroleum Engineering Software for the IBM PC and Compatibles*, Gulf Publishing Company, Houston, 1993.

Chin, W.C., "Eccentric Annular Flow Modeling for Highly Deviated Boreholes," Offshore Magazine, Aug. 1993, pp. 67-70.

Chin, W.C., *4D TurboView: Earth Imaging Software for the PC*, Oil & Gas Journal Energy Database, PennWell Publishing Company, Tulsa, OK, September 1993.

Chin, W.C., and Trevino, J.A., "Pressure Pulse Generator", U.S. Patent No. 4,785,300, November 15, 1988.

Churchill, R.V., *Fourier Series and Boundary Value Problems*, McGraw-Hill, New York, 1941.

Churchill, R.V., *Operational Mathematics*, McGraw-Hill, New York, 1958.

Claerbout, J.F., *Fundamentals of Geophysical Data Processing*, Blackwell Scientific Publishers, Oxford, 1985.

Claerbout, J.F., *Imaging the Earth's Interior*, Blackwell Scientific Publishers, Oxford, 1985.

Clark, R.A., and Reissner, E., "Bending of Curved Tubes," in *Advances in Applied Mechanics, Vol. II*, edited by Richard von Mises and Theodore von Karman, Academic Press, New York, 1951.

Clark, R.K., and Fontenot, J.E., "Field Measurements of the Effects of Drillstring Velocity, Pump Speed, and Lost Circulation Material on Downhole Pressures," SPE Paper No. 4970, 49th Annual Fall Meeting of the Society of Petroleum Engineers of AIME, Houston, Texas, Oct. 6-9, 1974.

Clay, C.S., *Elementary Exploration Seismology*, Prentice-Hall, Englewood Cliffs, New Jersey, 1990.

Clayer, F., Vandiver, J.K., and Lee, H.Y., "The Effect of Surface and Downhole Boundary Conditions on the Vibration of Drillstrings," SPE Paper No. 20447, presented at the 65th Annual Technical Conference and Exhibition of the Society of Petroleum Engineers, New Orleans, La., Sept. 23-26, 1990.

Clayton, R., and Engquist, B., "Absorbing Boundary Conditions for Acoustic and Elastic Wave Equations," Bulletin of the Seismological Society of America, 1977, Volume 67, pp. 1529-1540.

Close, D.A., Owens, S.C., and MacPherson, J.D., "Measurement of BHA Vibration Using MWD," IADC/SPE Paper No. 17273, presented at the 1988 IADC/SPE Drilling Conference, Dallas, Texas, Feb. 28 - Mar. 2, 1988.

Clough, R.W., and Penzien, J., *Dynamics of Structures*, McGraw-Hill, New York, 1975.

Codazzi, D., Till, P.K., Starkey, A.A., Lenamond, C.P., and Monaghan, B.J., "Rapid and Reliable Gas Influx Detection," IADC/SPE Paper No. 23936, presented at the 1992 IADC/SPE Drilling Conference, New Orleans, Louisiana, February 18-21, 1992.

Cohen, J.H., Maurer, W.C., and Westcott, P.A., "High-Performance TSD Bits Improve Penetration Rate," Oil & Gas Journal, Apr. 12, 1993, pp. 47-51.

Cole, J.D., *Perturbation Methods in Applied Mathematics*, Blaisdell Publishing, Waltham, Massachusetts, 1968.

Cook, R.L., Nicholson, J.W., Sheppard, M.C., and Westlake, W., "First Real Time Measurements of Downhole Vibrations, Forces, and Pressures Used to Monitor Directional Drilling Operations," SPE/IADC Paper No. 18651, presented at the 1989 SPE/IADC Drilling Conference, New Orleans, Louisiana, February 28 - March 3, 1989.

Coultrip, R.L., "High Accuracy Wavefront Tracing Traveltime Calculation,", Geophysics, Vol. 58, No. 2, Feb. 1993, pp. 284-292.

Courant, R., and Friedrichs, K.O., *Supersonic Flow and Shock Waves*, Springer-Verlag, Berlin, 1948.

Courant, R., and Hilbert, D., *Methods of Mathematical Physics, Volumes I and II*, John Wiley and Sons, New York, 1989.

Cunningham, R.A., "Analysis of Downhole Measurements of Drill String Forces and Motions," Journal of Engineering for Industry, Transactions of the ASME, May 1968, pp. 208-216.

Dablain, M.A., "The Role and Application of Entropy Terms for Acoustic Wave Modeling by the Finite Difference Method," Geophysics, Vol. 49, No. 9, Sept. 1984, pp. 1457-1465.

Dareing, D.W., "Vibrational Forces and Motions in Rotary Drill Strings," presented at the 1972 Rotary Drilling Conference, International Association of Drilling Contractors, Dallas, Texas, Feb. 23-25, 1972.

Dareing, D.W., "Drill Collar Length is a Major Factor in Vibration Control," SPE Paper No. 11228, presented at the 57th Annual Fall Technical Conference and Exhibition of the Society of Petroleum Engineers, New Orleans, Louisiana, Sept. 26-29, 1982.

Dareing, D.W., "Rotary Speed, Drill Collars Control Drill String Bounce," Oil & Gas Journal, June 6, 1983, pp. 63-68.

Dareing, D.W., "Vibrations Increase Available Power at the Bit," Oil & Gas Journal, March 5, 1984, pp. 91-96.

Dareing, D.W., "Vibrations Increase Available Power at the Bit," Journal of Energy Resources Technology, Transactions of the ASME, March 1985, pp. 138-141.

Dareing, D.W., and Livesay, B.J., "Longitudinal and Angular Drill String Vibrations with Damping," Journal of Engineering for Industry, Transactions of the ASME, October 1968, pp. 1-9.

Dareing, D.W., and Radzimovsky, E.I., "Effect of Dynamic Bit Forces on Bit Bearing Life," Society of Petroleum Engineers Journal, Dec. 1965, pp. 272-276.

Deily, F.H., Dareing, D.W., Paff, G.H., Ortloff, J.E., and Lynn, R.D., "Downhole Measurements of Drill String Forces and Motions," presented at the Petroleum Mechanical Engineering Conference, American Society of Mechanical Engineers, Philadelphia, Pa., Sept. 17-20, 1967.

Deily, F.H., Dareing, D.W., Paff, G.H., Ortloff, J.E., and Lynn, R.D., "Downhole Measurements of Drill String Forces and Motions," Journal of Engineering for Industry, Transactions of the ASME, May 1968, pp. 217-225.

De la Cruz, V., and Spanos, T.J.T., "Seismic Wave Propagation in a Porous Medium," Geophysics, Vol. 50, No. 10, Oct. 1985, pp. 1556-1565.

Den Hartog, J.P., *Advanced Strength of Materials*, McGraw-Hill, New York, 1952.

Denison, E.B., "Downhole Measurements Through Modified Drill Pipe," presented at the Petroleum Mechanical Engineering and Pressure Vessels and Piping Conference, of the American Society of Mechanical Engineers (ASME), Mexico City, Mexico, Sept. 19-24, 1976.

Denison, E.B., "Making Downhole Measurements Through Modified Drill Pipe," World Oil, Oct. 1976, pp. 86-89.

Denison, E.B., "High Data-Rate Drilling Telemetry System," Journal of Petroleum Technology, Feb. 1979, pp. 155-163.

Desbrandes, R., Bourgoyne, A.T., annd Carter, J.A., "MWD Transmission Data Rates Can Be Optimized," Petroleum Engineer International, June 1987, pp. 46-50.

Drumheller, D.S., "Acoustical Properties of Drill Strings," Sandia National Laboratory Report No. SAND-88-0502, August 1988.

Dufeyte, M.P., and Henneuse, H., "Detection and Monitoring of the Slip-Stick Motion: Field Experiments," SPE/IADC Paper No. 21945, presented at the SPE/IADC Drilling Conference, Amsterdam, Mar. 11-14, 1991.

Dunayevsky, V.A., Abbassian, F., and Judzis, A., "Dynamic Stability of Drillstrings Under Fluctuating Weight on Bit," SPE Drilling and Completion Journal, June 1993, pp. 84-92.

Dunayevsky, V.A., and Judzis, A., "Conservative and Nonconservative Buckling of Drillpipe," SPE Paper No. 11991, presented at the 58th Annual Technical Conference and Exhibition of the Society of Petroleum Engineers, San Francisco, Ca., Oct. 5-8, 1983.

Dunayevsky, V.A., Judzis, A., and Abbassian, F., "Use of Dynamic Drillstring Simulator for RPM and Bottomhole Assembly Optimization," OTC Paper 5902, presented at the 21st Annual Offshore Technology Conference, Houston, Texas, May 1-4, 1989.

Dunayevsky, V.A., Judzis, A., and Mills, W.H., "Onset of Drillstring Precession in a Directional Borehole," SPE Paper No. 13027, presented at the 59th Annual Technical Conference and Exhibition of the Society of Petroleum Engineers, Houston, Texas, Sept. 16-19, 1984.

Dunayevsky, V.A., Judzis, A., Mills, W.H., and Griffin, G., "Dynamic Stability of Drillstrings Under Fluctuating Weights-on-Bit," SPE Paper No. 14329, 60th Annual Technical Conference and Exhibition of the Society of Petroleum Engineers, Las Vegas, Nevada, Sept. 22-25, 1985.

Eisner, E., "The Design of Resonant Vibrators," in *Physical Acoustics, Vol. 1, Part B*, edited by W.P. Mason, Academic Press, New York, 1964.

Eronini, I.E., "On Dynamic Fracture of Rock Under Bit Impact Loads," Journal of Energy Resources Technology, Transactions of the ASME, June 1982, pp. 105-107.

Eronini, I.E., Somerton, W.H., and Auslander, D.M., "A Dynamic Model for Rotary Rock Drilling," Journal of Energy Resources Technology, Transactions of the ASME, June 1982, pp. 108-119.

Falconer, I.G., Burgess, T.M., and Wolfenberger, E., "MWD Interpretation Tracks Bit Wear," Oil & Gas Journal, Feb. 10, 1986, pp. 55-59.

Fear, M.J., Thorogood, J.L., Whelehan, O.P., and Williamson, H.S., "Optimization of Rock-Bit Life Based on Bearing Failure Criteria," SPE Drilling Engineering Journal, Sept. 1992, pp. 163-166.

Finnie, I., and Bailey, J.J., "An Experimental Study of Drill String Vibration," Journal of Engineering for Industry, Transactions of the ASME, May 1960, pp. 129-135.

Fontenot, J.E., and Clark, R.K., "An Improved Method for Calculating Swab/Surge and Circulating Pressures in a Drilling Well, SPE Paper No. 4521, 48th Annual Fall Meeting of the Society of Petroleum Engineers of AIME, Las Vegas, Nevada, Sept. 30 - Oct. 3, 1973.

Foster, M.R., and Patton, B.J., "Apparatus for Improving Signal-to-Noise Ratio in Logging-While-Drilling System," United States Patent No. 3,742,443, June 26, 1973.

Garrett, D.L., "Dynamic Analysis of Slender Rods," Journal of Energy Resources Technology, Transactions of the ASME, Dec. 1982, pp. 302-306.

Gelfand, I.M., and Fomin, S.V., *Calculus of Variations*, Prentice-Hall, Englewood Cliffs, New Jersey, 1963.

Geoltrain, S., and Brac, J., "Can We Image Complex Structures With First Arrival Traveltime?", Geophysics, Vol. 58, No. 4, April 1993, pp. 564-575.

Graff, K.F., *Wave Motion in Elastic Solids*, Ohio State University Press, 1975.

Gray, K.E., Armstrong, F., and Gatlin, C., "Two-Dimensional Study of Rock Breakage in Drag-Bit Drilling at Atmospheric Pressure," Journal of Petroleum Technology, Jan. 1962, pp. 93-98.

Grimes, R.E., Felderhoff, F.C., and Brown, H., "Heavy Weight Rock Bits Increase Penetration Rates in Hard Rock," Oil & Gas Journal, May 18, 1992, pp. 76-79.

Hamilton, T., Swanson, B., and Wand, P., "Use of New Kick Simulator Will Increase Wellsite Safety, World Oil, September 1992, pp. 71-77.

Hassanzadeh, S., "Acoustic Modeling in Fluid-Saturated Porous Media," Geophysics, Vol. 56, No. 4, April 1991, pp. 424-435.

Henneuse, H., "Surface Detection of Vibrations and Drilling Optimization: Field Experience," IADC/SPE Paper No. 23888, presented at the 1992 IADC/SPE Drilling Conference, New Orleans, Louisiana, Feb. 18-21, 1992.

Hildebrand, F.B., *Advanced Calculus for Applications*, Prentice-Hall, Englewood Cliffs, New Jersey, 1948.

Hildebrand, F., *Methods of Applied Mathematics*, Prentice-Hall, Englewood Cliffs, New Jersey, 1952.

Huang, T., and Dareing, D.W., "Buckling and Lateral Vibration of Drill Pipe," Journal of Engineering for Industry, Transactions of the ASME, Nov. 1968, pp. 613-619.

Hudson, J.A., *The Excitation and Propagation of Elastic Waves*, Cambridge University Press, Cambridge, 1980.

Jansen, J.D., "Whirl and Chaotic Motion of Stabilized Drill Collars," SPE Drilling Engineering Journal, June 1992, pp. 107-114.

Jeffryes, B.P., "Method and Apparatus for Obtaining Borehole Information Downhole," U.K. Patent Application GB 2 257 785 A, January 20, 1993.

Kalsi, M.S., Wang, J.K., and Chandra, U., "Transient Dynamic Analysis of the Drill String Under Jarring Operation Using Finite Element Method," SPE/IADC Paper No. 13446, presented at the 1985 SPE/IADC Drilling Conference, New Orleans, Louisiana, March 6-8, 1985.

Kane, J.E., "Dynamic Lateral Bottom Hole Forces Aid Drilling," Drilling Magazine, Aug. 1984, pp. 43-48.

Kanwal, R.P., *Generalized Functions: Theory and Technique*, Academic Press, New York, 1983.

Karal, F., "The Analogous Acoustical Impedance for Discontinuities and Constrictions of Circular Cross-Section," Journal of the Acoustical Society of America, Vol. 25, 1953, pp. 327-334.

Kennet, B.L.N., *Seismic Wave Propagation in Stratified Media*, Cambridge University Press, Cambridge, 1983.

Kinsler, L.E., and Frey, A.R., *Fundamentals of Acoustics*, John Wiley and Sons, New York, 1950.

Koch, W.M., "Analysis of the First Field Measurements of Downhole Drill Bit Motions and Forces," presented at the Twenty-Seventh Annual Meeting of the American Association of Oilwell Drilling Contractors, Oklahoma City, Oklahoma, Oct. 11-13, 1967.

Lal, M., "Surge and Swab Modeling for Dynamic Pressures and Safe Trip Velocities," IADC/SPE Paper No. 11412, presented at the IADC/SPE 1983 Drilling Conference, New Orleans, La., Feb. 20-23, 1983.

Lamb, H., *Hydrodynamics*, Dover Publications, New York, 1945.

Landahl, M.T., "Wave Mechanics of Breakdown," Journal of Fluid Mechanics, Vol. 56, Part 4, 1972, pp. 775-802.

Langeveld, C.J., "PDC Bit Dynamics," IADC/SPE Paper No. 23867, presented at the 1992 IADC/SPE Drilling Conference, New Orleans, Louisiana, Feb. 18-21, 1992.

Lear, W.E., and Dareing, D.W., "Effect of Drillstring Vibrations on MWD Pressure Pulse Signals," Journal of Energy Resources Technology, Transactions of the ASME, June 1990, pp. 84-89.

Leighton, W., *Ordinary Differential Equations*, Wadsworth Publishing Company, Belmont, California, 1967.

Lighthill, M.J., *An Introduction to Fourier Analysis and Generalized Functions*, Cambridge University Press, Cambridge, 1959.

Lin, Y.Q., and Wang, Y.H., "New Mechanism in Drillstring Vibration," OTC Paper No. 6225, presented at the 22nd Annual Offshore Technology Conference, Houston, Texas, May 7-10, 1990.

Longuet-Higgins, M.S., and Stewart, R.W., "Radiation Stresses in Water Waves: A Physical Discussion, with Applications," Deep Sea Research, Vol. 11, 1964.

Love, A.E.H., *A Treatise on the Mathematical Theory of Elasticity*, Dover Publications, New York, 1944.

Lutz, J., Raynaud, M., Gstalder, S., Quichaud, C., Raynal, J., and Muckleroy, J.A., "Instantaneous Logging Based on a Dynamic Theory of Drilling," Journal of Petroleum Technology, June 1972, pp. 750-758.

Ma, D., and Azar, J.J., "Dynamics of Roller Cone Bits," Journal of Energy Resources Technology, Transactions of the ASME, Dec. 1985, pp. 543-548.

Ma, D., and Azar, J.J., "A Study of Rock-Bit Interaction and Wellbore Deviation," Journal of Energy Resources Technology, Transactions of the ASME, Sept. 1986, pp. 228-233.

Mason, W.P., *Electromechanical Transducers and Wave Filters*, D. Van Nostrand Company, New York, 1942.

Maurer, W.C., "The 'Perfect-Cleaning' Theory of Rotary Drilling," Journal of Petroleum Technology, Nov. 1962, pp. 1270-1274.

Maurer, W.C., "Bit-Tooth Penetration Under Simulated Borehole Conditions," Journal of Petroleum Technology, December 1965, pp. 1433-1442.

McCann, C., and McCann, D.M., "A Theory of Compressional Wave Attenuation in Noncohesive Sediments," Geophysics, Vol. 50, No. 8, Aug. 1985, pp. 1311-1317.

Miklowitz, J., *The Theory of Elastic Waves and Waveguides*, North-Holland, Amsterdam, 1984.

Miles, J.W., "The Reflection of Sound Due to a Change in Cross Section of a Circular Tube," Journal of the Acoustical Society of America, Vol. 16, 1944, pp. 14-19.

Miles, J.W., "The Analysis of Plane Discontinuities in Cylindrical Tubes," Journal of the Acoustical Society of America, Vol. 17, 1946, pp. 272-284.

Miller, C.E., and Rollins, H.M., "Evaluation of a Vibration Damping Tool and of Drill-Stem Torque Requirements from Data Recorded by an Instrumented Drill-Stem Member," presented at the Petroleum Mechanical Engineering Conference of the ASME, Philadelphia, Pa., Sept. 17-20, 1967.

Mitchell, R.F., "Dynamic Surge/Swab Pressure Predictions," SPE Drilling Engineering Journal, Sept. 1988, pp. 325-333.

Mitchell, R.F., and Allen, M.B., "Lateral Vibration: The Key to BHA Failure Analysis," World Oil, March 1985, pp. 101-106.

Montaron, B.A., Hache, J.M.D., and Voisin, B., "Improvements in MWD Telemetry: The Right Data at the Right Time," SPE Paper No. 25356, 1993 SPE/IADC Drilling Conference, Amsterdam, Feb. 23-25, 1993.

Moore, P.L., *Drilling Practices Manual*, PennWell Publishing Company, Tulsa, Oklahoma, 1974.

Morse, P.M., and Feshbach, H., *Methods of Theoretical Physics, Parts I and II*, McGraw-Hill, New York, 1953.

Morse, P.M., and Ingard, K.U., *Theoretical Acoustics*, McGraw-Hill, New York, 1968.

Moser, T.J., "Shortest Path Calculation of Seismic Rays," Geophysics, Vol 56, No. 1, Feb. 1991, pp. 59-67.

Nakken, E.I., and Baltzersen, O., "Characteristics of Drill Bit Generated Noise," SPWLA 31st Annual Logging Symposium, June 24-27, 1990.

Nayfeh, A., *Perturbation Methods*, John Wiley and Sons, New York, 1973.

Nicholson, R.W., "Critical Rotary Speeds," presented at the 1975 Rotary Drilling Conference of the International Association of Drilling Contractors, March 5-7, 1975.

Nordgren, R.P., "On Computation of the Motion of Elastic Rods," Journal of Applied Mechanics, Transactions of the ASME, Sept. 1974, pp. 777-780

Outmans, H.D., "The Effect of Some Drilling Variables On the Instantaneous Rate of Penetration," 34th Annual Fall Meeting of the SPE, Dallas, Texas, Oct. 4-7, 1959.

Paslay, P.R., and Bogy, D.B., "Drill String Vibrations Due to Intermittent Contact of Bit Teeth," Journal of Engineering for Industry, Transactions of the ASME, May 1963, pp. 187-194.

Paslay, P.R., Jan, Y.M., Kingman, J.E.E., and Macpherson, J.D., "Detection of BHA Lateral Resonances While Drilling With Surface Longitudinal and Torsional Sensors," SPE Paper No. 24583, presented at the 67th Annual Technical Conference and Exhibition of the Society of Petroleum Engineers, Washington, D.C., Oct. 4-7, 1992.

Pierce, A.D., *Acoustics: An Introduction to Its Physical Principles and Applications*, McGraw-Hill, New York, 1981.

Podio, A., and Gray, K.E., "Single-Blow Bit-Tooth Impact Tests on Saturated Rocks Under Confining Pressure, I: Zero Pore Pressure," Society of Petroleum Engineers Journal, September 1965, pp. 211-224.

Qin, F., Luo, Y., Olsen, K.B., Cai, W., and Schuster, G.T., "Finite Difference Solution of the Eikonal Equation Along Expanding Wavefronts," Geophysics, Vol. 57, No. 3, March 1992, pp. 478-487.

Rappold, K., "Drilling Optimized With Surface Measurement of Downhole Vibrations," Oil & Gas Journal, Feb. 15, 1993, pp. 58-62.

Rappold, K., "Drillstring Vibration Measurements Detect Bit Stick-Slip," Oil & Gas Journal, Mar.1, 1993, pp. 66-70.

Rector, J.W., and Marion, B.P., "MWD VSP and Checkshot Surveys Using the Drill Bit as a Downhole Energy Source," OTC Paper No. 6024, presented at the 21st Annual Offshore Technology Conference, Houston, Texas, May 1-4, 1989.

Rector, J.W., and Marion, B.P., "Extending VSP to 3-D and MWD: Using the Drill Bit as a Downhole Seismic Source," Oil & Gas Journal, June 19, 1989, pp. 55-58.

Rewcastle, S.C., and Burgess, T.M., "Real-Time Downhole Shock Measurements Increase Drilling Efficiency and Improve MWD Reliability," IADC/SPE Paper No. 23890, presented at the 1992 IADC/SPE Drilling Conference, New Orleans, Louisiana, Feb. 18-21, 1992.

Rowley, D.S., Howe, R.J., and Deily, F.H., "Laboratory Drilling Performance of the Full-Scale Rock Bit," presented at the 35th Annual Fall Meeting of the Society of Petroleum Engineers, Denver, Co., Oct. 2-5, 1960.

Sananikone, P., Kamoshima, O., and White, D.B., "A Field Method for Controlling Drillstring Torsional Vibrations," IADC/SPE Paper No. 23891, presented at the 1992 IADC/SPE Drilling Conference, New Orleans, Louisiana, Feb. 18-21, 1992.

Schlichting, H., *Boundary Layer Theory*, McGraw-Hill, New York, 1968.

Schwarz, L., *Theorie des Distributions*, Hermann & Cie, Paris, 1950.

Sechler, E.E., *Elasticity in Engineering*, Dover Publications, New York, 1952.

Seeman, B., and Horowicz, L., "Vertical Seismic Profiling: Separation of Upgoing and Downgoing Acoustic Waves in a Stratified Medium," Geophysics, Vol. 48, No. 5, May 1983, pp. 555-568.

Spanos, P.D., and Payne, M.L., "Advances in Dynamic Bottomhole Assembly Modeling and Dynamic Response Determination," IADC/SPE Paper No. 23905, presented at the 1992 IADC/SPE Drilling Conference, New Orleans, Louisiana, Feb. 18-21, 1992.

Squire, W.D., and Whitehouse, H.J., "A New Approach to Drill-String Acoustical Telemetry," SPE Paper No. 8340, presented at the 54th Annual Fall Technical Conference and Exhibition of the Society of Petroleum Engineers of the AIME, Las Vegas, Nevada, Sept. 23-26, 1979.

Staff, "Antiwhirl PDC Bit Designs Reduce Vibrations," Oil & Gas Journal, Nov. 30, 1992, pp. 59-62.

Stulemeijer, I.P.J.M., "Analysis of the Longitudinal Vibrations in a Drill String in the NL/DST Test Well," KSEPL Contribution to the SOC Drilling Engineering Conference at Broadwater Beach, Biloxi, Mississippi, Feb. 6-8, 1984.

Timoshenko, S.P., *Strength of Materials, Part II: Advanced Theory and Problems*, Van Nostrand Reinhold Company, New York, 1958.

Timoshenko, S.P., and Gere, J.M., *Theory of Elastic Stability*, McGraw-Hill, New York, 1961.

Timoshenko, S.P., and Goodier, J.N., *Theory of Elasticity*, McGraw-Hill, New York, 1934.

Toksoz, M.N., and Johnston, D.H., *Seismic Wave Attenuation*, Geophysics Reprint Series No. 2, Society of Exploration Geophysicists, Tulsa, Oklahoma, 1981.

Tychonov, A.N., and Samarski, A.A., *Partial Differential Equations of Mathematical Physics, Vol. I*, Holden-Day, San Francisco, 1964.

Tychonov, A.N., and Samarski, A.A., *Partial Differential Equations of Mathematical Physics, Vol. II*, Holden-Day, San Francisco, 1967.

Vandiver, J.K., Nicholson, J.W., and Shyu, R.J., "Case Studies of the Bending Vibration and Whirling Motion of Drill Collars," SPE/IADC Paper No. 18652, presented at the 1989 SPE/IADC Drilling Conference, New Orleans, Louisiana, February 28-March 3, 1989.

van Dyke, M., *Perturbation Methods in Fluid Mechanics*, Academic Press, New York, 1964.

van Trier, J., and Symes, W.W., "Upwind Finite Difference Calculation of Traveltimes," Geophysics, Vol. 56, No. 6, June 1991, pp. 812-821.

Venkatraman, B., and Patel, S.A., *Structural Mechanics with Introductions to Elasticity and Plasticity*, McGraw-Hill, New York, 1970.

Vidale, J., "Finite Difference Calculation of Travel Times," Bulletin of the Seismological Society of America, Vol. 78, No. 6, Dec. 1988, pp. 2062-2076.

Vidale, J.E., and Houston, H., "Rapid Calculation of Seismic Amplitudes," Geophysics, Vol. 55, No. 11, Nov. 1990, pp. 1504-1507.

Walker, B.H., and Friedman, M.B., "Three-Dimensional Force and Deflection Analysis of a Variable Cross-Section Drill String," Journal of Pressure Vessel Technology, May 1977, pp. 367-373.

Wang, J.K., Kalsi, M.S., Chapelle, R.A., and Beasley, T.R., "A Practical Approach to Jarring Analysis," SPE/IADC Paper No. 16155, 1987 SPE/IADC Drilling Conference, New Orleans, March 15-18, 1987.

Warren, T.M., "Drilling Model for Soft Formation Bits," Journal of Petroleum Technology, June 1981, pp. 963-970.

Warren, T.M., "Factors Affecting Torque for a Tricone Bit," SPE Paper No. 11994, presented at the 58th Annual Technical Confrence and Exhibition, San Francisco, Ca., Oct. 5-8, 1983.

Warren, T.M., "Penetration Rate Performance of Roller Cone Bits," SPE Paper No. 13259, presented at the 59th Annual Technical Conference and Exhibition, Houston, Texas, Sept. 16-19, 1984.

White, J.E., "Seismic Waves in Fluid-Saturated Rocks: An Examination of the Biot Theory," Colorado School of Mines Quarterly, Vol. 1, 1984, pp. 1-32.

Whitham, G.B., *Linear and Nonlinear Waves*, John Wiley and Sons, New York, 1974.

Wolf, S.F., Zacksenhouse, M., and Arian, A., "Field Measurements of Downhole Drilling Vibrations," SPE Paper No. 14330, 60th Annual Technical Conference and Exhibition of the Society of Petroleum Engineers, Las Vegas, Nevada, Sept. 22-25, 1985.

Yang, J.H., and Gray, K.E., "Single-Blow Bit-Tooth Impact Tests on Saturated Rocks Under Confining Pressure, II: Elevated Pore Pressure," Society of Petroleum Engineers Journal, December 1967, pp. 389-408.

Yew, C.H., and Jogi, P.N., "Study of Wave Motions in Fluid-Saturated Porous Rocks," Journal of the Acoustical Society of America, Vol. 60, No. 1, July 1976, pp. 2-8.

Zukas, J.A., Nicholas, T., Swift, H.F., Greszczuk, L.B., and Curran, D.R., *Impact Dynamics*, John Wiley and Sons, New York, 1982.

Index

A

Absolute stability, 169
AC/DC interaction, 142-146, 155
Accordion model, 149
Acoustic impedance, 290
Acoustics, 8, 91, 93, 120, 282, 292
Aeroelasticity, 107
Amplification factor, 169
Amplitude law, 74-75
Analytical continuation 25, 74, 174
Anecdotal stories, 126
Angle of twist, 240
Anisotropic media, 85
Annular flow, 301-306, 334
Ansatz, 11
Antisymmetry. *See* Symmetry.
Apparent viscosity, 303
Area changes, 174-175, 181-182, 232, 283, 285, 289, 296-298
Artificial viscosity, 361
Asymptotic analysis, 3, 21, 58, 78
Attenuation, 62, 64, 137, 223
Axial vibration, 7, 121, 136

B

Backward differencing, 163
Backward whirl. *See* Whirling.
Banded matrix solver, 227
Beam. *See* Lateral vibration.
Bending. *See* Lateral vibration.
Bending method, 357
Bending moment, 90, 212, 227
Bernoulli equation, 104, 292
Bit bounce, 139, 149, 206
Blow-out, 298
Borehole flow modeling, 112-113, 301-315
Bottomhole assembly, 119
Boundary conditions, 10, 12
Boundary layers, 114
Boundary value problem, 10
Bromwich integration, 23. *Also see* Complex variables.
Buckley-Leverett condition, 355
Buckling (static), 273
Bulk modulus, 120, 282
Burger's equation, 354

C

Canonical variables, 31
Capillary waves, 102
Causality, 5, 165, 230
Center frequency, 70
Center wavenumber, 70, 74
Central differencing, 163, 229
Chain rule, 5, 31
Channel parameter, 306
Characteristic plane, 6
Characteristic values, 13
Characteristics, 6, 31, 60, 299
Circle theorem, 110
Classical wave equation, 5, 139
Cole-Hopf solution, 354
Completeness, 14, 91
Complex exponential, 15, 17, 21
Complex variables, 23
Compressibility, 9, 99, 120, 294
Computational solution. *See* Numerical integration.
Conservation form, 64, 74, 78
Constructive interference, 329
Continuous spectrum, 217
Continuous wave telemetry, 196
Convective derivative, 77
Convergence, 169
Convolution theorem, 25-26, 30, 42, 161
Couple. *See* Dipole.
Coupled vibrations, 123, 255
Courant-Friedrichs-Lewy (CFL) condition, 170, 231
Critical Reynolds number, 111
Cross-structure, 92, 94, 104
Crosswell seismics, 349, 356
Crosswell tomography, 349, 356
Cumulative errors, 55
Cutoff frequency, 96

D

D'Alembert's solution, 7
DC, 25, 126, 142
Delay line method, 195
Delta-p, 301, 321-323, 326
Delta-velocity, 324, 326
De Moivre's theorem, 17
Desktop experiment, 122, 256, 271
Destructive interference, 329
Detecting lateral vibrations, 224
Diagonal dominance, 168, 175-176, 298
Difference molecule, 198
Differential technique, 196-198
Digital analysis, 198
Dimension (Fortran), 168
Dipoles, 36-37, 46, 48
Dirac delta function, 13, 27, 31, 33, 40, 44-46, 100, 225
Directional drilling, 128, 272
Directional filtering, 31-32, 49, 195, 331
Discontinuities, 14, 34-35, 43, 101, 120, 141, 284-287, 289
Dispersion, 6-7, 19, 57, 96, 214
Dispersion relation, 19-21, 57, 96, 102, 104, 210, 214, 338
Displacement source, 14, 101, 141, 149, 153-154, 173
Dissipation, 15-16, 33, 62
Distortion, 8, 25, 27, 29
Distributed forces, 12
Distributions, 34, 37
Doglegs, 182
Domains of dependence and influence, 5
Doppler shift, 366
Double-valued, 34, 37, 298
Downgoing waves, 6
Drill bit kinematics, 152-153, 158

Drill collar design, 222
Drill pipe acoustic telemetry, 320
Dual bending modes, 212, 256, 272
Dynamically steady, 15, 68, 80, 124, 180, 232, 352

E

Eccentricity, 123, 183
Eigenfunctions, 10, 13, 90, 96
Eigenvalues, 13, 90, 108-109
Eikonal equation, 53, 74, 85, 339, 343
Elastic impact, 191
Elastic line, 130, 180, 240, 272
Elastic membrane, 106-110
Electromagnetic wave telemetry, 320
Elliptic equations, 5, 102
Energy balance, 155
Energy variation, 84
Entropy condition, 355
Equation of state, 94
Equipartition of energy, 106
Euler variational equations, 342
Eulerian description, 92, 107, 290
Evanescent mode, 97, 120
Even functions, 9
Existence, 10
Exponential shift, 47
Extended eikonal equation, 350
External forces, 33

F

Fan, 357
Fermat's principle of least time, 341-345
Finite difference method, 163-170, 197, 229-231, 245, 259-260, 296, 358, 361
First arrival, 356
First-order equations, 50-53
Fishing, 206
Focus, 210, 215
Forcing function, 35
Formation imaging, 129
Fortran source code, 176-180, 233-234, 247-248, 263-269, 274
Forward difference, 164
Forward whirl. *See* Whirling.
Fourier series, 9, 13, 97
Fourier transform, 29
Free-fall limit, 23-24, 142
Free surface, 102-103
Frequency, 19-20, 57, 84, 210
Frequency filter, 195

G

Galilean transformation, 291
Gaussian elimination, 167
Generalized function, 37
GEODYN, 129-130
Gradient, 99
Gravity-capillary waves, 102
Gravity waves, 62, 102, 366
Grid size, 169
Group velocity, 20, 59-60, 64, 98, 105, 204, 211

H

Hertz, 20
Heterogeneous media, 7
High frequency limit, 338
Homogeneous media, 7
Hooke's law, 7, 94
Horizontal wells, 274
Hydraulics, 9, 93, 120, 282, 292
Hydrodynamic loading, 367
Hydrodynamic stability, 107, 111-114
Hyperbolic equations, 3, 5

I

Imaginary frequency, 21, 213-214
Imaging, 129, 184, 188
Impedance mismatch, 120, 141, 174
Impulse response, 40, 42
Incompressible, 9
Inelastic impacts, 189
Initial conditions, 10, 12, 146, 228
Initial value problem, 7
Inverse scattering theory, 354
Inviscid flow, 99
Irrotationality, 59, 82, 99, 114, 292

J

Jarring, 182-183
Joints, 57
Jump conditions. *See* Discontinuities.

K

Kinematic barrier, 215-219, 368-370
Kinematic boundary conditions, 103
Kinematic shock, 67
Kinematic viscosity, 365
Kinematic wave theory, 6, 21, 55, 59, 204, 337, 364
Kirchoff equations, 272
Kirchoff migration, 70, 357
Korteweg deVries equation, 354

L

Lagrangian (average), 81-82
Lagrangian description, 282
Laplace equation, 103
Laplace transform, 14, 22-23, 46
Lateral vibrations, 88, 122
Least squares, 14
Left-going waves, 6
Lithology sensor, 319
Longitudinal vibrations. *See* Axial vibration.
Lumped mass, 125

M

Manifold, 205
Marine riser, 205
Mass conservation, 92
Mass-spring-damper, 24, 137, 146, 157
Material derivative, 77
Mathematical induction, 72
Measurement-While-Drilling, 49, 57, 97-100, 124, 128, 196-197, 241, 246-247, 262, 281, 289-291, 318, 334
Method of characteristics, 6, 299
Migration, 337
Minima, 18
Mixed boundary conditions, 13-14, 24, 27, 121, 161, 189
Modal waves, 92, 94
Mode shapes, 13
Modified drillpipe telemetry, 319
Modulator, 319
Momentum equation, 92
Monochromatic wave, 19, 70, 338
Monopole, 35
Moving flow, 80, 365
Mud motor, 128
Mud pulse telemetry, 97, 246, 321
Mud siren, 97, 325-326
Multiple point forces, 39
Multiple scaling method, 55, 64, 71
Multiplexing, 241

N

Natural frequency, 13
Navier-Stokes equations, 92
Negative damping, 243
Negative pressure pulser, 97, 100, 323-325
Network theory, 357
Neutral point, 80, 122, 145, 205-206, 211-212, 215
Neutral stability (curve) , 111
Newtonian flow, 111
Nodes, 18
Noether's theorem, 77
Nonlinearity, 22
Non-Newtonian flow, 302
Nonuniformities, 21, 60
Normal modes, 13, 125
Numerical integration, 61, 79
Numerical stability, 168

O

Odd functions, 10
Operational calculus, 34
Orr-Sommerfeld equation, 111
Orthogonality, 14, 91
Outgoing wave condition, 33

P

P-waves, 338
Packaging constraints, 327
Packed assembly, 273
Parameterization, 51, 274
Parametric resonance, 272
Particular solution, 41
Passbands, 320
Pentadiagonal matrix, 229-230, 234, 259-260
Perfect cleaning, 152
Perturbation method, 55
Phase, 16, 18, 29, 69, 339
Phase angle, 20
Phase function, 58, 74-75
Phase velocity, 20, 57, 98
Polar moment of inertia, 240
Polycrystalline diamond compact (PDC) bits, 128, 239, 255
Poppet valve, 321-323
Position sensor, 319
Positive pressure poppet valve, 97, 100, 321
Potential function, 99
Power requirements, 328
Power system, 319
Precession, 271-274
Pressure gradient, 101
Principle of least time, 341-345
Probability of bit error, 196, 332
Production system, 205
Propagating wave, 18-19, 63, 96

Q

Q-model, 62, 346-347
Quality factor. *See* Q-model.

R

Radiation condition, 33, 35, 49
Radiation stress, 214, 365
Rate-of-penetration (ROP), 119, 128, 147-152, 159-162
Ray convergence, 68, 360
Rayleigh equation, 107-108, 111
Rays, 6, 60, 83, 85, 220, 339, 341
Recirculating vortex flow, 305
Recorded downhole data, 320
Rectangular pulse, 27
Reflection properties, 7
Refraction, 210

Regularity condition, 30
Resonance, 125, 147, 208, 287
Reynolds number, 111
Rheology, 93, 183, 292, 301
Right-going waves, 6
Rigid termination, 8
Rock-bit interaction, 121, 141, 147-152, 161, 170, 186, 189, 267-271
Rotary drilling, 254
Round-off error, 168

S

Saddle point method, 58
Seismic limit, 351-352
Seismology, 37, 62-63, 154, 351
Separation of variables, 10, 12, 89
Shear flow, 99, 107, 114
Shear force, 90, 212, 227, 231
Shear modulus, 240
Shock-fitting, 356
Shocks, 124, 208, 355
Shooting method, 357
Sifting property, 41, 47
Signal strength, 49
Signal-to-noise (S/N) ratio, 196, 241, 319, 332
Single tooth impact, 151
Singularity, 56, 67, 353
Slick assembly, 273
SNAP log, 186-188
Solid body rotation, 241
Sound speed, 5-7
Splitting, 49
Spotting fluids, 183
Stability, 169
Stability envelope, 112, 273
Stable open (or close), 329
Standing waves, 18, 63, 96, 124
Static indeterminacy, 231
Stationary phase, 58
Stick-slip, 123, 142, 184, 242-244
Stiffness, 122, 212
Streamfunction, 108, 305
Stress free end, 8
Stress-strain relationship, 63, 93
Stuck pipe, 182
Substantive derivative, 77
Surface instrumentation, 137
Surface receiver, 319
Surface signal processor, 319
Surface tension, 102
Swab-surge, 298-301
Symbolic manipulation, 24
Symmetries, 40, 100
Synthetic data, 33

T

Taylor series, 9, 71, 74, 163-169
Telemetry package, 319
Telemetry scheme, 327, 332
Tension leg platform, 205
Three-wave formulation, 198
Top drive, 273
Torque-on-bit, 246
Torque reversal, 123, 206
Torsional rigidity, 240
Torsional stiffness, 240
Torsional vibration, 122, 239
Torsional wave telemetry, 320
Trajectories, 6
Transfer function, 18, 194
Transition frequency, 347
Transmission channel, 330-331
Transparency to waves, 35-36
Transport equation, 339
Transverse vibrations, 5, 33, 122. *See also* Lateral vibrations.
Travel time modeling, 356
Traveling block, 137
Traveling salesman problem, 357

Tridiagonal equations, 23, 167, 229
Tripping speed, 299
Truncation error, 168
Turbine, 97
Turbodrill, 123, 254
Turning point, 122
Twist-off, 204

U

Unconditional stability, 169
Uniform plane wave, 21
Uniqueness, 10
Unwinding, 242
Upgoing waves, 6
Upwind differencing, 361

V

Variation of parameters, 41
Variational principles, 77, 81, 342
Vector calculus, 94, 99
Vector potential, 99, 103, 294-295
Velocity model, 356
Vertical seismic profiling, 185
Vibration theory, 3
Vibroseis, 185
Viscometer, 304
Volume velocity, 283
von Neumann stability, 168
Vorticity vector, 99

W

Water hammer, 287
Wave action, 81
Wave crest, 19
Wave crest conservation, 59, 72, 76
Wave energy, 20, 60, 77, 84, 219, 364
Wave group, 6
Wave momentum, 77, 219, 364
Wave propagation, 3
Wave trapping, 77, 209
Waveguide, 6, 25, 91, 120
Wavelength, 20, 105
Wavenumber, 19, 57, 98, 210
Weight-on-bit, 122, 144-145, 158, 246
Whirling, 123, 239, 248-250
WKB solution, 21, 55, 71, 78

Y

Young's modulus, 120